GW00370562

GLACIOTECTONICS
FORMS AND PROCESSES

PROCEEDINGS OF VARIOUS MEETINGS OF THE GLACIOTECTONICS WORK GROUP: FIELD MEETING, MØN, DENMARK (1986) / INQUA CONGRESS, OTTAWA, CANADA (1987) / FIELD MEETING, NORFOLK, UK (1988)

Glaciotectonics

Forms and processes

Edited by
DAVID G.CROOT
Department of Geographical Sciences, Plymouth Polytechnic

A.A.BALKEMA / ROTTERDAM / BROOKFIELD / 1988

The texts of the various papers in this volume were set individually by typists under the supervision of each of the authors concerned.

Published by

A.A.Balkema, P.O.Box 1675, 3000 BR Rotterdam, Netherlands

A.A.Balkema Publishers, Old Post Road, Brookfield, VT 05036, USA

ISBN 90 6191 848 0

Printed in the Netherlands

Glaciotectonics: Forms and Processes, Croot (ed.), © 1988 Balkema, Rotterdam. ISBN 90 6191 848 0

Contents

Glaciotectonics: Forms and Processes, Croot (ed.), © 1988 Balkema, Rotterdam. ISBN 90 6191 848 0

Introduction

David G.Croot
Department of Geographical Sciences, Plymouth Polytechnic, Plymouth, Devon, UK

This collection of papers represents a cross-section of research work conducted by members of the Work Group on Glaciotectonics (WGGT) over the last five years. The Work Group, part of the International Quaternary Association's Commission on the Formation and Properties of Glacial Deposits, was formed in 1982 following an initiative by Dr A Berthelsen of Copenhagen, Denmark. The papers have been presented at three field meetings, workshops and conferences in 1986, 1987 and 1988. Two were Work Group Meetings (Mon, Denmark October 1986 and Norfolk, UK October 1988) whilst the third was the quinquennial INQUA Congress in Ottawa, Canada (July/August 1987). The aim of producing the papers in this single volume is to present the material for consumption by a much wider audience than those who were able to attend the various meetings. The format has also enabled researchers to produce much more complete evidence of their results and findings than would normally be possible in other media.

The papers deal with a wide range of glaciotectonic topics: some are 'theoretical' or conceptual, others re-assess previously reported and differently interpreted sections, yet others provide entirely new research results from hitherto undescribed areas or sections. I hope the reader finds all of the contributions enlightening and thought-provoking at the very least.

I have of necessity edited the English of the non-English-speaking contributors. I trust that I have made amendments to their manuscripts which maintains the spirit of the paper, rather than the letter. This has not been an easy task, and in some places the result is an unwieldy compromise. I offer my apologies to the grammatical purists. The greatest care has been taken to minimise typographical and spelling errors, but since each author was responsible for his/her camera-ready copy manuscript, and worked successfully to a very strict deadline, I have no doubt that errors have crept in.

Papers appear in alphabetical order of the author's surname, with one exception; The Glaciotectonic Bibliography. This is included as an aid to current and future glaciotectonic research workers, because the literature on the subject is so diverse. As with any academic publication, this appears last in the order.

As editor on behalf of the Work Group I express my thanks to all the contributors for working hard to meet deadlines; to my wife, and colleagues who have been most supportive; and finally to Debbie Petherick who not only typed my contributions, but repeatedly corrected and re-typed many other contributions.

Glaciotectonics: Forms and Processes, Croot (ed.), © 1988 Balkema, Rotterdam. ISBN 90 6191 848 0

Ice-shoved hills of Saskatchewan compared with Mississippi Delta mudlumps – Implications for glaciotectonic models

James S. Aber
Earth Science Department, Emporia State University, Emporia, Kans., USA

ABSTRACT: The Dirt Hills and Cactus Hills of southern Saskatchewan are outstanding examples of large ice-shoved hills. Glaciotectonic models for the genesis of these and other ice-shoved hills fall into two groups: (1) models in which the glacier thrusts permafrozen strata and (2) models in which permafrost is not considered a prerequisite. Mudlumps of the prograding Mississippi Delta are deforming in a style and magnitude much like ice-shoved ridges. Mudlumps may be regarded as modern nonglacial analogs to large ice-shoved hills of Pleistocene age.

Both situations share several traits in common; the only significant difference is the nature of the advancing load--glacier ice or delta sand. The increased load, in either case, caused compaction and overpressuring of incompetent strata and led to deformation and uplift of ridges in front of the advancing load. Loading by glacier ice during deformation of the Dirt Hills is estimated to have been triple the magnitude of loading associated with Mississippi mudlumps. Permafrost is not necessary for this style of deformation in either glacial or nonglacial settings.

1. INTRODUCTION

The Dirt Hills and Cactus Hills, located near Regina in southern Saskatchewan (Fig. 1), are outstanding examples of ice-shoved hills, because of their great size, superb development as arcuate ranges, and excellent geomorphic expression. Upper Cretaceous strata have been uplifted as much as 200 m into imbricately thrust and folded ridges (Aber 1985a). Similar, impressive ice-shoved hills are found elsewhere in the Canadian Plains (Kupsch 1962), in North Dakota (Bluemle and Clayton 1984), in northern Europe (Aber 1985b), in New England (Oldale and O'Hara 1984), and in the Arctic (Mackay 1959). These and other glaciotectonic features testify to the power of glaciation to modify the landscape by deformation of pre-existing bedrock and sediment.

Models for glaciotectonic genesis of such large ice-shoved hills fall into two groups: (1) models in which the glacier deforms permafrozen strata and (2) models in which permafrost is not considered a prerequisite. Deformed sedimentary floes are mostly composed of poorly consolidated to unconsolidated material. Thus, many geologists have assumed that floes were deformed while frozen and that their thickness is an indication of the former depth of permafrost (Gry 1940; de Jong 1952, 1967; Rutten 1960; Mathews and Mackay 1960; Kupsch 1962; Kaye 1964; Clayton and Moran 1974; Banham 1975; Berthelsen 1979; Moran et al. 1980; Aber 1982).

Other geologists, however, have rejected the necessity for permafrost. Pre-existing lithologic contrasts or stratigraphic boundaries in combination with elevated hydrostatic pressure during ice advance are considered adequate conditions for glacier thrusting (Mackay and Mathews 1964; Moran 1971; Rotnicki 1976; Bluemle and Clayton 1984; van der Wateren 1985). Large ice-shoved hills comparable to the Dirt Hills or Cactus Hills are not forming around existing glaciers. However, smaller ice-shoved ridges have been described from modern glaciers in both frozen (Klassen 1982) and unfrozen (Humlum 1983; Croot 1987, this volume) settings.

Large thrusts and diapirs are forming today in certain nonglacial sedimentary environments, in particular the Mississippi Delta. The similarity between delta "mudlumps" and glaciotectonic deformation has

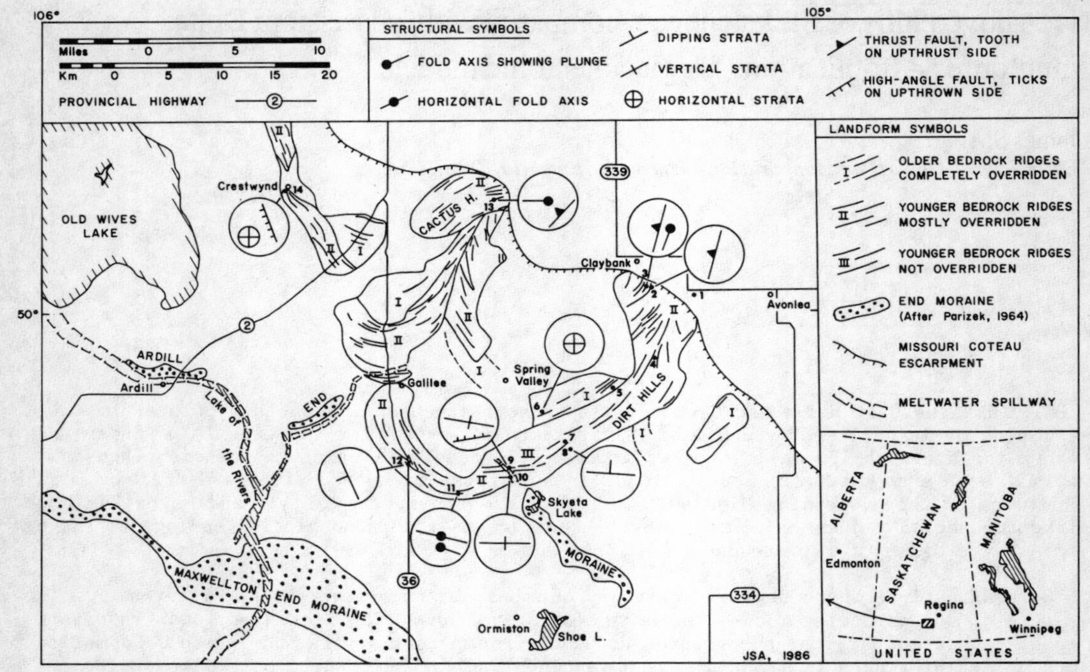

Fig. 1. Map of Dirt Hills and Cactus Hills vicinity showing pattern of ice-shoved ridges, locations of sites, and structural features at each site (circle diagrams). Based on interpretation of satellite images, aerial photographs and topographic maps plus field work during 1984 and 1986.

been noted before (Ruszczynska-Szenajch 1976; Aber 1982). Thrust blocks, folds, and diapirs are comparable in style to ice-pushed structures and display similar structural uplifts (>100 m). In both settings, the deformations were created by increased loading, either by advancing glacier ice or by prograding delta sand. Mississippi Delta mudlumps are here regarded as modern, non-glacial analogs to large ice-shoved hills of Pleistocene age.

The Dirt Hills and Cactus Hills have received considerable study (Byers 1959; Kupsch 1962; Parizek 1964; Christiansen 1971; Christiansen and Whitaker 1976) as have Mississippi Delta mudlumps (Morgan 1961; Andersen 1961; Morgan et al. 1968). The structural development of each will be described briefly, based on previous work, only insofar as necessary to draw comparisons between the two settings.

2. DIRT HILLS AND CACTUS HILLS

The Dirt Hills and Cactus Hills are located on the Missouri Coteau, a major northeast-facing escarpment that marks the boundary between the Saskatchewan and Alberta Plains. Higher elevation of the Alberta Plain to the south and west is a reflection of more resistant terrestrial sandstone interbedded with mudstone and lignite. The Saskatchewan Plain, in contrast, is underlain by softer marine shale. The regional geologic structure consists of essentially flat-lying strata which dip very gently to the east or northeast. Steeply dipping, folded, and faulted bedrock structures are common, however, in ice-shoved hills along the Missouri Coteau (Fig. 1).

During the maximum Lostwood (late Wisconsin) glaciation, the ice margin lay to the south near the United States-Canada border. Ice movement at this time was from the northeast. As the Laurentide ice sheet began to shrink, the Weyburn ice lobe developed immediately north of the Missouri Coteau, and it flowed southeastward parallel to the Coteau escarpment (Christiansen 1956).

Three upper Cretaceous bedrock units can be seen deformed in the Dirt Hills and Cactus Hills. These three in ascending order are (Fig. 2): Eastend (Ke), Whitemud (Kw),

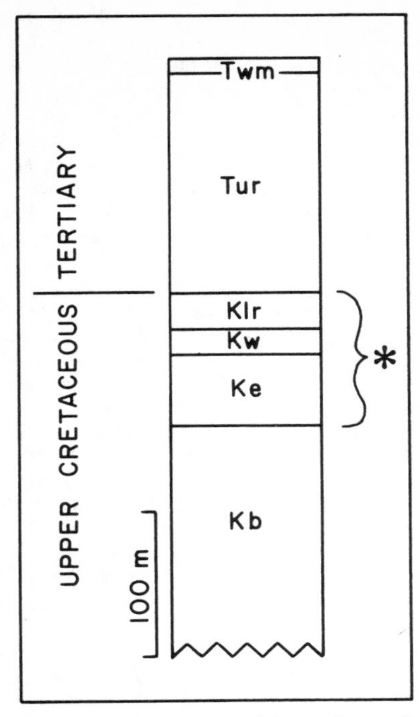

Fig. 2. Stratigraphic column for bedrock of Dirt Hills and Cactus Hills vicinity. Deformed bedrock units indicated by asterisk. Based on Fraser et al. (1935) and Parizek (1964).

and lower Ravenscrag (Klr). The deformed strata are predominately terrestrial, bentonitic or kaolinitic sandstone and mudstone, having a total thickness up to 90 m. Claystone makes up the lower Eastend Formation, and lignite beds are common within the Whitemud and lower Ravenscrag Formations. The deformed strata are underlain by presumably undisturbed marine shale of the Bearpaw Formation (Kb).

Folded and thrust bedrock floes stacked in an imbricated pattern comprise the overall structure of the Dirt Hills and Cactus Hills (Fig. 1). At most sites, drift covers the bedrock but is not involved in the deformations. However, at a few sites older masses of drift are deformed along with bedrock. Not only have bedrock blocks been displaced horizontally, but considerable vertical uplift has also occurred. Maximum structural uplift is documented in the southern Dirt Hills (site 9), where a block of Eastend Formation is found standing vertically 200 m above its normal stratigraphic position.

Remarkable agreement exists between orientations of bedrock structures, trends of individual ridges, and overall outlines of the Dirt Hills and Cactus Hills. This amply confirms Kupsch's (1962) conclusion that ice-shoved ridges on the Missouri Coteau are direct or first-order morphologic expressions of bedrock structures produced by ice pushing. Bedrock structures at most sites are related to a single episode and direction of ice pushing, but some sites show evidence for multiple phases of deformation. Variations in bedrock competence clearly influenced structural development. Thrust faults are usually located within lignite or claystone beds; conversely, thicker sandstone layers comprise the larger folds and fault blocks.

The Dirt Hills and Cactus Hills form two large loop-shaped ranges which along with nearby hills define the outlines of three ice tongues that caused thrusting of bedrock ridges. These three ice tongues were: (1) Galilee, west of Cactus Hills, (2) Spring Valley, between Cactus Hills and Dirt Hills, and (3) Avonlea, east of Dirt Hills. The ice tongues pushed up three classes of ridges (Fig. 1). Subdued, class I ridges were presumably created during an earlier glacier advance of unknown age. Prominent, class II and III ridges were thrust up by the last ice advance to push onto the Missouri Coteau.

Thrusting of the Dirt Hills and Cactus Hills did not happen during initial advance of the Lostwood ice sheet, which reached its maximum position near the United States border about 17,000 BP (Christiansen 1979; Dyke and Prest 1986). At that time, the Dirt Hills and Cactus Hills did not yet exist. Thrusting of the hills occurred later, during a strong readvance of the Weyburn lobe. This lobe generated lateral offshoots or ice tongues that pushed into embayments of the Coteau. Thrusting of bedrock occurred around the margins of these ice tongues due to rapid loading and forward movement.

Christiansen (1979, Fig. 12) interpreted the age of Weyburn readvance and Dirt Hills thrusting at 15,500 BP. However, other geologists favored younger dates in the vicinity of 13,000 BP (Teller et al. 1980; Fenton et al. 1983; Dyke and Prest 1986).

All three ice tongues caused thrusting of class I bedrock ridges during an earlier advance. The main thrusting of class II and III ridges occurred during a readvance of the Galilee and Spring Valley ice tongues. The Avonlea ice tongue also readvanced at this time, but without thrusting up any new ridges. The Galilee and Avonlea ice tongues overran ice-pushed ridges reaching positions marked by the Ardill end moraine and Lake of the Rivers spillway. The Spring Valley ice tongue, however, stopped on the inner (northern) side of the Dirt Hills, from where a spillway was cut across class III ridges toward Skyeta Lake. Class III ridges, thus, formed a nunatak between active ice to the north and older stagnant ice lying on the Coteau to the south.

Building of the Ardill end-moraine system and cutting of associated spillways were related to the same ice advances that caused the main phase of thrusting in the Dirt Hills and Cactus Hills. Following thrusting of the Dirt Hills and Cactus Hills, building of the Ardill end moraine, and cutting of spillways, the ice tongues stagnated and downwasted leaving an irregular accumulation of hummocky moraine over much of the area north of the ice margin.

Rapid loading of competent sandstone bedrock over saturated, incompetent mudstone strata (lower Ravenscrag, Whitemud, and upper Eastend Formations over lower and middle Eastend Formation) caused thrusting around the margins of the ice tongues. The fact that bedrock was most likely thawed and saturated is confirmed by abundant evidence for meltwater spillways, large proglacial lakes, and wasting ice masses throughout southern Saskatchewan during deglaciation. Collapse structures associated with melting stagnant ice (McDonald and Shilts 1975) are common in ice-contact drift of the region. There is no evidence that permafrost existed at the time of glacier thrusting. Ubiquitous meltwater means that thrust blocks could not be moved by freezing onto the undersides of the ice tongues, but were displaced by squeezing out from under the ice margin.

3. MISSISSIPPI MUDLUMPS

The Mississippi Delta is a classic, large, bird-foot style delta (Fig. 3). One of its lesser known features are mudlumps--small islands or shoals formed over uplifted clay structures near bar-finger sand at the mouths of major distributary channels. Mudlumps are active features with lifespans measured in decades before becoming buried by prograding delta sand. The zone of mudlump uplift has migrated outward during the past century as the delta has grown (Morgan et al. 1968).

Mudlump islands range in size from small pinnacles to maximum areas of about 8 hectares. They are mostly oval in form, with a length usually 3-4 times the width. The surficial portion of a typical mudlump is formed by an anticline trending parallel to the mudlump's long axis. A narrow, shallow graben marked by many small normal faults runs along the crest of the anticline. Strata within the graben are highly irregular and confused. Active mudlumps are known to rise at rates >60 cm/month (Morgan 1961). Extrusion of mud volcanoes and venting of methane-rich gas during uplift indicate that excess hydrostatic pressure must be developed within the clay core of the mudlump.

Test drilling near South Pass distributary has revealed the subsurface stratigraphy and structure of several mudlumps (Fig. 4). Beneath bar-finger sand, which may reach up to 120 m in thickness, a sequence of prodelta clay units rests on a late Wisconsin algal reef zone 180 m deep. The clay units have been deformed into a series of asymmetrically thrust anticlines or diapirs with the algal reef zone acting as a structural basement. Clay strata exposed in mudlump islands contain formaminifera derived from at least 120 m or more in depth (Andersen 1961).

The mudlumps of South Pass distributary have developed during the past 100 years. Thick sand has built up in the subsiding synclines between mudlumps during the same time interval. Assuming distributary sand accumulates at sea level, then synclines have subsided at average rates >1 m/year (Morgan et al. 1968). This subsidence has compensated for uplift in mudlumps, and the asymmetry of folds, faults, and diapirs reflects differential loading by the accumulating sand mass.

Mudlumps have formed only at the mouths of major distributaries; they have not developed in connection with lesser or shallow distributaries. Mudlumps undergo three developmental stages: (1) initial uplift as a submarine shoal, (2) growth into an island, and (3) erosion and truncation by waves. Many mudlumps experience episodic uplift, which invariably coincides with river flooding, when rapid sedimentation occurs at the distributary mouth. It is therefore concluded that mudlump development is intimately connected with the sediment load created by seaward growth of distributaries (Morgan et al. 1968).

4. MODELS FOR DEFORMATION

Structural development of the Dirt Hills and Cactus Hills and the Mississippi Delta mudlumps is remarkably similar in style and size. These two situations share several traits in common; the only significant dif-

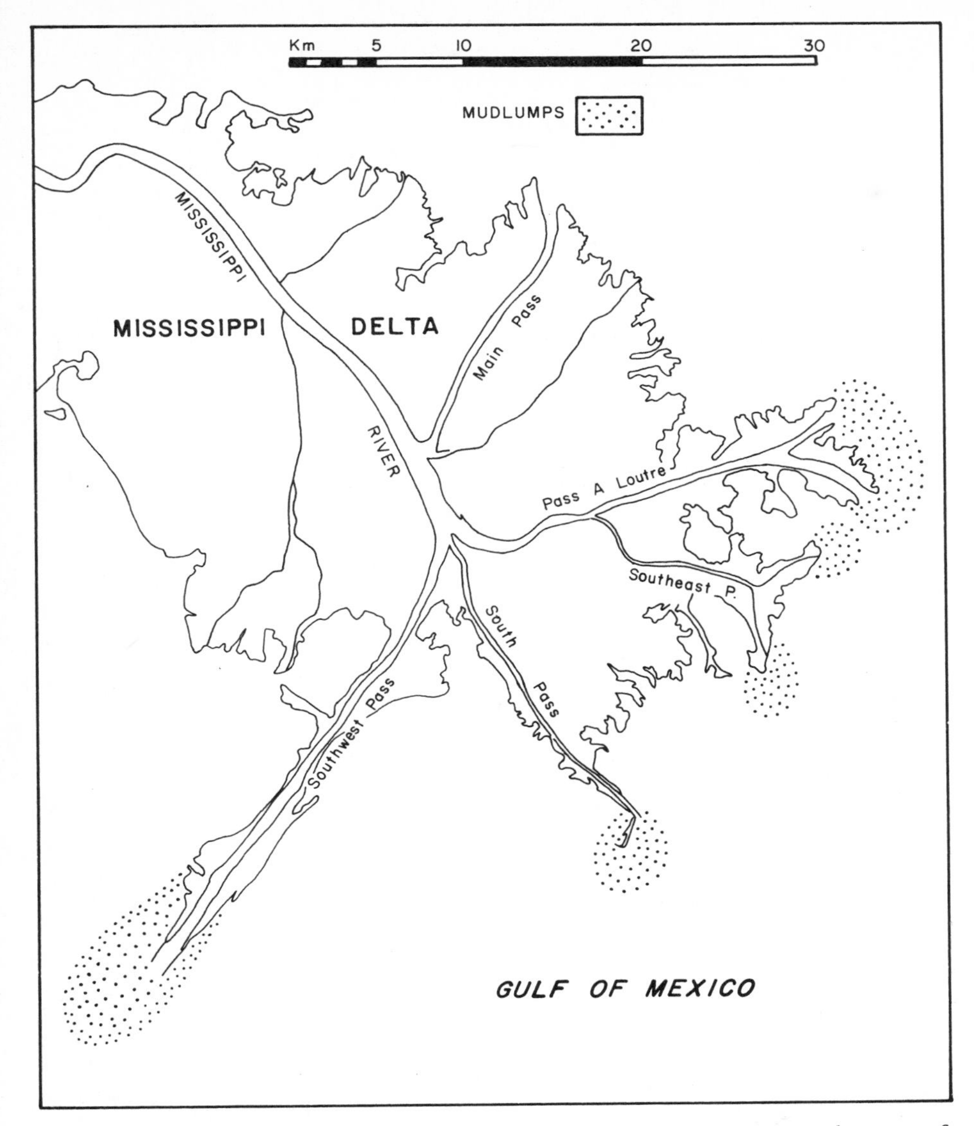

Fig. 3. Map of Mississippi Delta showing larger distributaries and zones of mudlump development (dotted). Based on Morgan et al. (1968, Fig. 1).

ference is the nature of the advancing load --glacier ice or distributary sand.

Hubbert and Rubey (1959) explained the mechanism for this style of deformation in their classic analysis of overthrusting. As hydrostatic pressure approaches or equals lithostatic pressure, shear strength of the sediment is greatly reduced and becomes effectively zero. The necessary fluid pressure may be developed when porous, impermeable, mechanically weak sediment is subjected to increased loading. This raises fluid pressure within the compacted sediment, so the overburden essentially floats on a high-pressure fluid cushion.

In the Mississippi Delta, clay units above the algal bed form the zone of failure (Fig. 5). Loading, compaction, and over-pressuring of prodelta clay units occur as thick sand accumulates at the delta front. Clay begins to flow outward and upward toward the zone of lower pressure. Asymmetric diapirs and

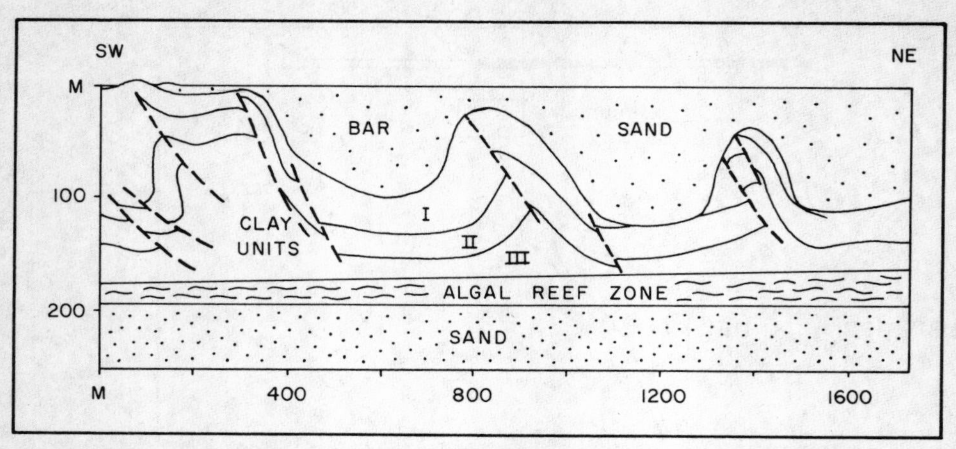

Fig. 4. Subsurface section located southwest of South Pass distributary. Thrusts and overturned folds developed in prodelta clay strata are present beneath mudlumps. Vertical exaggeration = 2.5 X; adapted from Morgan et al. (1968, Fig. 9).

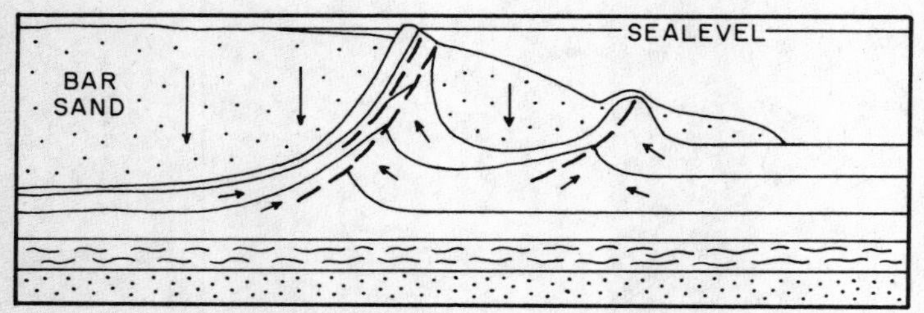

Fig. 5. Schematic model for structural development of mudlumps during advance of delta sand from left to right. Short arrows show direction of clay flowage; long arrows indicate differential loading by sand. Symbols same as Fig. 4; adapted from Morgan et al. (1968, Fig. 23F).

thrusts develop in the clay, and uplifted anticlines emerge at the surface. Flowage of clay into the diapir core creates a subsiding basin behind the mudlump that is filled by accumulating sand. The mudlump stablizes when the supply of clay in the basin is depleted, and continued delta growth eventually buries the old mudlump as new mudlumps begin to develop in a seaward direction.

The pressure conditions at the base of distributary sand are easily calculated, following the method of Bouwer (1978), assuming maximum sand thickness of 120 m, grain density of 2.6 g/cc, sediment porosity of 50 percent, and no overpressuring of the sand. Total pressure equals 21.6 kg/square cm; hydrostatic pressure equals 12.0 kg/square cm; intergranular pressure equals 9.6 kg/square cm. The intergranular pressure represents the minimum increased load imposed on the underlying clay. Some additional load is created by greater hydrostatic pressure due to subsidence. Therefore, the increased load on clay below syncline troughs is probably >10 kg/square cm. This load has resulted in collapse and overpressuring of the underlying clay units.

For the Dirt Hills, claystone in the lower Eastend Formation was the primary zone of failure below each major thrust block (Fig. 6). Failure also occurred along lignite beds within the Whitemud and lower Ravenscrag Formations. It is not necessary to invoke permafrost to explain ice pushing in this situation. It is necessary that ice advance and loading occurred quickly, so that groundwater could not escape from compacted formations. The most attractive mechanism is fast advance or possible surging of an ice tongue over a layer of basal water. This provides for the rapid loading

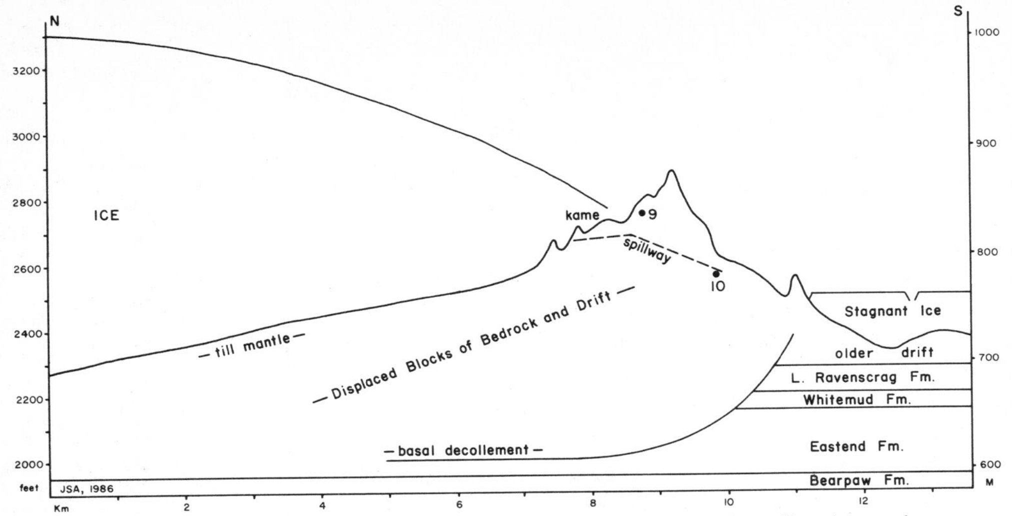

Fig. 6. Schematic model for thrusting of southern Dirt Hills. North-south profile of present land surface through high ridges west of Skyeta Lake spillway (shown by dashed line). Basal decollement is located in claystone beds of the lower Eastend Formation; at least 200 m of structural uplift has occurred at site 9. Profile of ice tongue based on minimum thickness of 300 m at northern end of section; ice may have been thicker.

and temporary trapping of groundwater to facilitate thrusting of large bedrock blocks.

The minimum pressure (in kg/square cm) developed at the base of a glacier is simply calculated: ice thickness (in m) x .09. This does not, however, account for any hydrostatic pressure from water trapped within or beneath the glacier. Assuming the ice tongue was at least 300 m thick along its axis in the Spring Valley vicinity gives a minimum increased load on underlying strata of 27 kg/square cm.

The Spring Valley ice tongue advanced into an embayment of the Missouri Coteau, so it can be assumed that ground water was locally trapped resulting in elevated hydrostatic pressure. The total load increase is, therefore, estimated to have been >30 kg/square cm, triple the load increase associated with Mississippi Delta mudlumps. This load was transmitted by relatively competent sandstone layers to underlying claystone and lignite beds, where compaction and overpressuring took place.

Shear stress at the base of a temperate (sliding or surging?) glacier is considered to be insignificant for ice pushing of this magnitude (van der Wateren 1985). Rather the horizontal pressure gradient created by differential loading seems to be of primary importance for both the Dirt Hills and the Mississippi Delta. Thus, mudlumps are squeezed out in front of the advancing delta distributaries, and bedrock blocks were squeezed out in front of advancing ice tongues. Distributaries of the Mississippi Delta are analogous to ice tongues of the Weyburn lobe in their effects on deformation of sedimentary strata.

On this basis, I conclude that permafrost is not a prerequisite for this style of deformation in either glacial or nonglacial settings. Nonetheless, permafrost is not ruled out in all situations. Permafrost almost certainly played a major role for ice thrusting in some northern areas, such as the Canadian Arctic (Rampton 1982; Klassen 1982). However, permafrost was probably not an important factor for ice thrusting of large hills along the southern margin of Laurentide glaciation.

ACKNOWLEDGEMENTS

This study began in 1984, while I was a Visiting Exchange Professor in the Department of Geography, University of Regina, Saskatchewan. Financial support was provided by a grant from the U.S. Information Agency through the Center for Great Plains Studies, Emporia State University, Kansas. Additional financial support has since been provided by Emporia State University. I have benefitted by many discussions, particularly with H.V. Andersen, M.M. Fenton, H. Ruszczynska-Szenajch, and D.J. Sauchyn. D.G. Croot read and criticized the manuscript.

REFERENCES

Aber, J.S. 1982. Model for glaciotectonism. Geological Society Denmark, Bulletin 30: 79-90.

Aber, J.S. 1985a. Ice-shoved hills of southern Saskatchewan, Canada. Kansas Academy Science, Abstracts 4:1.

Aber, J.S. 1985b. The character of glaciotectonism. Geologie en Mijnbouw 64:389-395.

Andersen, H.V. 1961. Genesis and paleontology of the Mississippi River mudlumps: Part II, Foraminifera of the mudlumps, lower Mississippi River delta. Louisiana Geological Survey, Bulletin 35.

Banham, P.H. 1975. Glacitectonic structures: a general discussion with particular reference to the contorted drift of Norfolk. In, Wright, A.E. and Moseley, F. (eds.), Ice ages: ancient and modern. Geology Journal, Special Issue 6:69-94.

Berthelsen, A. 1979. Recumbent folds and boudinage structures formed by sub-glacial shear: an example of gravity tectonics. In, van der Linden, W.J.M. (ed.), Van Bemmelen and his search for harmony. Geologie en Mijnbouw 58:253-260.

Bluemle, J.P. and Clayton, L. 1984. Large-scale glacial thrusting and related processes in North Dakota. Boreas 13:279-299.

Bouwer, H. 1978. Groundwater hydrology. McGraw-Hill Book Company, New York, 480 p.

Byers, A.R. 1959. Deformation of the Whitemud and Eastend Formations near Claybank, Saskatchewan. Transactions Royal Society Canada 53, series 3, sect. 4:1-11.

Christiansen, E.A. 1956. Glacial geology of the Moose Mountain area Saskatchewan. Saskatchewan Department Mineral Resources, Report 21.

Christiansen, E.A. 1971. Tills in southern Saskatchewan, Canada. In, Goldthwait, R.P. (ed.), Till/a symposium, p. 167-183. Ohio State University Press, Ohio.

Christiansen, E.A. 1979. The Wisconsinan deglaciation of southern Saskatchewan and adjacent areas. Canadian Journal Earth Sciences 16:913-938.

Christiansen, E.A. and Whitaker, S.H. 1976. Glacial thrusting of drift and bedrock. In, Legget, R.F. (ed.), Glacial till. Royal Society Canada, Special Publ. 12: 121-130.

Clayton, L. and Moran, S.R. 1974. A glacial process-form model. In Coates, D.R. (ed.), Glacial geomorphology, p. 89-119. SUNY-Binghamton Publ. in Geomorphology, Binghamton, New York.

Croot, D.G. 1987. Glacio-tectonic structures: a mesoscale model of thin-skinned thrust sheet? Journal Structural Geology 9:797-808.

Dyke, A.S. and Prest, V.K. 1986. Wisconsinan and Holocene retreat of the Laurentide Ice Sheet. Geological Survey Canada, Map 1702A, scale = 1:5,000,000.

Fenton, M.M., Moran, S.R., Teller, J.T. and Clayton, L. 1983. Quaternary stratigraphy and history in the southern part of the Lake Agassiz Basin. In, Teller, J.T. and Clayton, L. (eds.), Glacial Lake Agassiz. Geological Association Canada, Special Paper 26:49-74.

Fraser, F.J., McLearn, F.H., Russell, L.S., Warren, P.S. and Wickenden, R.T.D. 1935. Geology of Saskatchewan. Geological Survey Canada, Memoir 176, 137 p.

Gry, H. 1940. De istektoniske forhold i moleromraadet. Meddelelser Dansk Geologisk Forening 9:586-627.

Hubbert, M.K. and Rubey, W.W. 1959. Role of fluid pressure in mechanics of overthrust faulting. Geological Society America, Bulletin 70:115-166.

Humlum, O. 1983. Dannelsen af en dislocеret randmoraene ved en avancerende isrand, Hofdabrekkujökull, Island. Dansk Geologisk Forening, Arsskrift for 1982:11-26.

Jong, J.D. de 1952. On the structure of the pre-glacial Pleistocene of the Archemerberg (Prov. of Overijsel, Netherlands). Geologie en Mijnbouw 14:86.

Jong, J.D. de 1967. The Quaternary of the Netherlands. In, Rankama, K. (ed.), The Quaternary, vol. 2, p. 301-426. J. Wiley, New York.

Kaye, C.A. 1964. Illinoian and early Wisconsin moraines of Martha's Vineyard, Massachusetts. United States Geological Survey, Professional Paper 501-C:140-143.

Klassen, R.A. 1982. Glaciotectonic thrust plates, Bylot Island, District of Franklin. Geological Survey Canada, Current Research, part A, Paper 82-1A:369-373.

Kupsch, W.O. 1962. Ice-thrust ridges in western Canada. Journal Geology 70:582-594.

Mackay, J.R. 1959. Glacier ice-thrust features of the Yukon coast. Geographical Bulletin 13:5-21.

Mackay, J.R. and Mathews, W.H. 1964. The role of permafrost in ice-thrusting. Journal Geology 72:378-380.

Mathews, W.H. and MacKay, J.R. 1960. Deformation of soils by glacier ice and the influence of pore pressures and permafrost. Transactions Royal Society Canada 54, series 3, sect. 4:27-36.

McDonald, B.C. and Shilts, W.W. 1975. Interpretation of faults in glaciofluvial sediments. In, Jopling, A.V. and McDonald, B.C. (eds.), Glaciofluvial and glaciolacustrine sedimentation. Society Economic Paleontologists Mineralogists, Special Publ. 23:123-131.

Moran, S.R. 1971. Glaciotectonic structures in drift. In, Goldthwait, R.P. (ed.), Till/a symposium, p. 127-148. Ohio State University Press, Ohio.

Moran, S.R., Clayton, L., Hooke, R.LeB., Fenton, M.M. and Andriashek, L.D. 1980. Glacier-bed landforms of the Prairie region of North America. Journal Glaciology 25:457-476.

Morgan, J.P. 1961. Genesis and paleontology of the Mississippi River mudlumps: Part I, Mudlumps at the mouths of the Mississippi River. Louisiana Geological Survey, Bulletin 35.

Morgan, J.P., Coleman, J.M. and Gagliano, S.M. 1968. Mudlumps: diapiric structures in Mississippi Delta sediments. In, Braunstein, J. and O'Brien, G.D. (eds.), Diapirism and diapirs. American Association Petroleum Geologists, Memoir 8:145-161.

Oldale, R.N. and O'Hara, C.J. 1984. Glaciotectonic origin of the Massachusetts coastal end moraines and a fluctuating late Wisconsinan ice margin. Geological Society America, Bulletin 95:61-74.

Parizek, R.P. 1964. Geology of the Willow Bunch Lake Area (72-H) Saskatchewan. Saskatchewan Research Council, Geology Division, Report 4, 46 p.

Rampton, V.N. 1982. Quaternary geology of the Yukon Coastal Plain. Geological Survey Canada, Bulletin 317, 49 p.

Rotnicki, K. 1976. The theoretical basis for and a model of the origin of glaciotectonic deformations. Quaestiones Geographicae 3:103-139.

Ruszczynska-Szenajch, H. 1976. Glacitectonic depressions and glacial rafts in mid-eastern Poland. In, Rozycki, S.Z. (ed.), Pleistocene of Poland, p. 87-106. Studia Geologica Polonica, vol. L, Warsaw.

Rutten, M.G. 1960. Ice-pushed ridges, permafrost, and drainage. American Journal Science 258:293-297.

Teller, J.T., Moran, S.R. and Clayton, L. 1980. The Wisconsinan deglaciation of southern Saskatchewan and adjacent areas: Discussion. Canadian Journal Earth Sciences 17:539-541.

Wateren, D.F.M. van der 1985. A model of glacial tectonics, applied to the ice-pushed ridges in the central Netherlands. Geological Society Denmark, Bulletin 34:55-74.

Glaciotectonics: Forms and Processes, Croot (ed.), © 1988 Balkema, Rotterdam. ISBN 90 6191 848 0

A time-transgressive kinetostratigraphic sequence spanning 180° in a single section at Bradtville, Ontario, Canada

Katharine Albino & Aleksis Dreimanis
Department of Geology, University of Western Ontario, London, Ontario, Canada

ABSTRACT: Glacial tectonic and soft-sediment deformation structures in a new exposure of Tyrconnell Formation sediments at Bradtville, Ontario, record three local directions of glacial advances during the Late Wisconsinan. Prominent NW-dipping reverse faults and shear planes formed during an early advance of the Huron Lobe from the NW. Later inter-lobate flow of the Huron and Erie lobes produced N-dipping shear planes, various soft-sediment deformational features, associated extensional faults and boudinaged layers. Finally, during advance of the Erie lobe from SE, the Catfish Creek till was deposited on already-deformed sediments of the Tyrconnell Formation, and a few SE-dipping shear planes formed; an older NW-dipping fault zone was filled with sediments and transformed into a extension fracture at this time.

1 INTRODUCTION

Structural analyses of orientations of many small-to medium-scale glacial tectonic and spatially related soft-sediment deformation structures can provide useful information on local directions of glacial movement. Such data are especially useful from sediments which underlie former zones of interlobate ice flow, where the exact kinetostratigraphy is uncertain. This paper describes a new exposure of deformed Middle Wisconsinan sediments from an area where the Huron-Georgian Bay lobe locally advancing from the northwest, met the Erie Lobe locally advancing from the southeast, during the Nissouri Stadial at the beginning of Late Wisconsinan.

During 1985 intense erosion of the Lake Erie shoreface at "Bradtville", Ontario, uncovered a new section of deformed sediments of the Middle Wisconsinan Tyrconnell Formation, overlain by till. "Bradtville", the type locality of the Tyrconnell Formation, is an informal name for a cluster of cottages located 35km south-southwest of London, Ontario, on the north shore of Lake Erie (Fig. 1); the new exposure described here is 420m SW of the main Bradtville gully (Fig. 2). The purpose of this study was to determine the direction of ice movement and the resulting stresses that deformed these sediments, and to determine whether the

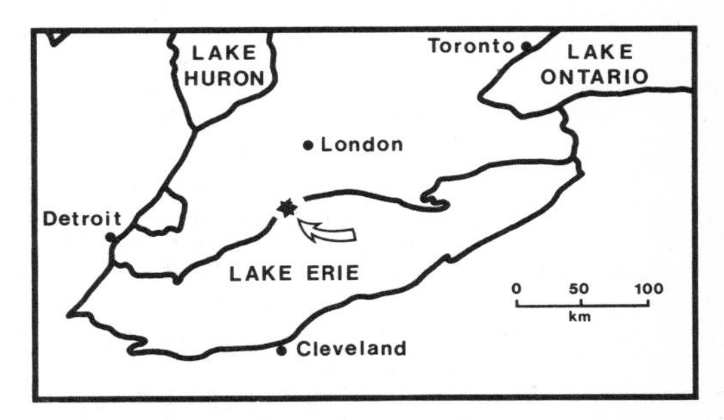

Fig. 1. Location of the study area (after Dreimanis 1985). Star shows approximate location of the section studied.

immediately overlying till is correlative with the little-exposed Late Wisconsinan Dunwich till, deposited by the Huron-Georgian Bay lobe and now considered to be the lowermost member of Catfish Creek Drift, or with the next younger Erie Lobe member of the Catfish Creek Drift.

At a nearby site, 0.5 km NE of the present one (Fig. 2), Hicock and Dreimanis (1985) documented two local directions for the initial Erie lobe ice-advance, based on striae on boulders and the orientation of clasts at the base of Catfish Creek till, and of glaciotectonic structures in both till and underlying sediments of the Tyrconnell Formation. Their data suggest an initial Late Wisconsinan Erie lobe

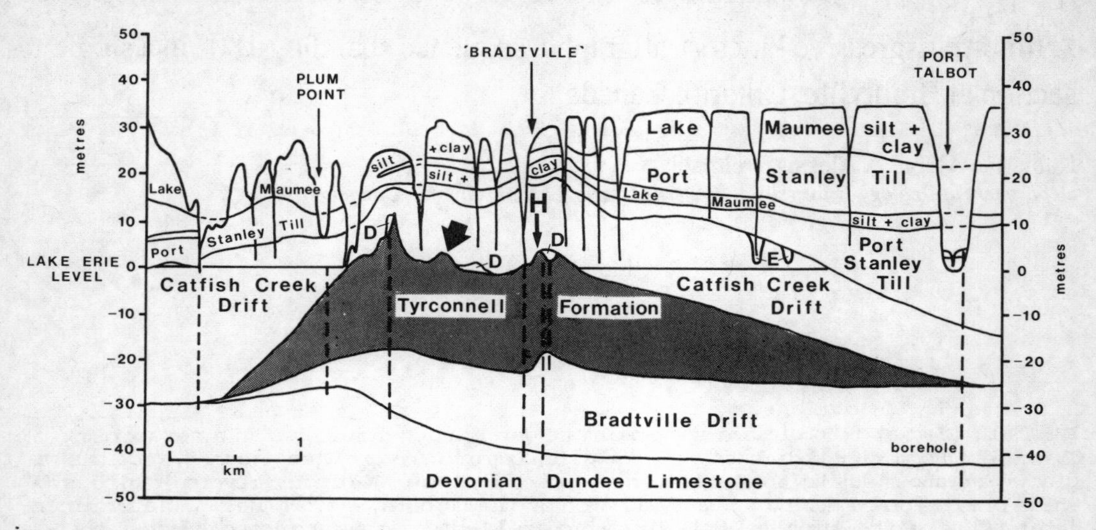

Fig. 2. Lake Erie bluffs stratigraphy in the Bradtville area (after Dreimanis and Barnett 1985). Large arrow points to section studied in this paper; H = section studied by Hicock and Dreimanis (1985); D = Dunwich drift; E = Erie Interstadial beach sediments.

advance from the NE, followed by one from the SE.

The Wisconsinan deposits of the Bradtville area have been described in many papers. The latest summaries, with references, may be found in Dreimanis (1987a and 1987b); the genetic, structural and sedimentologic complexity of the lower part of the Erie lobe Catfish Creek Drift at Bradtville is discussed by Dreimanis, Hamilton and Kelly (1987).

It should be mentioned here that the deformed Tyrconnell Formation has been encountered in several sections of the Lake Erie Bluffs between Plum Point and Bradtville (Dreimanis and Barnett, 1985; updated in Dreimanis, 1987a: Fig. 39), but most of them have been covered by slump and seldom available for detailed investigation. Our present section is in the middle of the approximately 2 km wide portion of the lake bluffs, where Tyrconnell Formation appears above lake level (Fig. 2).

2 Methods

Lithologic units and structural features of Tyrconnell sediments in the exposure were mapped on a photographic base during a two week period in the fall of 1985. Measurements were made of 87 structural elements in the Tyrconnell sediments, and the azimuth and plunge of 54 stone a-axes in the overlying till at the NE end of the section (Fig. 3:31-1, referring to grid coordinates in Fig. 3 of 31m across and 1m up). Five samples were collected for laboratory analysis: two from till on either side of the fabric site, and one each of the massive clayey silt of unit B, white silt of Member C, and rhythmically laminated clayey silt of Member D, to compare them with published data on various local stratigraphic units. Particle-size distributions were determined by the hydrometer method (Day, 1965), and carbonate content using the Chittick apparatus (Dreimanis 1962).

3 Stratigraphy

The lake bluffs expose the following two visually distinguishable units (Fig. 3): (1) a 25 m long and up to 4 m high deformed fine-textured Tyrconnell sediment body, capped by; (2) an apparently undeformed 18m thick Catfish Creek Drift layer. The bluff section is oriented SW-NE, obliquely to most structural elements, and therefore the dips as seen in Fig. 3 are all apparent dips. As a result the pattern seen in Fig. 3 is misleading, as main shear and thrust planes seem to rise southwestward in the section. The true dips and strikes of the structures had to be deciphered by digging into the section, at every place of measurement.

3.1 Tyrconnell Formation

The nature and stratigraphy of the Tyrconnell Formation sediments is well-established from drill-core and limited exposures along the Lake Erie bluffs (Dreimanis et al. 1966; Quigley and Dreimanis 1972; Dreimanis 1987a) and consists of the following four members (Dreimanis 1987a), from youngest to oldest:

D - massive to rhythmically laminated, carbonate-rich (30-45%) glaciolacustrine clayey silt; up to 25 m thick;

C - massive to finely laminated lacustrine silt, carbonate-rich and more dolomitic than B or D; contains organic remains radiocarbon dated 42-48 ka old; 2-20 m thick;

B - massive to rhythmically laminated glaciolacustrine clayey silt, similar in appearance and composition to Member D, except for its lowermost carbonate poorer part; 5-18 m thick, and;

A - accretion gley soil, noncalcareous; 0.5-3.5 m thick; encountered in test drillings only.

The upper three members, B, C and D, are exposed at the study site (Fig. 3). It is uncertain whether the laminated and massive clayey silt at the NE end of the section belongs to Member D or B, as those units are similar, according to Dreimanis et al. (1966: Glaciolacustrine units I and II). We have assigned it to Member D.

Member B consists mostly of hard, massive light gray clayey silt, which overlies and is interbedded with rhythmically laminated clayey silt. Member C consists of massive white and slightly finer-grained light gray silts which have a loose consistence. Member D consists dominantly of rhythmically laminated clayey silt, with alternating light gray and gray laminae that average 4mm in thickness. Of the three units exposed in this outcrop, Member C is the coarsest-grained (silt/clay ratio of 5.6:1) and contains the greatest percentage of total carbonate as well as the highest amount of dolomite (Table 1). Both Members B and D contain similar amounts of carbonate (Table 1), but the massive clayey silt of Member B is coarser-grained than laminated clayey silt of Member D, with silt/clay ratios of 2.8:1 and 1.2:1, respectively.

The lowermost clayey silts of Member B are exposed at the SW end of the outcrop, and are overlain by massive silts of Member C, along a contact paralleled by a zone of NW-dipping reverse faults (Fig. 3). Member C occupies the central portion of the outcrop, and is overlain by laminated clayey silts of Member D to the NE,

Table 1. Grain size distribution and carbonate content of Members of the Tyrconnell Formation

Members	B	C	D
Grain sizes			
Sand %	0	1	0
Silt %	74	84	54
Clay %	26	15	46
Carbonate content			
Dolomite %	29	38	27
Calcite %	14	10	16
Total carbonate %	43	48	43
Calcite/dolomite	0.5	0.3	0.6

along a NW-dipping sediment-filled reverse fault. Member D is overlain by till at the NE end of the outcrop, along a contact closely paralleled by a zone of N-dipping shear planes within clayey silts of Member D; the till itself is unsheared. The contact between silts of Member C and overlying till in the central portion of the outcrop is convex-up, and parallel to a contact between white and light gray silts immediately below.

3.2 Till

The till which overlies deformed sediments of the Tyrconnell Formation in this exposure (Fig. 3) is a light gray to light brownish gray diamicton, containing subrounded carbonate pebbles with subordinate igneous/metamorphic clasts; limestone pebbles are slightly more abundant than dolostone pebbles (Table 2). The till matrix is poorly-sorted silty sand, and contains 43% total carbonate with a calcite/dolomite ratio of 0.5 (Table 3).

Table 2. Lithologies of till pebbles

Till pebbles	Percentage
Limestone	39
Dolostone	30
Igneous & metamorphic	24
Sandstone	5
Chert	1
Shale & siltstone	1

These data on stone lithologies, and grain size and carbonate content of the till matrix correlate best with those for Catfish Creek till as reported by May and Dreimanis (1976) and May et. al. (1980),

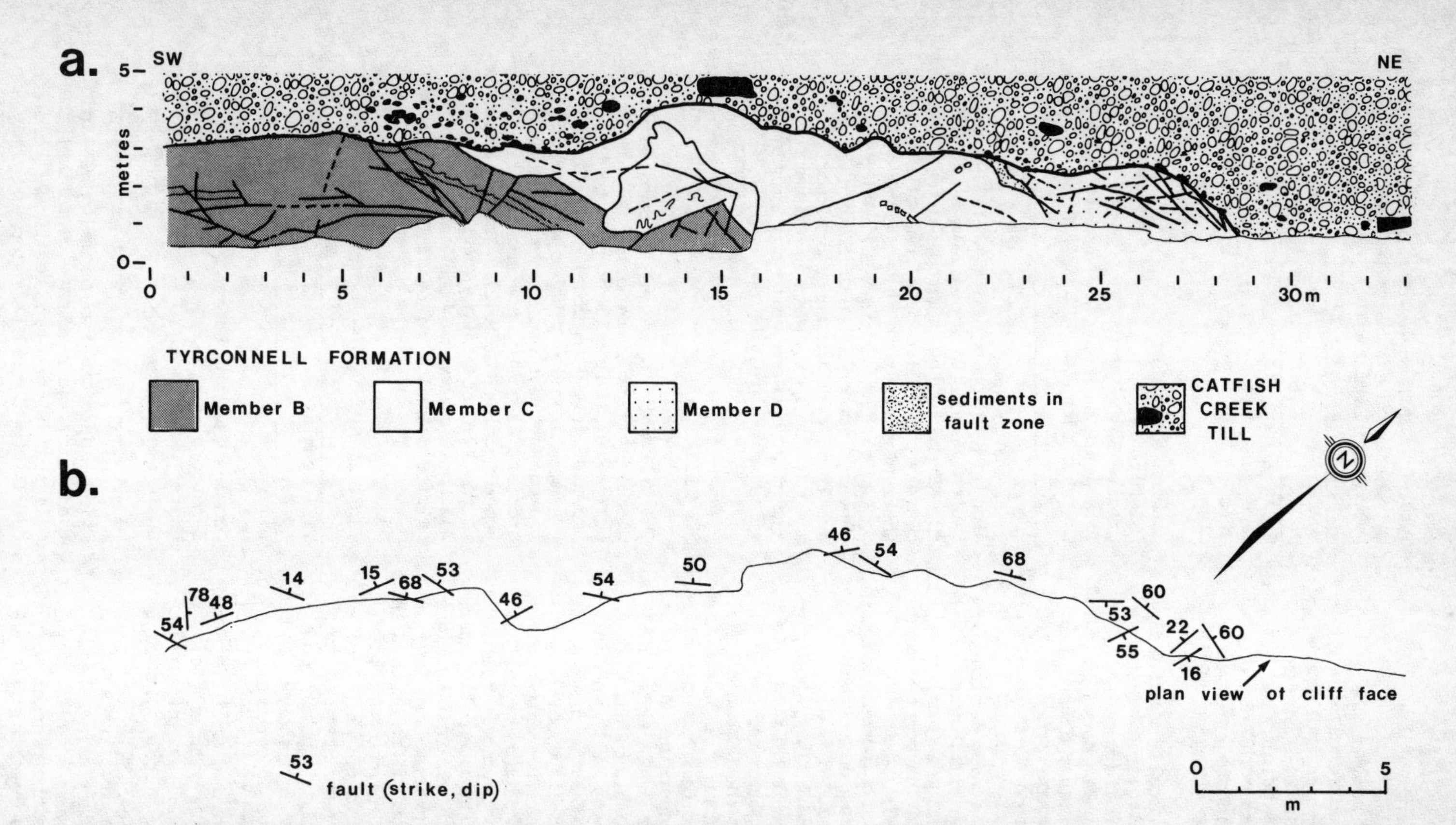

Fig. 3. (a) Sketch of cliff section. Heavy lines represent major faults and shear planes; thin lines represent bedding, convoluted bedding, ball and pillow structures and boudinaged layers (explained in text); (b) Plan view of study site, on which a selection of major faults are plotted.

Table 3. Grain size distribution and carbonate content of till matrix.

Grain sizes		Carbonate content	
Sand	50-54 %	Dolomite	28 %
Silt	34-35 %	Calcite	15 %
Clay	12-15 %	Total carbonate	43 %
		Calcite/dolomite	0.5:1

suggesting that the till was deposited by the Erie lobe from the southeast, rather than the Huron-Georgian Bay lobe from the north. The calcite/dolomite ratio of this till is lower than the expected 1:1, perhaps due to incorporation of some underlying Tyrconnell Formation sediments which have lower calcite/dolmite ratios.

The correlation of this till with Catfish Creek till deposited by the Erie lobe is supported by till fabric measurements. Stone a-axes from the basal till layer at the NE end of the outcrop exhibit a strong SE-NW alignment (Fig. 4), with axes plunging both NW and SE. A similar till fabric is reported by G. Brown (pers. comm. 1988) from stones immediately above the contact with Member D in a different location near the NE end of the outcrop.

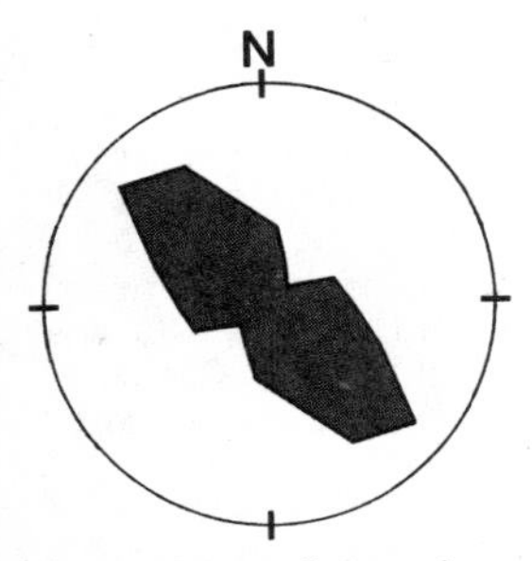

Fig. 4. Two-dimensional rose diagram of 54 measurements of till stone a-axes.

4 STRUCTURES

The new exposure at Bradtville contains both glacial tectonic and soft-sediment deformation structures. Glacial tectonic structures are produced by deformation and dislocation of substrata, due to glacier-ice movement or loading (definition modified from Aber 1985). Berthelsen (1973, 1978) defined the concepts of kineto-stratigraphy, in which orientations of structures in deformed sequences are used to determine stress directions relative to glacial movement, and serve as valuable stratigraphic indicators. Glacial tectonic case-studies are abundant in the literature; two recent examples include those by Brodzikowski and Van Loon (1985) and Åmark (1986). The nature and origin of soft-sediment deformation structures have been recently reviewed by Allen (1982).

4.1 Glacial tectonic structures formed by compression

Glacial tectonic structures are restricted to sediments of the Tyrconnell Formation: the till overlying the deformed sediments is unsheared. Compressional structures are predominant, and include curviplanar thrust faults, reverse faults, shear planes, fractures, and minor folds (Fig. 3). A minority of glacial tectonic structures formed as a result of extension and are discussed in conjunction with soft-sediment deformation structures in the next section. It was only possible to directly determine the direction of displacement for a few of the faults in the outcrop, using offset bedding planes. The term ´shear plane´ is therefore used to designate prominent faults which are occasionally lined with fault gouge, for which the sense of offset is unknown. The term ´fracture´ is used to describe minor planar structures along which neither brecciation nor displacement has occurred, which are usually parallel to major faults but may be randomly oriented. The following abbreviation will be used when discussing the orientation of faults: strike/dip, e.g. 252/53°NW means strike 252°, dip 53°NW.

Reverse faults and shears that strike NE-SW to E-W dominate among structures measured in the outcrop; the two most prominent are NW-dipping reverse faults which occur at or parallel to lithologic contacts within the Tyrconnell Formation. One is found in the NE portion of the outcrop (Fig. 3:23-2; Fig. 5:closed squares), where Member D has been thrust to the SE over Member C along a fault with an average orientation of 232/68° NW. In detail this reverse fault is a zone bounded by both NE-SW and E-W striking shear planes (azimuths ~229 and 269°), filled with lenses of well-sorted sediment that are elongated down-dip. The other prominent fault lies below and roughly parallel to the contact between Members B and C (Fig. 3:7-2; Fig. 5:open squares), and is a half-metre-wide zone of sub-parallel curviplanar reverse faults with an average orientation of 252/53° NW. Southeast-directed movement along this

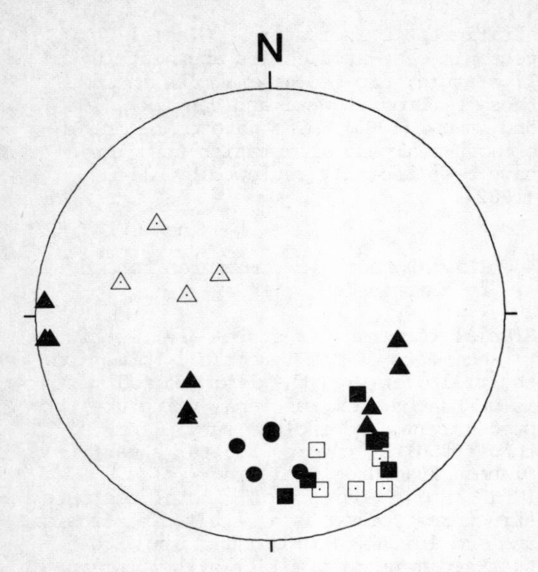

Fig. 5. Lower hemisphere equal-area stereogram of poles to a selection of representative faults. Squares = poles to sediment-filled reverse fault zone in Fig. 3:23-2; open squares = poles to reverse fault zone in Fig. 3:7-2; dots = poles to shear planes in Fig. 3:27-2; triangles = poles to syn-diapiric faults in Fig. 3:from 9-1.5 to 14-1); open triangles = poles to late SE-dipping faults.

fault is shown by offset bedding planes, and some individual fault planes are discontinuously lined with fault gouge.

Away from the two prominent NE-SW faults, the outcrop may be divided into domains, each of which is characterized by a time sequence of structures of different orientations. Deformational structures in Member D at the NE end of the outcrop (Fig. 3:27-2; Fig. 5:dots), are dominated by closely-spaced N-dipping shear planes (average orientation 268/60°N), cut by less prominent sub-horizontal faults (180/22°W, 010/16°E) which are themselves cut by ESE- to SE-dipping shear planes (016/55°ESE, 040/53°SE). In contrast, deformational structures at the SW end of the outcrop in Member B (Fig. 3:1.5-1.5) are dominated by subhorizontal shear planes (200/15°WNW) which are cut by a succession of mostly extensional faults (listed from older to youngest): 306/78°NE; 246/54°NW, normal offset; 202/48°W-NW, normal offset.

4.2 Soft-sediment and glacial tectonic structures formed as a result of extension

Soft-sediment deformation structures are restricted to the central portion of the outcrop, where they are found in silts of Member C. They include a diapir, convolute bedding and ball-and-pillow structures. The diapir (Fig. 3:13.5-3) is visible as a convex-up contact between lower light gray and upper white silts, below and parallel to the arched-up contact between white silt of Member C and overlying Catfish Creek till. Immediately below the diapir, interbedded light gray and white silts exhibit convolute bedding (Fig. 3:13.5-1.5). Ball-and-pillow structures occur NE of the diapir (Fig. 3:21.5-2.5), and consist of balls of coarser-grained white silt in a finer-grained light gray silt matrix.

A minority of the glacial tectonic structures in the outcrop were formed as a result of extension. These include small-scale normal faults and associated horst structures, as well as two boudinaged layers. One set of late normal faults, already mentioned above, offsets subhorizontal shear planes in Member B at the SW end of the outcrop. As well, a conjugate set of late, small-scale normal faults occurs in the central portion of the outcrop in Member B (Fig. 3:from 9-1.5 to 14-1; Fig. 5:closed triangles). The orientation of the principal stress (σ_1) of this conjugate set of normal faults is 180/45°S, directed from the north. The close spatial association of the diapir and these extensional faults suggests that diapirism in sediments of Member C may have been accompanied by displacement of underlying sediments of Member B, along high-angle faults. The other small-scale extensional features present in the outcrop are two boudinaged clayey silt layers (248/54°NNW and 260/60°NNW; Fig. 3:19.5-1.5 and 26.5-2), one of which is found in Member C, a unit otherwise dominated by soft-sediment deformation structures. The orientations of the boudinaged layers provide evidence for extension directed upward toward the SSE, which compliments the S-directed orientation of the principal stress responsible for syn-diapiric faulting.

5 DISCUSSION

Evidence for relative ages of faults suggests that glaciodynamic stress was initially directed from the NW as a result of the Huron Lobe advance, and

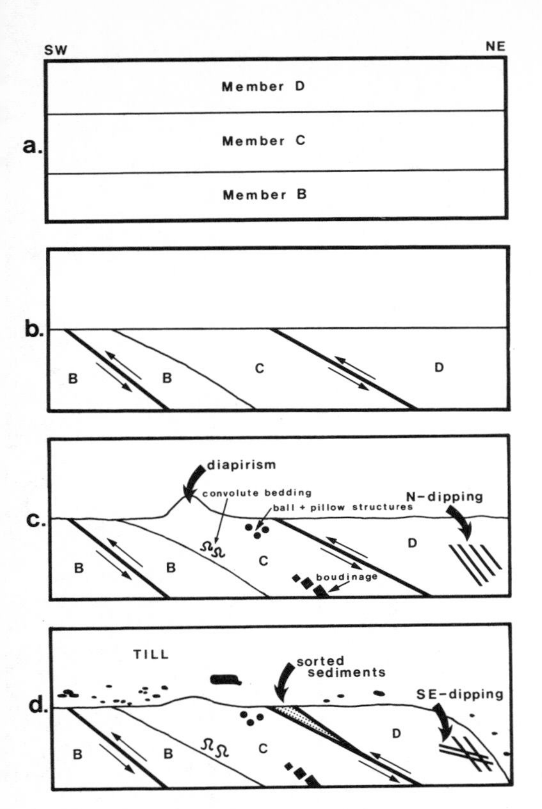

Fig. 6. Schematic diagrams showing sequence of formation of glacial tectonic and soft-sediment deformation structures in the section at Bradtville, and their cross-cutting relationships: (a) undisturbed lacustrine sediments; (b) during advance of the Huron lobe from NW-NNW, a major reverse fault between Members D and C and minor parallel shear planes form; (c) during interlobate flow of both Huron and Erie lobes from the N, N-dipping shear planes form, clayey silt layers are boudinaged, water escape structures such as the diapir, convolute bedding and ball-and-pillow structures are formed, and syn-diapiric faulting occurs, and; (d) during advance of the Erie Lobe from the SE, Catfish Creek till is deposited on top of previously-deformed sediments, a few SE-dipping shear planes form, and the major reverse fault zone may have been opened and transformed into a sediment-filled extension fracture. Note that dips of faults as shown are apparent, because the cliff section is oriented oblique to the directions of glacial advance.

subsequently from the N due to interlobate flow after the Huron and Erie lobes met, and later from the SE due to the advance of the Erie Lobe. This sequence of advances resulted in the formation of: (1) early, steep NW-dipping faults; (2) later, N-dipping shear planes and soft-sediment deformation structures, and; (3) lastly, SE-dipping shear planes (Fig. 6).

During the Huron Lobe advance from the NW, under conditions of compressive stress in the proglacial zone, detachment took place along the contact between Members D and C (Fig. 6b). Clayey silts of Member D were thrust over silts of Member C along a major NW-NNW-dipping reverse fault (Fig. 3:23-2), at the same time as minor reverse faults, shear planes and fractures formed within Members D and B, sub-parallel to the major fault. The lack of till deposited by the Huron Lobe at this site is understandable: our section is in the middle of the large deformed zone of Tyrconnell Formation, which is perhaps a topographic high created during ice-marginal thrust-faulting by Huron Lobe ice (Fig. 2).

When the Erie lobe, locally advancing from the east, reached the SE edge of the Huron lobe, an interlobate southward to southwestward-directed flow first developed along this contact, as suggested by the oldest deformation structures under Catfish Creek Drift or in its basal part at Bradtville, 0.3-0.5 km to the NE (Hicock and Dreimanis 1985; Dreimanis, Hamilton and Kelly 1987) and 1.3 km to the SW (Dreimanis 1982). The E-W striking, steeply N-dipping shear planes just below the contact with the Catfish Creek till at the NE end of the outcrop (Fig. 3:27-2) as well as the large diapir, associated convolutions, ball and pillow structures, syn-diapiric faults, and spatially related boudinage structures (Fig. 3:from 9-1.5 to 21.5-2.5), were formed during this inter-lobate phase (Fig. 6c). The coexistence of a NW-dipping boudinaged layer and nearby undeformed (i.e. post-thrusting) soft-sediment deformation structures in Member C, argues that water escape through more permeable silts was initiated during, and continued after, deformation directed toward the S.

Next, the Erie lobe advanced over the Bradtville area from the SE, and deposited Catfish Creek till upon the earlier-deformed sediments of the Tyrconnell Formation (Fig. 6d), imparting a strong SE-NW fabric to the till pebbles (Fig. 4). The major NW-dipping reverse fault (Fig. 3:23-2) may have been opened and

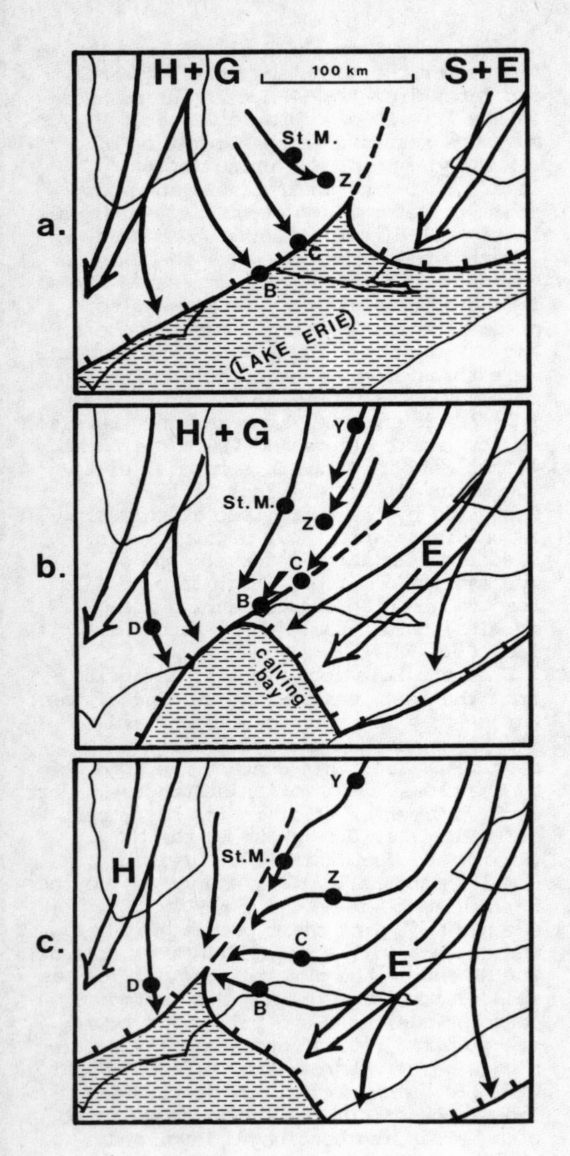

Fig. 7. Regional sequence of glacial advances during the Late Wisconsinan, based on measurements of fabrics, deformations and/or striae in a number of locations (references cited in text). B = Bradtville; C = New Sarum area of Catfish Creek valley; D = Dresden; E = Erie lobe; H = Huron lobe; H + G = Huron and Georgian Bay lobe; S + E = Lake Simcoe and Erie lobe; St.M. = St.Mary's; Y = Yatton; Z = Zorra.

In the sections at Bradtville: (a) the oldest, NW-dipping faults record the initial advance of the Huron (and Georgian Bay) lobe; (b) the N- to NE-dipping faults (see also Hicock and Dreimanis 1985), and soft-sediment deformation structures related to stress directed from the N and NE, record the interlobate flow of Huron and Erie lobe ice from the NNE; (c) the Catfish Creek till lithology, till fabric and a few SE-dipping shear planes reflect the advance of the Erie lobe from the SE.

transformed into a sediment-filled extension fracture at this time. Only a few SE-dipping shear planes (Fig. 3:25.5-2) developed in underlying sediments during this advance, perhaps because a mantle of 'wet' Catfish Creek till on top of previously-deformed sediments of the Tyrconnell Formation acted to attenuate stress, sparing the substrate much further deformation. In addition, the area where our section is located was covered by a proglacial lake at the time when the Huron and Erie lobes advanced on it (Evenson et al. 1977; Gibbard 1980), which would also have acted to attenuate stress applied to the substrate.

5.1 Regional correlations

The orientations of glaciotectonic deformations in the section investigated and their cross-cutting relationships, suggest three directions of glacial movement that followed each other in a clockwise sequence: (1) from NW; (2) from N and NE, and; (3) from E and SE. The oldest glacial advance from NW is related to the Huron-Georgian Bay lobe (Fig. 7a), as confirmed also by the lithological and geochemical composition (Mason, 1986) of Dunwich till (the lowermost member of Catfish Creek Drift) exposed at the base of the bluffs adjoining the NE end of our section of Fig. 3; also a NNW-SSE pebble fabric was determined in this till by Mason (1986) subsequent to our investigation.

The other two directions, first NNE, and later from E and SE, have been related to the Erie lobe, as proven by the Catfish Creek till lithology, pebble fabrics and orientation of structures (Dreimanis 1982; Hicock and Dreimanis 1985; Dreimanis, Hamilton and Kelly 1987). The local NE-SW glacial movement was hypothetically related to a possible calving bay (Fig. 7b) along the front of the Erie lobe by Hicock and Dreimanis (1985), but it could also be interpreted as an interlobate confluence flow of the Erie and Huron lobes (Fig. 7c:dashed line). The youngest Erie lobe movement towards W and NW is well recorded as parallel sets of striae on boulder pavements in the lower part of Catfish Creek till (Dreimanis, Hamilton and Kelly 1987). Such a direction of glacial movement in association with the lodged boulders would suggest a strong local glacial advance of a grounded Erie lobe towards NW, which was not obstructed by the Huron lobe. Thus the edge of the Huron lobe must have retreated by this time, while the Erie lobe was advancing (Fig. 7c).

If the interpretation of the three successive directions of glacial movements (Fig. 7) is correct, then it should be supported by directional evidence from other sections of Catfish Creek Drift, particularly in the area towards the NE. It is an area of thick cover of younger surface tills. Deep sections exposing the base of Catfish Creek Drift have been encountered only in the Catfish Creek valley (Dreimanis 1987), the Canada Cement Plant quarry at Zorra (Westgate and Dreimanis 1967), at Yatton (Feenstra 1975) and at St Mary´s Cement Company quarry (Winder and Dreimanis 1957). Three different sets of fabrics, deformations and/or striae have been encountered in Catfish Creek till at the Catfish Creek and Zorra sections (Fig. 7), two at St. Marys and just one at Yatton. In addition, two directions of local advances are recorded on boulder pavements at Dresden (Dreimanis, unpublished data). All of these data are in agreement with our interpretation of the sequence of events at Bradtville.

6 CONCLUSIONS

The kinetostratigraphic evidence deciphered from glacial tectonic deformations in our section, fits well into the regional sequence of glacial advances at the beginning of the Nissouri Stadial. What is more important, this exposure produces the most complete evidence for the interaction of the Huron and Erie lobes on the north shore of Lake Erie, having kept a record of at least three directions of glacial stresses in a single 4x30m section.

ACKNOWLEDGEMENTS

This study was partially supported by an NSERC Research Grant to Aleksis Dreimanis, and was presented in a poster session at the XII INQUA Congress in Ottawa, Canada in August, 1987. We are grateful to George Albino for field assistance, to Gregory Brown for advice in the laboratory, and to Stephen Hicock for help with computer plotting of structural data. Our interpretations benefited through discussions with George Albino, Brian Hart, Gregory Brown and David Croot.

REFERENCES

Aber, J.S. 1985. The character of glaciotectonism. Geol. Mijnbow 64:389-395.

Allen, J.R.L. 1982. Soft-sediment deformation structures. In J.R.L.Allen (ed.), Sedimentary structures, their character and physical basis, Volume II, p. 343-393. Amsterdam: Elsevier.

Åmark, M. 1986. Glacial tectonics and deposition of stratified drift during formation of tills beneath an active glacier - examples from Skane, southern Sweden. Boreas 15:155-171.

Berthelsen, A. 1973. Weichselian ice advance and drift successions in Denmark. Bull.Geol.Inst.Univ. Uppsala, New Series 5:21-29.

------ 1978. The methodology of kinetostratigraphy as applied to glacial geology. Bull.Geol.Soc. Denmark 27:25-38.

Brodzikowski, K. & A.J. Van Loon 1985. Inventory of deformational structures as a tool for unravelling the Quaternary geology of glaciated areas. Boreas 14:175-188.

Day, P.R. 1965. Particle fractionation and particle-size analysis. In C.A.Black (ed.), Methods of soil analysis. Agronomy, No. 9, Part 1, p. 545-567. Madison, Wisconsin: American Soc. Agron.

Dreimanis, A. 1962. Quantitative gasometric determination of calcite and dolomite by using chittick apparatus. J.Sed.Pet. 32:520-529.

------ 1982. Two origins of the stratified Catfish Creek Till at Plum Point, Ontario, Canada. Boreas 11:173-180.

------ 1985. Genetic complexity of a subaquatic till tongue at Port Talbot, Ontario, Canada. Geol.Surv. of Finland, Spec.Pap. 3:23-38.

------ 1987a. The Port Talbot Interstadial site, southwestern Ontario. Geol.Soc.Am. Centennial Field Guide-Northeastern Section, 1987, p.345-348.

------ 1987b. Day 6:London to Port Stanley area. In P.J.Barnett & R.I.Kelly (eds.), Quaternary history of southern Ontario, XIIth INQUA Congress field excursion A-11, p. 44-53. Ottawa: National Res.Counc. Canada.

Dreimanis, A. & P.J.Barnett 1985. Quaternary geology of the Port Stanley area, southern Ontario. Ont.Geol. Survey May P2827. Geol. Series-Prelim. Map, scale 1:50,000. Geology 1964-84.

Dreimanis, A. & R.P.Goldthwait 1973. Wisconsin glaciation of the Huron, Erie and Ontario Lobes. In R.F.Black et al. (eds.), Geol.Soc. America Mem. 136, p. 71-106.

Dreimanis, A., J.Hamilton & P.Kelly 1987. Complex subglacial sedimentation of Catfish Creek till at Bradtville, Ontario, Canada. In J.J.M. van der Meer (Ed.), Till and Glaciotectonics, p. 73-87. Rotterdam: Balkema.

Dreimanis, A., J.Terasmae & G.D.McKenzie 1966. The Port Talbot Interstade of the Wisconsin Glaciation. Canadian J.Earth Sci. 3:305-325.

Evenson, E.B., A. Dreimanis & J.W.Newsome 1977. Subaquatic flow tills: a new interpretation for the genesis of some laminated till deposits. Boreas 6:115-133.

Feenstra, B. 1975. Late Wisconsin stratigraphy in the northern part of the Stratford-Conestigo area, southern Ontario. Unpublished M.Sc. Thesis, University of Western Ontario, London, Canada.

Gibbard, P. 1980. The origin of stratified Catfish Creek Till by basal melting. Boreas 9:71-85.

Hicock, S.R. & A. Dreimanis 1985. Glaciotectonic structures as useful ice-movement indicators in glacial deposits: four Canadian case studies. Canadian J. Earth Sci. 22:339-346.

Mason, E.D. 1986. Genesis and stratigraphic significance of a new exposure of Dunwich Till and associated drift in the north shore of Lake Erie. Unpublished B.Sc. thesis, Univ. Western Ontario, London, Canada, 57 p.

May, R. W. & A.Dreimanis 1976. Compositional variability in tills. In R.F.Legget (ed.), Glacial till, an interdisciplinary study. Royal Soc. Canada, Spec.Pub. 12, p. 99-120.

May, R.W., A.Dreimanis & W.Stankowski 1980. Quantitative evaluation of clast fabrics within the Catfish Creek Till, Bradtville, Ontario. Canadian J.Earth Sci. 17:1064-10

Quigley, R.M. & A.Dreimanis 1972. Weathered interstadial green clay at Port Talbot, Ontario. Canadian J. Earth Sci. 9:991-1000.

Westgate, J.A. & A.Dreimanis 1967. The Pleistocene section at Zorra, southwestern Ontario. Can.J. Earth Sci. 4:1127-1143.

Winder, C.G. & A.Dreimanis 1957. Limestone quarry of St. Mary's Cement Company Ltd. The Geology of Canadian Industrial Mineral Deposits, p. 152-153.

Glaciotectonics: Forms and Processes, Croot (ed.), © 1988 Balkema, Rotterdam. ISBN 90 6191 848 0

Thin-skinned glaciotectonic structures

P.H.Banham
Department of Geology, Royal Holloway & Bedford New College, London, UK

ABSTRACT: After discussing the grounds on which structures may be argued to have a glaciotectonic origin, a survey of the literature reveals that for some time interpretations consistent with the complete model of thin-skinned tectonics have arisen from studies of glaciotectonic structures.

1 INTRODUCTION

Although its very distinctive terminology has not been widely used, a close approach to thin-skinned tectonics has for some time formed the basis for the interpretation of many glaciotectonic structures (Banham 1975 and references therein). Indeed, some of the earliest theoretical models of complete, thin-skinned tectonics, ie with a trailing-end, extensional regime characterised by listric normal faults as well as a front-end, compressional regime dominated by listric thrusts, were evolved in the glacial context (eg Weertman 1961, Boulton 1972; for a full, recent account of thin-skinned tectonics see Butler 1987).

Despite this lead in the development of a complete model, interpretations of glacio-tectonic structures in the field have over-whelmingly concentrated on the front-end, compressional regime. Until recently, structures related to a complementary, up-ice, extensional regime have been noted only in passing. True, from a kinetostrati-graphic point of view Berthelsen (1979) recognised the need for a complete model and the significance of glaciotectonically induced "hill-and-hole" topography has been realised in Poland (eg Ruszczynska-Szenajch 1976, 1985, Ber 1987) and elsewhere (eg Moran et al. 1980, Humlum 1985). However, only the most recent field-based studies have sought to integrate these compressional and extensional components (Croot 1987, Eybergen 1987, van der Wateren 1987).

The present paper therefore re-examines earlier published accounts of glaciotectonic deformation to reveal that several probable thin-skinned and especially extensional, structures have been described even if not recognised as such. First, however, it is appropriate briefly to address the broader question as to how glaciotectonic structures as a whole may be identified with some confidence.

2 THE RECOGNITION OF GLACIOTECTONIC STRUCTURES

The narrow definition that glaciotectonics is the "study of major structural features found in ice masses" (Drewry 1986) is in-appropriate even from a purely glaciological point of view, given that the deformability of sub-glacial sediments etc may exercise an important control on the mobility of ice-sheets at mid-latitudes (Boulton et al. 1985). However, as structures can be proved to be glaciotectonic in the sense of this volume (ie induced within both ice and rock - including sediments - by surface ice masses) only when seen in relation to existing or historically recorded ice-sheets, it would be wise to be cautious when claiming specific structures as glacio-tectonic. (See also Ruszczynska-Szenajch, this volume).

Unfortunately for those attempting this discrimination, glaciotectonic structures show clear geometrical similarities not only with tectonic structures proper, but also with soft-sediment structures. Factors considered to be important in the formation of glaciotectonic structures, particularly the presence of ice, water and water-saturated sediments, also facilitate the development of such structures as slumps and mud diapirs. All these structures are generally considered to have formed as a consequence of the rapid imposition of a load and/or the existence of pore-water and/or viscosity/ductility/density contrasts within layered sequences. Much of the large-scale deformation of the Pleistocene Contorted Drift of Norfolk is considered by Banham (1975 and this volume) to have been of non-glaciotectonic, soft-sediment type.

Although an existing or historically documented ice-mass is needed for a conclusive proof, several factors may be taken to indicate a likely glaciotectonic origin:

1. Upslope vergence of structures, especially where this is symmetrically displayed on both slopes of a valley or around a depression.
2. The existence of a stratigraphic hole where ice has melted out.

Other factors are thought to be less strongly indicative:

3. A time/space association with till and other glacigenic deposits.
4. Development immediately below (usually with 30 m) the contemporary land surface.
5. A regularity of orientation of structures, eg parallelism of fold hinges, thrust faults etc and a dominant sense of vergence.
6. A laterally progressive development of structural style and/or intensity of deformation.
7. Absence of evidence for a structural prime-mover apart from ice.

Even allowing for changes in ice regimes since the Pleistocene, it is of concern that most reports of "glaciotectonic" structures come from areas where contemporaneous, contiguous ice-sheets can only be deduced by arguments which sometimes perilously approach the circular. As far as possible, the cases examined here have been considered in the light of the qualifying statements given above.

3 THIN-SKINNED GLACIOTECTONIC STRUCTURES

3.1 The compressional regime

Although isoclinal folding is important in some Pleistocene cases eg the Bride Moraine, Isle of Man (Thomas 1984, Dackombe and Thomas 1985), a stack of thrust slices, arguably a duplex, is the most characteristic structural element within compressional glaciotectonic regimes (eg Babcock et al. 1978, Barefoot 1978, Höfle and Lade 1983, Maarleveld 1983, Sjorring 1983, Wilke and Ehlers 1983, Oldale and O'Hara 1984, Thomas and Summers 1984, Lundquist 1985, Boulton 1986, McCabe and Hirons 1986). In terms of thin-skinned tectonics, such thrust stacks can normally be conceived (Butler 1987) as having formed between a hangingwall above, consisting of an ice-sheet and/or glacigenic sediments, and a footwall below usually consisting of glacigenic sediments and/or bedrock. Note that the duplex develops distally by progressive thrusting of the footwall ramp. A fine example of a likely ramp and culmination is figured by Andrews (1980) from a glaciotectonised shale and limestone sequence in New York State and the "strike slip faults" of Schwan and van Loon (1981) from Funen, Denmark, may perhaps represent lateral ramps or possibly hangingwall cross-faults within the thin-skinned model.

These and similar points need not be laboured, however, for there appears to be little doubt that many of the finest developments of glaciotectonic structures can be interpreted in terms of the compressional, front-end, thin-skinned regime.

3.2 The extensional regime

Perhaps not surprisingly it is only in the best early accounts of glacial geology that descriptions can be found of possible extensional thin-skinned structures. Dreimanis (1935) had been among the first to recognise the influence of ice on bedrock and as early as 1964 MacClintock and Dreimanis reported from a till in the St. Lawrence valley "curved shear planes" several meters in length which are "concave upwards", which "resemble gouges" and which dip in the down-ice direction as determined from other evidence. That these listric surfaces are normal faults is not claimed because the homogeneous nature of the till prevented a determination of their sense of displacement.

In several other areas, however, similar shear planes and fractures can clearly be demonstrated as normal faults which commonly dip in the down-ice direction and which displace glacigenic sediments and/or bedrock by up to several meters (eg Bijn-Duval et al. 1974, Broster et al. 1979, Nielsen 1981, Kruger et al. 1984). Sometimes antithetic normal faults also seem to be present (eg Mills and Wells 1974, Hicock and Dreimanis 1985), although frequently, especially where there is evidence for more than one phase of deformation, the fault patterns described are too complicated for (re-) interpretation at a distance (eg Shotton 1965, Stone and Koteff 1979, McMillan and Brown 1983, Pedersen and Petersen 1985). Intriguingly, in the glacial sediments of Newfoundland, one of the best developed sets of listric normal faults has been interpreted by Eyles (1977) as the consequence of melting ground-ice. In view of their possible significance for regional glaciotectonic interpretations these and similar structures elsewhere merit positive vetting from an extensional, thin-skinned tectonic standpoint.

If these suggested thin-skinned tectonic (re-)interpretations from the older literature are taken together with some recent studies which demonstrate at least a large part of the complete thin-skinned tectonic model (eg Humlum 1985, Croot 1987, Eybergen 1987, van der Wateren 1987) then there is now a better basis in field studies to justify an attempt to outline a general model for the formation of glaciotectonic structures.

4 TOWARDS AN IDEALISED GLACIOTECTONIC MODEL

Although the glacial context imposes unusual constraints on the thin-skinned model, for present purposes these can be qualitatively guaged from recent advances in understanding of ice-loading stresses, ice-sheet movements and glacial hydrology (eg Boulton et al. 1985, Kamb et al. 1985, Beget 1986, Iken and Bindschandler 1986, Meeking and Johnson 1986, Shoemaker 1986). From all these considerations an idealised model would apparently need to incorporate the following simplified features:

1. A pre-existing scarp topography with a dip-slope in the down-ice direction and a scarp-slope facing the ice-sheet.
2. Ponding of water against the scarp.
3. A ground surface formed of highly porous, wet, soft sediment eg till from a slightly earlier ice advance.
4. At least a thin bed of highly porous, wet, soft sediment should be present within 30 m below the scarp ground surface.
5. If present at all, frozen ground should be shallow (under 30 m).
6. Rapid arrival of ice.
7. Temperate ice, melting to contribute further water to the sediment below the ice-sheet.

Then, it may be envisaged that the rapid arrival and ponding of an ice-sheet against the scarp will cause rapid loading of the underlying, wet sediments so that successive listric normal fault/thrust fault slices will be progressively detached down-ice from the natural ramp of the scarp.

As many individual natural examples of glaciotectonic structures depart significantly from this generalised and supposedly idealised model, there is a great need for locally-based, precise modelling. Only then may the true importance of thin-skinned tectonics be seen in the glaciotectonic context.

5 SUMMARY OF CONCLUSIONS

1. Thinking of the thin-skinned tectonic type has influenced the interpretation of glaciotectonic structures for many years even although its distinctive modern terminology has not been applied until recently.
2. More fundamentally, there should be concern that structures are not accepted as glaciotectonic without supporting argument for this conclusion.
3. Many good accounts in the literature provide a basis for a re-assessment of an idealised model for the formation of glaciotectonic structures in the light of the geometry of thin-skinned tectonics.
4. An idealised, thin-skinned tectonic model may provide useful insights, but, especially in view of the common departure of the natural from the supposedly ideal, there is a great need for locally-based, precise modelling of glaciotectonic processes.

REFERENCES

Andrews, D.E. 1980. Glacially thrust bedrock - an indication of late Wisconsin climate in West New York State. Geology 8:97-101.

Babcock, E.A., M.M.Fenton & L.D.Andriashek 1978. Shear phenomena in ice-thrust gravels, Central Alberta. Can.J.Earth Sci. 15:277-283.

Banham, P.H. 1975. Glaciotectonic structures: a general discussion with particular reference to the Contorted Drift of Norfolk. In A.E.Wright & F. Moseley (eds.), Ice ages: ancient and modern, p.69-94. Geological J.Spec. Issue no.6. Liverpool, Seel House Press.

Barefoot, M.J. 1978. Geology applied to sub-surface mining. In J.L.Knill (ed.), Industrial geology, p.65-77. Oxford, OUP.

Beget, J.E. 1986. Modelling the influence of till rheology on the flow and profile of the Lake Michigan lobe, S. Laurentide ice-sheet, USA. J.Glaciol. 32:235-241.

Ber, A. 1987. Glaciotectonic deformation of glacial landforms and deposits in the Suwalki Lakeland (NE Poland). In J.J.M. van der Meer (ed.), Tills and Glaciotectonics, p.135-143. Rotterdam, Balkema.

Berthelsen, A. 1979. Recumbent folds and boudinage structures formed by subglacial shear: an example of gravity tectonics. Geol. en Mijnbou, 58:253-260.

Bijn-Duval, B., M.Deynoux & P.Ragnon 1974. Essai d'interpretation des "fractures en graduis" observees dans les formations glaciaires precambriennes et ordoviciennes du Sahara. Revue de Geog. Physique et de Geol. dynamique XVI, 5:503-512.

Boulton, G.S. 1972. The role of thermal regime in glacial sedimentation. In R.J. Price & D.E.Sugden (eds.), Polar geomorphology. Spec.Publ.Inst.Brit.Geogr. 4:1-19.

Boulton, G.S. 1986. Push moraines and glacial contact fans in marine and terrestrial environments. Sedimentology 33:677-698.

Boulton, G.S., G.D.Smith, A.S.Jones & J. Newsome 1985. Glacial geology and glaciology of the last mid-latitude ice-sheets. J.Geol.Soc. Lond. 142:447-474.

Broster, B.E., A.Dreimanis & J.C.White 1979. A sequence of glacial deformation, erosion and deposition at the ice-rock interface during the last glaciation: Cranbrook, B.C., Canada. J.Glaciol. 23:283-295.

Butler, R.W.H. 1987. Thrust sequences. J. Geol.Soc.Lond. 144:619-634.

Croot, D.G. 1987. Glacio-tectonic structures: a meso-scale model of thin-skinned thrust sheets? J.Struct.Geol. 9: 797-808.

Dackombe, R.V. & G.S.P.Thomas 1985. Field Guide to the Quaternary of the Isle of Man. QRA 1985.

Eybergen, F.A. Glacier snout dynamics and contemporary push moraine formation at the Turtmannglacier, Wallis, Switzerland. In J.J.van der Meer (ed.), Tills and Glaciotectonics, p.217-231. Rotterdam: Balkema.

Eyles, N. 1977. Late Winconsinan glaciotectonic structures and evidence of post-glacial permafrost in north-central Newfoundland. Can.J.Earth Sci. 14:2797-2806.

Hicock, S.R. & A.Dreimanis 1985. Glaciotectonic structures as useful ice-movement indicators in glacial deposits: four Canadian case studies. Can.J.Earth Sci. 22:339-346.

Humlum, O. 1985. Genesis of imbricate push moraine, Hofdabrekknjokull, Iceland. J. Geol. 93:185-195.

Hofle, H.C. & U.Lade 1983. The stratigraphic position of the Lamstedter Moraine within the Younger Drenthe substage (Middle Saalian). In J.Ehlers (ed.), Glacial deposits of North-West Europe, p. 343-346. Rotterdam, Balkema.

Iken, A. & R.A.Bindschadler 1986. Combined measurements of subglacial water pressure and surface velocity of Findelengletscher, Switzerland: conclusions about drainage system and sliding mechanism. J.Glaciol. 32:101-119.

Kamb, B., C.F.Raymond, W.D.Harrison, H. Engelhardt, K.A.Echelmeyer, N.Humphrey, N.M.Brugman & T.Pfeffer 1985. Glacial surge mechanism: 1982-83 surge of Variegated glacier, Alaska. Science 227: 469-479.

Kruger, J. & H.H.Thomsen 1984. Morphology, stratigraphy and genesis of small drumlins in front of the glacier Myrdalsjokull, S. Iceland. J.Glaciol. 30:94-105.

Lundquist, J. 1985. Glaciotectonic and till or tillite genesis: examples from Pleistocene glacial drift in central Sweden. Palaeogeog., Palaeoclim. & Palaeoecol. 51: 389-395.

Maarleveld, G.C. 1983. Icepushed ridges in the central Netherlands. In J.Ehlers (ed.), Glacial deposits of North-West Europe, p.393-397. Rotterdam, Balkema.

MacClintock, P. & A.Dreimanis 1964. Reorientation of till fabric by over-riding glacier in the St. Lawrence valley. Amer. J.Sci. 262:133-142.

McCabe, A.M. & K.R.Hirons 1986. South-East Ulster Field Guide. QRA & IAQS.

McMillan, A.A. & A.E.Brown 1983. Glaciotectonic structures at Belshill, east of Glasgow. Quatern. Newsletter 40:1-6.

Meeking, R.M. & R.E.Johnson 1986. On the mechanics of surging glaciers. J.Glaciol. 32:120-132.

Mills, H.C. & P.D.Wells 1974. Ice-shove deformation and glacial stratigraphy of Port Washington, Long Island, N.Y. Bull. Geol.Soc.Amer. 85:357-364.

Moran, S.R., L.Clayton, R.leB.Hooke, M.M. Fenton & L.D.Andriashek 1980. Glacier-bed landforms of the prairie region of north America. J.Glaciol. 25:457-476.

Nielsen, P.E. 1981. Till fabric re-orientation by subglacial shear. Geol For.i Stockholm Forhand. 103:383-388.

Oldale, R.N. & C.J.O'Hara 1984. Glaciotectonic origin of the Massachusetts coastal end moraines and a fluctuating late Wisconsinan ice margin. Bull.Geol. Soc.Amer. 95:61-74.

Pedersen, S.A.S. & K.S.Petersen 1985. Sandkiler i moler pa Fur. Dansk Geol. Unders.Intern. rapport 32.

Ruszczynska-Szenajch, H. 1976. Glaciotectonic depressions and glacial rafts in mid-eastern Poland. Studia Geol.Pol. 50: 1-106.

Ruszczynska-Szenajch, H. 1985. Origin and age of the large-scale glaciotectonic structures in central and eastern Poland. Ann.Soc.Geol.Pol. 55:307-332.

Schwan, J. & A.J.van Loon 1981. Structure and genesis of a buried ice-pushed zone near Rold (Funen, Denmark). Geol. en Mijnbouw 60:385-394.

Shoemaker, E.M. 1986. Subglacial hydrology for an ice-sheet resting on a deformable

aquifer. J.Glaciol. 32:20-31.
Shotton, F.W. 1965. Normal faulting in British Pleistocene deposits. Q.J.Geol. Soc.Lond. 121:419-434.
Sjorring, S. 1983. Ristinge Klint. In J. Ehlers (ed.), Glacial deposits of north-west Europe, p.219-226. Rotterdam, Balkema.
Stone, B.D. & C.Koteff 1979. A late Wisconsinan ice re-advance near Manchester, New Hampshire. Amer.J.Sci. 279:590-601.
Thomas, G.S.P. 1984. The origin of the glacio-dynamic structure of the Bride Moraine, Isle of Man. Boreas 13:355-364.
Thomas, G.S.P. & A.J.Summers 1984. Glacio-dynamic structures from the Blackwater Formation, County Wexford, Ireland. Boreas 13:6-12.
van der Wateren, D. 1987. Structural geology and sedimentology of the Dammer Berge push moraine, FRG. In J.J.M.van der Meer (ed.), Tills and glaciotectonics, p. 157-182. Rotterdam, Balkema.
Weertman, J. 1961. Mechanism for the formation of inner moraines found near the edge of cold ice caps and ice sheets. J. Glaciol. 3:965-978.
Wilke, H. & J.Ehlers 1983. The thrust moraine of Hamburg-Blankenese. In J. Ehlers (ed.), Glacial deposits of north-west Europe, p.331-333.

Glaciotectonics: Forms and Processes, Croot (ed.), © 1988 Balkema, Rotterdam. ISBN 90 6191 848 0

Polyphase glaciotectonic deformation in the Contorted Drift of Norfolk

P.H.Banham
Department of Geology, Royal Holloway & Bedford New College, London, UK

ABSTRACT: New observations indicate five main phases of deformation. The three advances of the Cromer (North Sea) ice produced three phases of till-related shear foliations, lineations and intrafolial and other folds (D1-D3). The first Cromer advance from the NNE also probably caused the initial emplacement of most if not all of the large chalk and sand rafts. West of Cromer these structures have been progressively re-worked into a later shear foliation (D4) which is shared by the Marly Drift above and to the west. An expansion of Lowestoft ice from the west is most likely to have been responsible for these structures. Finally, rapid deposition of gravel and sand outwash induced the formation of non-glaciotectonic diapiric domes and basins (D5).

1 INTRODUCTION

Polyphase deformation in the glacial context is well-known and is perhaps best documented from Denmark Houmark-Nielsen 1987). Within the Contorted Drift of Norfolk at least two phases of deformation have been recognised for some time (eg Reid 1882, Slater 1926, Solomon 1932, Dhonau & Dhonau 1963), and Banham (1975) put forward arguments for four main phases of deformation during the Anglian Stage. The first three phases were considered to be glaciotectonic and associated with each of the three known advances of the Cromer (North Sea) icesheet (D1-D3). The numerous chalk and sand rafts were thought to have been emplaced initially during the first advance of the Cromer ice (D1). The fourth and final phase then recognised (D4, Banham 1975) comprises large domes and basins formed not glaciotectonically, but by diapirism induced by the rapid deposition of a thick overburden of outwash.

These conclusions still appear essentially correct for the southeastern portion of the Contorted Drift which lies between Overstrand and Mundesley (Figure 1). However, there is now further structural evidence from the section between Cromer and Weybourne to the northwest which suggests a greater eastward extent of the Lowestoft ice than has been recognised in recent years. A new phase of deformation (D4 of this paper) post-dates the three Cromer till-related phases, but pre-dates the load induced diapiric domes and basins (now D5).

In general terms it is argued here that the Contorted Drift is a rather complicated glaciotectonite which developed in a zone of interaction between the Cromer ice to the east and the Lowestoft ice to the west. This paper sets out the structural evidence in particular, but the lithological data is summarised as appropriate.

2 THE STRUCTURAL EVIDENCE

The structural evidence is conveniently considered in three contiguous, largely coastal sections. Two sections outside the Contorted Drift provide the background (viz Corton-Mundesley and Weybourne-Fakenham) while the intervening section between Mundesley and Weybourne displays the Contorted Drift itself (Figure 1). As several recent accounts are readily available, the first two sections are dealt with quite briefly (see especially Banham 1975,1977, Shotton et al.1977, Perrin et al.1979, Boulton et al.1984).

2.1 Corton-Mundesley, the Cromer and Lowestoft Tills

Recent geological mapping in the Corton area (Hopson & Bridge 1987) has confirmed that the eastern margin of the clay- and chalk-rich Lowestoft Till overlies three sandy Cromer Tills which can be traced north into those recognised by Reid (1882) and

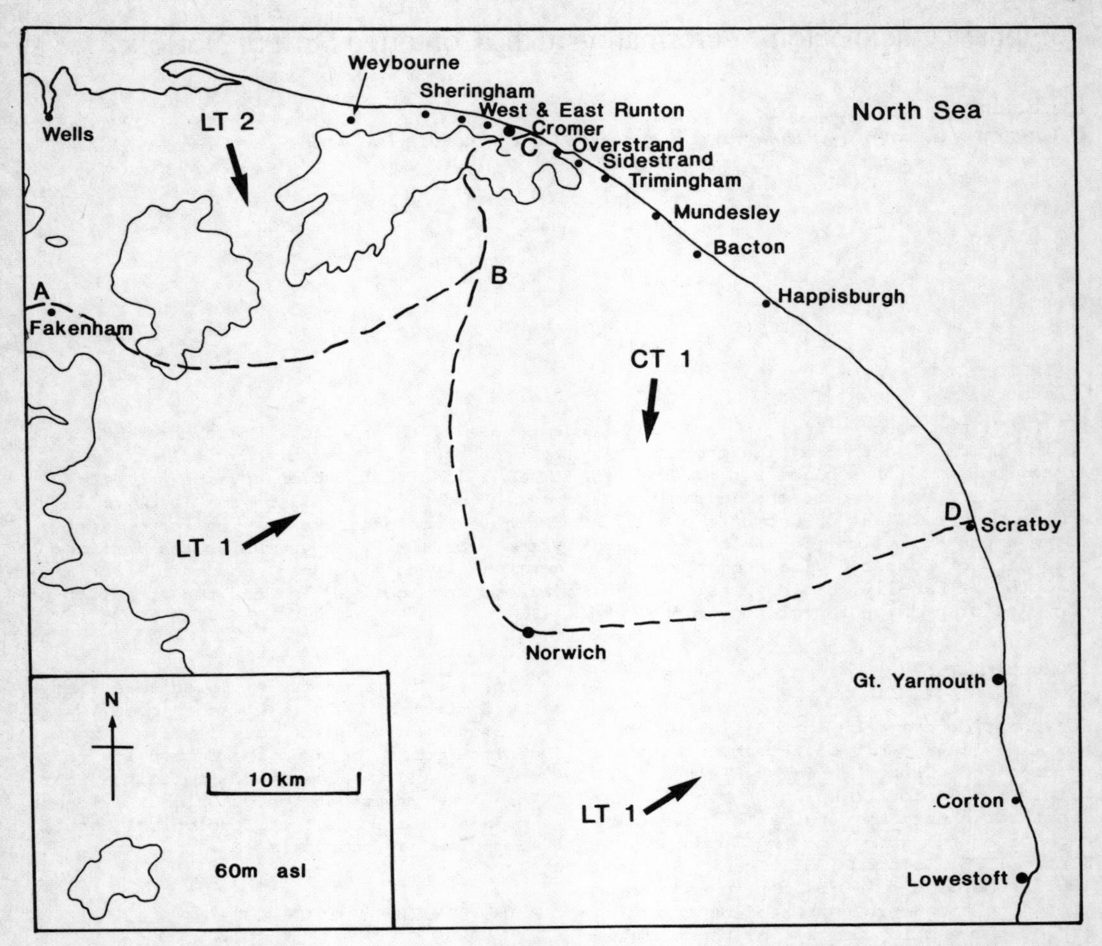

Fig. 1 Locality map; A-B-C - approx. southeastern limit of the Marly Drift; C-B-D - approx eastern limit of the Lowestoft Till and ice and approx. western limit of the Cromer Tills at the surface (NB the Cromer Tills continue under the Lowestoft Till to the southwest); CT1 - direction of advance of the First Cromer ice; LT1 - probable direction of advance of the main (first) Lowestoft ice; LT2 - possible direction of later advance of the Lowestoft ice.

Banham (1975). Within all three Cromer Tills the structures comprise shear foliations with intrafolial folds and pebble lineations. Small thrust faults and upright, asymmetric to overturned folds also affect both the tills and the sediments immediately below each till. Consideration of all the structural evidence led Banham (1975) to the conclusion that whereas the First Cromer Till had been emplaced from the NNE, the Second and Third Cromer Tills had been emplaced from the WNW and W respectively. This is surprising since all three of these tills have a similar, sandy, "North Sea" character. As the next, Lowestoft ice apparently advanced from the west (see below) it is possible that it deflected the two later Cromer advances from an original direction approximately from the NE.

For other reasons Cox & Nickless (1972) and Perrin et al. (1979) also argue for an abutment of the Cromer and Lowestoft ice-sheets in the Norwich and north Norfolk areas respectively. The deep (15m), post-Cromer Tills, pre-Lowestoft Till erosion in the Corton area stressed by Hopson & Bridge (1987) could perhaps have occurred quite quickly in the glacial context in soft sediments and not have required the "considerable period of erosion" they deduce.

2.2 Weybourne-Fakenham, the Marly Drift (= Lowestoft Till)

This section is well-exposed only at its eastern end near Weybourne. From there it passes inland (Figure 1) across the wide outcrop of the Marly Drift, a complex of very chalky tills and sand and gravel outwash. According to Banham et al.(1975) and Perrin et al.(1979) the Marly Drift is lithologically of Lowestoft type in the west, where it passes laterally into Lowestoft Till of normal type, and of Cromer type in the east where it has a transitional, but generally overlying relationship with the Cromer Tills-based Contorted Drift farther east. Banham(1977) argued that this compositional range within the Marly Drift reflected variability in the substrate during its formation and that the whole of the Marly Drift should be considered as a chalky facies of the Lowestoft Till. Perrin et al. (1979) and Ehlers et al. (1987) broadly concur, although the former prefer to retain the possibility of an influence by Cromer ice in the east (ie the abutment referred to above).

Measurements of pebble lineations (West & Donner 1956)and fold hinges (Banham & Ranson 1965), although few in number were consistent in indicating movement of the Lowestoft ice from the SW as it deformed and deposited the Marly Drift. However, arguments involving mineral provenance (Perrin et al.1979) and geomorphology (Straw 1979) have suggested ice movement from a more northerly direction. A regional re-examination of the structural evidence in particular (Ehlers et al.1987) has led to the conclusion that the Marly Drift consists of a lower unit transported from the SW and an upper unit derived from the NNW. Whatever it is clearly agreed that the Marly Drift lies above and to the west of the Contorted Drift.

2.3 Mundesley-Weybourne, the Contorted Drift

As generally described the Contorted Drift is a glaciotectonite defined by the development of structures (eg Reid 1882, Banham 1968). Although Boulton et al.(1984: p 111) imply that this structural unit also has an individual lithological character, other workers are agreed that the Contorted Drift consists of the three Cromer Tills with their intervening waterlaid beds, plus the Cromer Forest Bed Series and Chalk below all now variably structurally re-worked. North of Mundesley it is possible to trace the layer-cake sequence of Cromer Tills etc northwards into the Contorted Drift. Reid (1882) recognised the Cromer Tills as far north as Sidestrand and to the northwest of Overstrand Banham(1968) described a relatively undeformed partial sequence of Cromer Tills etc below a sharp plane of decollement at the base of the Contorted Drift tectonite. Even where deformation has been so intense that a lithostratigraphical sequence can no longer readily be recognised, field measurements and laboratory studies have consistently indicated that the bulk of the Contorted Drift comprises Cromer Till-like lithologies (reviewed in Banham 1975; see also Banham et al. 1975 and Perrin et al. 1979).

Farther to the northwest, between Cromer and Weybourne, the Contorted Drift consists of a matrix of highly deformed, sandy, Cromer-type till containing lensoid rafts of chalk, sand etc up to at least 200x50x 10m in size. In every case studied, Peake & Hancock (1961) found that the faunal zones of the chalk rafts matched those of the solid chalk directly below. Deformation of the Contorted Drift appears to have occurred without either marked lateral transport (except possibly along the N-S strike of the faunal zones of the Chalk) or the addition of new material.

High in the Contorted Drift two further lithostratigraphical units occur. First, in the northwest, between Overstrand and Weybourne, lensoid erosional remnants of Marly Drift of Cromer type rest upon the lower, more sandy Contorted Drift. This is the "Chalky Boulder Clay (C7)" of Solomon (1932) and Baden-Powell (1948). Secondly, the Gimingham Sands are frequently present; they show their greatest thickness (15m+) in synclinal sand basins (Figure 2).

North of Mundesley the lateral passage between the Cromer Tills to the southeast and the Contorted Drift to the northwest is marked by a rapid increase in the amplitude of anticlinal domes and complementary synclinal basins which have no clear preferred orientation. Structures of this type are in fact present farther southeast, but there they are normally developed only in the Gimingham Sands and Third Cromer Till at the top of the sequence. From Mundesley nothwestward these structures progressively increase in amplitude and their influence therefore deepens into lower beds. Southeast of Cromer, for example, the domes and basins normally bottom-out above the Cromer Forest Bed Series, but farther northwest, between Cromer and Weybourne these structures penetrate down into solid chalk. The greatest downward penetration is likely to have been facilitated by the prior detachment of thrust rafts of Cromer Forest Bed Series and Chalk (see below). As the dome and basin folds increase in amplitude,the more ductile tills

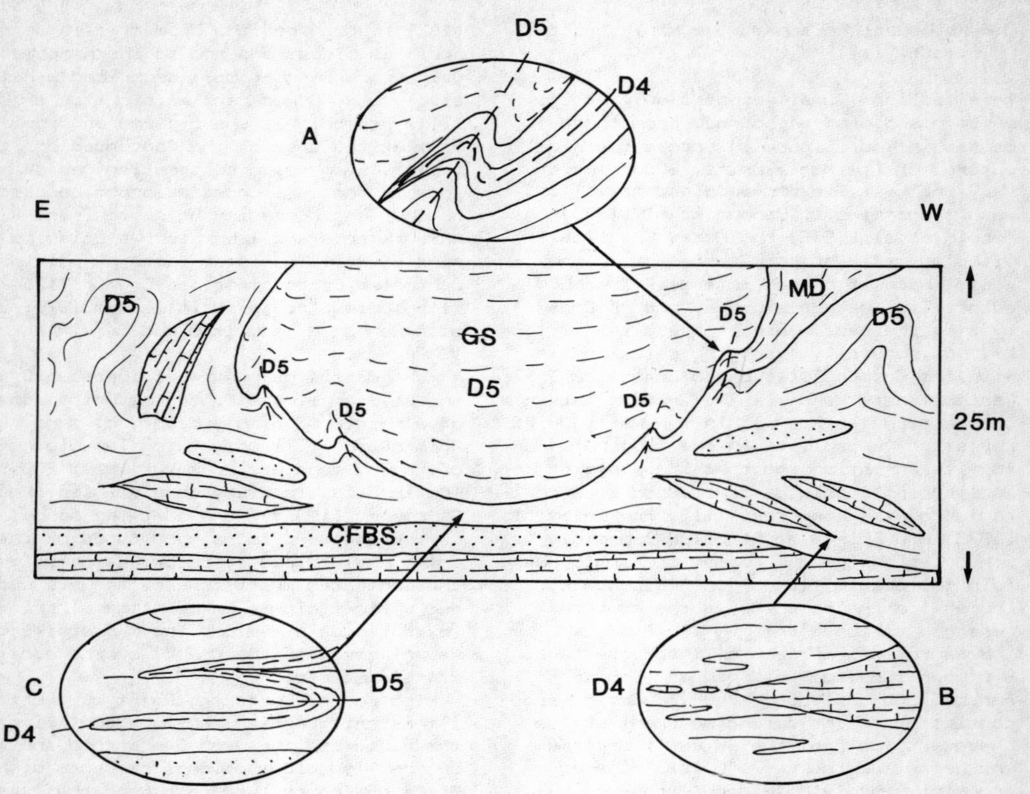

Fig. 2 Idealised, true-scale section through the Contorted Drift near West Runton. A major D5 synclinal basin ("sand basin") is shown downfolding the Gimingham Sands (GS) and Marly Drift (MD, = Lowestoft Till). The D5 minor folds (inset A) were the consequence of upward diapiric flow of the Cromer Tills etc into the adjacent, complementary anticlinal domes. Pre-existing (D1+?) thrust rafts of chalk and Cromer Forest Bed Series (CFBS) were also rotated to inversion during D5. Marginal shearing (D4 +?) of the chalk rafts in particular caused the development of chalky laminae and associated intrafolial folds (inset B) which were re-folded by D5 intrafolial folds in domains of extreme attenuation eg at the crests of major D5 synclinal basins (inset C).

etc in particular are folded into near isoclinal, meso- and minor-folds which are congruent with the major, often diapired domes and basins (Figure 2).

In places detached rafts of chalk dominate the Contorted Drift. At Marl Point north of Mundesley Peake & Hancock (1961) found that several, large thrust and folded rafts trend consistently ESE and are overturned to the SSW. As these rafts are apparently little-affected by the dome and basin folds, are still partially enveloped in First Cromer Till and, in common with these rafts in general, exhibit the same faunal zone as that in the solid chalk directly below, it may be concluded that they were detached during the first advance of the Cromer ice and transported only a short distance from the NNE. Rafts of a similar type and trend are also found at Sidestrand. In the simplest interpretation, the large rafts which dominate the Contorted Drift farther northwest, especially between East and West Runton, would also be considered to have been detached and initially emplaced during an (the first ?) advance of the Cromer ice. This cannot be convincingly demonstrated however, because since emplacement they have suffered intense shearing as indicated by their fragmented, granulated and streaked-out condition. They have also been rotated and even inverted during the dome and basin folding (Figure 2).

Apart from the initial emplacement of the rafts, Banham (1975) attributed all structures in the Contorted Drift to various stages in the development of the dome and basin folds which were regarded as a conse-

Table 1. A summary of the geographical distribution of deformation phases.

SE → NW

Deformation phase (D)	Corton - Mundesley	Mundesley - Overstrand	Overstrand - Weybourne	Weybourne - Fakenham
D5	weak	strong	strongest	weak (?)
D4	not found	weak	strong	strongest
D1-D3	weak	strong	strongest	strong

Table 2. Sequence of structural and other events in NE Norfolk during the Anglian; CFBS - Cromer Forest Bed Series; CTs 1-3 - First to Third Cromer Tills; Int. Beds - Intermediate Beds; Mund. Sands - Mundesley Sands; LT - Lowestoft Till; MD - Marly Drift; Gim. Sands - Gimingham Sands; BKD Grav. - Brick Kiln Dale Gravels; BL Grav. - Briton's Lane Gravels.

Deformation phase (D)	Structures	Movement direction	Sed/erosion events	Regional events
D5	diapiric dome and basin folds	vertical loading	Gim. Sands, BKL & BL Gravs.	melting of Cromer & Lowestoft ice sheets
D4	foliation, lineation, folding	from W?	LT/MD	Lowestoft ice advance
			Corton Sands	lake/sea
D3	foliation, lineation, folding	from W (finally)	CT3	Cromer ice re-advance
			Mund. Sands	lake/sea
D2	foliation, lineation, folding, (thrusting)	from NW (finally)	CT2	Cromer ice re-advance
			Int. Beds	lake/sea
D1	foliation, lineation, folding, thrusting	from NNE	CT1	Cromer ice advance
			CFBS	estuary/sea

quence of the rapid deposition of a thick sequence (45m+) of overlying outwash (ie the Gimingham Sands plus the Brick Kiln Dale and Briton's Lane Gravels). That this conclusion is an oversimplification is shown by a consideration of three further observations.

First, near West Runton especially, the overprinting of structures suggests the following sequence of events (Figure 2):

1. Deposition of the Cromer Tills and detachment of rafts of Cromer Forest Bed Series and Chalk.

2. Shearing to form chalky and sandy laminae with tight intrafolial folds which trend weakly N-S.

3. Re-folding of these intrafolial folds by the dome and basin folds:

a). on the limbs by relatively open, upright, meso- and minor-folds, and

b). on the crests by further, tight, intrafolial folds leading to the transposition of the earlier fabric (Figure 2).

Secondly, between Overstrand and Sidestrand, particularly, there are several large, ENE-trending, upright folds which fold already sheared Cromer Tills and chalk rafts and which are consistently overturned to the SSE. These folds are difficult to relate to the dome and basin folds; rather, they are consistent with a post-Cromer Tills, Lowestoft ice advance from the NNW (Ehlers et al. 1987).

Thirdly, the Marly Drift (=Lowestoft Till) can be found in the upper part of the Contorted Drift nearly as far east as Overstrand (Figure 1) where, as elsewhere, it contains streaked-out, sandy laminae with intrafolial folds.

Although much work remains to be done, these new observations now suggest the following structural sequence for the Contorted Drift:

1. Three advances of the Cromer ice, deposition of the Cromer Tills and detachment of rafts of Cromer Forest Bed Series and Chalk(D1-D3).

2. Advance of the Lowestoft ice generally from the west, deposition of the Marly Drift and deformation of the Marly Drift and the pre-existing stack of Cromer Tills etc to form a shear foliation and tight intrafolial folds in the west and more open, upright (ice-marginal ?) folds in the east (eg at Sidestrand; D4).

3. Melting of the Lowestoft(and Cromer?) ice, rapid deposition of Gimingham Sands and Briton's Lane etc Gravels as outwash to cause the formation of dome and basin load structures and associated meso- and minor-folds (D5).

3 SYNOPSIS

The distribution of the phases of deformation throughout the area is summarised in Table 1 and the sequence of structures is placed within a stratigraphical context in Table 2. Generalising, it seems reasonably clear that three early phases of deformation (D1-D3) associated with the advances of the Cromer ice from the NNE were followed by later phases associated with the advance of the Lowestoft ice generally from the west(D4) and with the rapid build-up of load-inducing outwash consequent on the melting of the icesheet(s) (D5).

REFERENCES

Baden-Powell, D.F.W. 1948. The chalky boulder clay of Norfolk and Suffolk. Geol. Mag. 85: 279-296.

Banham, P.H. 1968. A pre liminary note on the Pleistocenestratigraphy of northeast Norfolk. Proc. Geol. Assoc. 79: 507-512.

Banham, P.H. 1975. Glacitectonic structures: a general discussion with particular reference to the Contorted Drift of Norfolk. In A.E.Wright & F.Mosely (eds.), Ice Ages: Ancient and Modern, p.69-94.Geol.J.Spec. Issue No.6, Liverpool, Seel House Press.

Banham, P.H. 1977. Glacitectonites in till stratigraphy. Boreas 6: 164-174.

Banham, P.H., H.Davies & R.M.S.Perrin 1975. Short field meeting in north Norfolk. Proc. Geol. Assoc. 86: 251-258.

Banham, P.H. & C.E.Ranson 1965. Structural study of the Contorted Drift and disturbed chalk at Weybourne, north Norfolk. Geol. Mag. 102: 164-174.

Boulton,G.S., F.Cox, J.Hart & M.Thornton 1984. The glacial geology of Norfolk.Bull. Geol. Soc. Norfolk 34: 103-122.

Dhonau, T.J. & N.B.Dhonau 1963. Glacial structures on the north Norfolk coast. Proc. Geol.Assoc. 74: 433-439.

Ehlers, J. P.L.Gibbard & C.A.Whiteman 1987. Recent investigations of the Marly Drift of northwest Norfolk, England. In J.J.M. Van Der Meer (ed.) Tills and Glaciotectonics, p.39-54. Rotterdam, Balkema.

Hopson, P.M. & D.McC.Bridge 1987. Middle Pleistocene stratigraphy in the lower Waveney valley, east Anglia. Proc. Geol. Assoc. 98: 171-186.

Houmark-Nielsen, M.1987. Pleistocene stratigraphy and glacial history of the central part of Denmark. Bull. Geol. Soc. Denmark 36:(1-2) : 1-189.

Peake,N.B. & J.M.Hancock 1961. The Upper Cretaceous of Norfolk. Trans. Norfolk & Norwich Nat. Soc. 19: 293-339.

Perrin, R.M.S., J.Rose & H.Davies 1979. The distribution, variation and origins of pre-Devensian tills in eastern England. Phil. Trans. Roy. Soc. Lond. B 287: 535-570.

Reid, C. 1882. The geology of the country around Cromer. Mem. Geol. Surv. GB.

Shotton, F.W., P.H.Banham & W.W.Bishop 1977. Glacial-interglacial stratigraphy of the Quaternary in midland and eastern England. In F.W.Shotton (ed.) British Quaternary studies p.267-282. Oxford, OUP.

Slater, G. 1926. Glacial tectonics as reflected in disturbed drift deposits. Proc. Geol. Assoc. 37: 392-400.

Solomon, J.D. 1932. The glacial succession on the north Norfolk coast. Proc. Geol. Assoc. 43: 241-271.

Straw, A. 1979. The geomorphological significance of the Wolstonian glaciation in eastern England. Trans. Inst. Brit. Geogs. 4(NS): 540-549.

West, R.G. & J.J.Donner 1956. The glaciations of east Anglia and the east Midlands: a differentiation based on stone-orientation measurements of the tills. Q. J. Geol. Soc. Lond. 112: 69-91.

Glaciotectonics: Forms and Processes, Croot (ed.), © 1988 Balkema, Rotterdam. ISBN 90 6191 848 0

Morphological, structural and mechanical analysis of neoglacial ice-pushed ridges in Iceland

David G.Croot
Department of Geographical Sciences, Plymouth Polytechnic, Plymouth, Devon, UK

ABSTRACT: Morphological, structural and mechanical analysis of neoglacial push-ridges in Iceland. The stratigraphy, morphology, external and internal structures and mechanical characteristics of a series of ice-pushed ridges in eastern Iceland are described. The ridges were formed by a glacier advance into penecontemporaneous sediments in 1890. The structural elements can be divided into two groups: subglacial extension and proglacial compression. A stage-by-stage reconstruction of the evolution of the structures is modelled, and demonstrates clear relationships between the resultant structures and style of deformation, the mechanical properties of the sediments and meltwater production.

1 INTRODUCTION

Glaciotectonic structures vary in style and scale from simple single syngenetic folds and faults to very large scale cogenetic features up to several hundred kilometres long which involve massive displacement of bedrock. (For discussion see Aber, Croot and Fenton 1988, Berthelsen 1978 and Banham 1975, 1977). Classification of features and structures is at an early stage and each author approaches the topic in a slightly different manner, depending on his/her aim.

The features described in this paper, are most similar in topographic form to those described by Gripp (1929, 1975) and Kalin (1971), who used the terms "Stauchendmoraine" and "push-moraine" respectively. The same term, however, is used by Rabassa et al. (1979) to describe a feature of quite different form at the snout of a surging glacier in Argentina.

The aim of this paper is to describe the form, stratigraphy, structure and mechanical properties of a series of ice pushed ridges created in Iceland at the snout of a surging glacier (Eyjabakkajokull) in 1890. This data is then used to restore the section to its original state and subsequently model its evolution.

1.1 Characteristics of the glacier and foreland

The ridge complexes described and analysed below have been formed by the advance of Eyjabakkajokull, a north eastern outlet glacier of the Vatnajokull ice cap, Iceland, into pre-existing valley floor sediments (Fig. 1).

The thermal characteristics of ground into which a glacier advances, and in which glaciotectonic activity takes place are believed by most workers to be critical to the style of deformation (Banham 1975, Kalin 1971, Thomas 1984, Van der Wateren 1985). Unfortunately there are no climatic records available for the vicinity of the snout of Eyjabakkajokull, and extrapolation both spatially and temporally from the nearest present-day station at Hallormstadur would be imprecise. During the summer when fieldwork for this paper was carried out the top 10cm of ground in the proglacial area was unfrozen, but below that depth the ground was frozen to depths varying from 0.5m to 2.0m. In his discussion of the distribution of permafrost landforms in Iceland, Thorarinsson (1951) describes a number of such features occurring in the vicinity of Eyjabakkajokull. Although these two pieces of evidence suggest that the ground at the time of the 1890 advance may have been frozen to a depth of up to 2m, it is not possible to prove that fact. Later in this paper it is shown that the

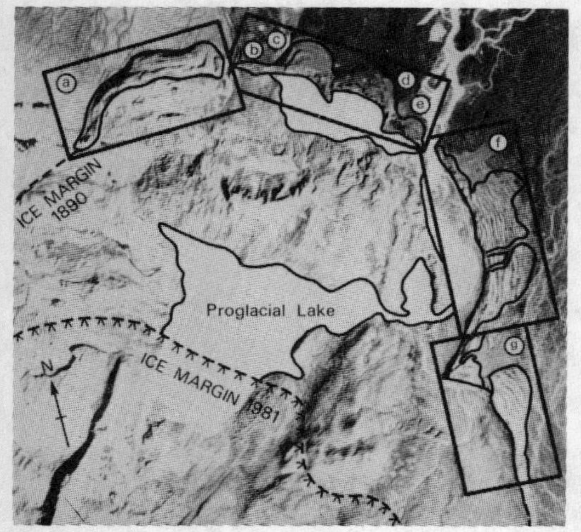

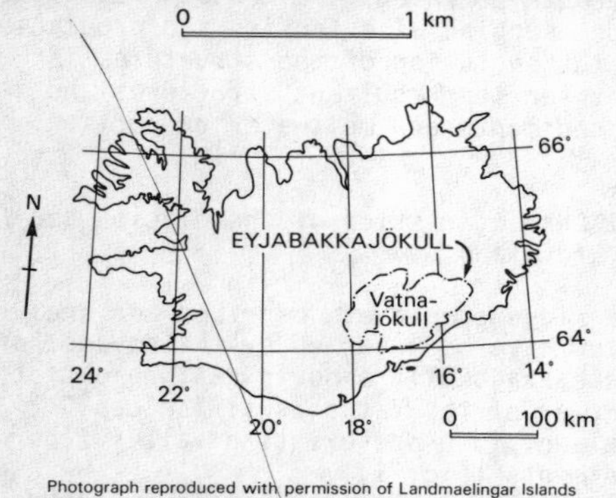

Photograph reproduced with permission of Landmaelingar Islands

Figure 1. Air photograph of part of the proglacial area of Eyjabakkajokull and location map of Iceland. Areas of push-ridges formed adjacent to the 1890 ice margin are labelled a-g. (See text for description). Photograph reproduced with permission of Landmaelingar Islands.

presence of permafrost is not an essential element in the evolution of the push-ridges described.

The current glacier foreland is divisible into two major morphological units, separated by the push-ridges under discussion (Fig. 1). The recently deglaciated area between the push-ridges and glacier margin comprises a suite of landforms and sedimentary units which are products of the stagnation of the ice following the 1890, 1931 and 1972 surge advances. Outside the limits of the 1890 advance, the valley is generally broad and flat, comprising a range of fluvioglacial stratigraphic units with an extensive, well developed herbaceous cover (Fig 2). Standing above this vegetated sandur are low (10m or less) isolated basaltic hills with little or no vegetation cover. The meltstreams from Eyjabakkajokull and isolated perennial snow patches meander and braid across the sandur and eventually converge into a single channel at the north eastern neck of the valley before plunging down a gorge into the deeply entrenched valley beyond.

2 PUSH RIDGE MORPHOLOGY

Morphological mapping and topographic surveys of the push-ridges were carried out in the field at a scale of 1:2,500 using enlargements of air photographs (Landmaelingar Islands 1967) and the results are shown in Fig 3.

Although describing a broad arc across the valley floor representing the ice limit achieved in 1890, the push-ridges can be subdivided into seven groups or complexes, each one representing a discreet unit of deformation ('a'-'g' Figs 1 and 3). These complexes are sometimes separated by meltwater channels which have exploited the topographically low areas at the junctions of adjacent groups of ridges. The western part of the valley floor is devoid of ridges, being an area of very active meltwater erosion and accretion. There may have been push-ridges in this area immediately after the 1890 surge, but any evidence of such occurrence is now missing. Elsewhere, the form of the ridges does not appear to have changed at all since their creation in 1890. Consequently morphological and structural details are excellently preserved. The single exception is ridge complex 'a' (Fig 1, Fig 3), which is adjacent to the active sandur. These ridges have been and are being eroded by meltwater activity and their original morphology and dimensions are impossible to ascertain. Despite this erosion, however, they are the bulkiest of all the ridge complexes, and bear an even closer resemblance to push-ridges described by Kalin (1971) and Gripp (1929), than those described below.

Each ridge complex is lobate or crescentic in plan form, and overall slopes downward from the proglacial margin (usually ice-contact) to the sandur plain beyond (Fig 3). However, the size and shape of each group varies across the area (compare for example ridge groups 'a', 'e' and 'g'). In all cases the former ice margin position is very clearly marked on the air photographs (change of tone) and

Figure 2. Panorama of part of push-ridge complexes c, d and e. Glacier advanced from right to left, terminating part-way up the glacier - proximal slope of the push-ridges. Continuity of vegetation cover (pre-1980) from deformed to undeformed sandur.

on the ground (by the presence of a thin layer of ablation till). The upice area of all the ridge groups is a topographic low, either occupied by a shallow lake ('c', 'd', 'e') or filled with post-deformation outwash gravels ('a', 'f' and 'g').

Although the dimensions of ridge groups 'b'-'g' vary (see Table 1), each group is similar in that it comprises individual ridge forms aligned parallel (or nearly so) to the former ice margin. These individual ridges can often be traced across the lobate complex although some rise and are pinched out again within the lateral margins. Invariably, the ridge nearest the former ice margin is the highest, and crest heights fall successively towards the glacier distal limit. Ridges comprising groups 'f' and 'g' are quite different in form. These ridges are spatially more extensive, but of low amplitude above the sandur level (Fig 4, Profile 12).

Table 1. Dimensions of the ridge complexes (See Fig. 5 for definitions)

Ridge complex	H (m)	L (m)	W (m)	θ (°)	Aspect ratio*	Volume† ($\times 10^5 m^3$)	Mass‡ ($\times 10^8$ kg)
a	40	220	360	7	5.5	15.8	26.9–37.9
b	7	50	111	9	7.1	00.19	0.3–0.4
c	15	230	330	3.5	15.3	5.69	9.7–13.6
d	8	200	305	4	25	2.44	4.15–5.9
e	9§	110	200	8	12.2	0.99	1.7–2.4
f	5§	280	527	2	56	3.68	6.3–8.8
g	5§	270	270	2	54	1.82	3.1–4.4

* Aspect ratio = L/H.
† Volume estimated to be $V = ((L/2)\, wh)\ m^3$.
‡ Density values probably range from 1700 (silt) to 2400 kg m^{-3} (gravel); min.–max. range of mass values is therefore given.
§ These values are more representative of height than values read from profiles.

Other landforms are found within the push-ridge complexes. These include deeply incised blind-head gullies or canyons orthogonal to the trend of the ridge crests, and occasional topographic flats. Interlobate areas are frequently masked by thin gravel and sand deltas.

It is possible, on the basis of morphology alone, to classify the groups of ridges into 'a' (incomplete, but bulky features); 'b' (complete, small scale, simple anticlinal features); 'c', 'd', 'e' (extensive, complete complex-morphology features); 'f' and 'g' (extensive, simple symmetrical anticlinal structures of low amplitude). More detailed analysis of ridge groups 'c', 'd' and 'e' was carried out as a section through ridge complex 'e' provided data concerning the internal structure of this type of ridge complex. This analysis included detailed mapping of hydrogeomorphic and structural features; stratigraphic and sedimentological analysis, and detailed plotting of the large section exposed at the eastern end of ridge group 'e'.

The ridges and intervening swales of ridge groups 'c', 'd' and 'e' are raised above the adjacent sandur level. Pea-sized gravel and sand in the bottom of some of the swales and in the blind-head gullies which truncate the ridges (Fig. 4)

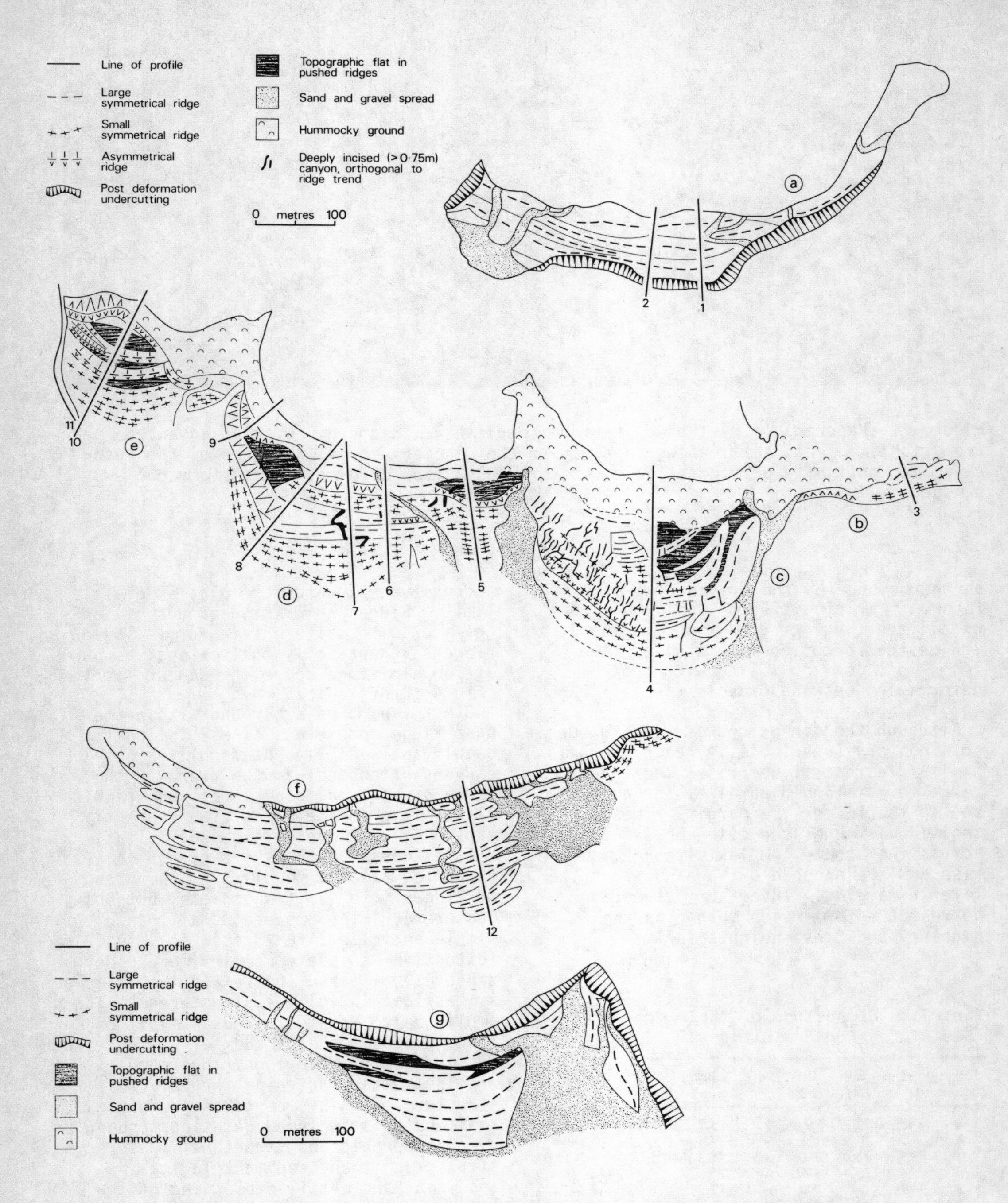

Figure 3. Morphology of push-ridge complexes a-g. (See text and Figure 1 for location).

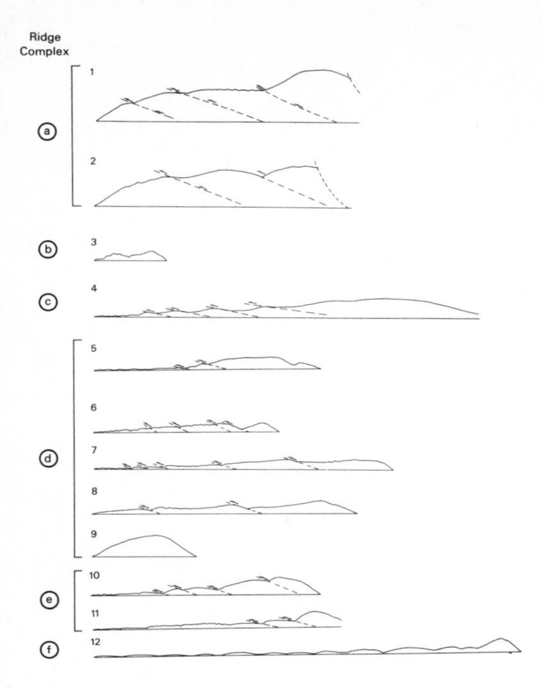

Figure 4. Profiles 1-12, surveyed across ridge complexes a-f. (See Fig 3. for location).

indicate occupation by water at some time in the past. There is no water source evident in any of these features at present and vegetation overgrowth suggests that they have not been used for some time. Occasional deeply entrenched gullies oriented normal to the ridge crests are cut through to the lake-filled depressions on the proximal side of the ridges and their heads 'hang' 2-3 metres above the general sandur and lake level. The

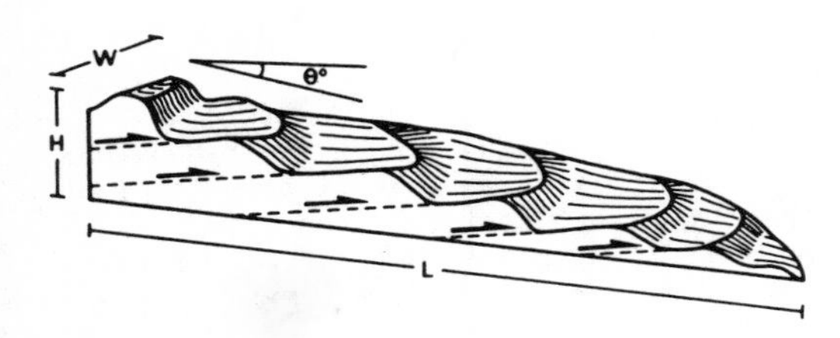

Figure 5. Idealised model of push-ridge complex, showing derivation of dimensions given in Table 1.

relationships of these gullies to other features in the area suggest that they were cut contemporaneously with the evolution of the ridges, and are products of meltwater discharging from the former ice-margin. Water appears to have sometimes discharged through the body of proximal ridges and emerged at the surface as springs mid-way through the ridge complex, or alternatively originated at the junction of the ice and push-ridges and incised downwards across the structure of the ridge complex. Wherever these channels discharge on to the sandur downvalley, the meltwater deposited a thin cover of gravel and sand, up to 10 cm in thickness, on top of the pre-existing turf layer.

3 STRATIGRAPHY

Riverbank sections along Profile 11 demonstrate the stratigraphic continuity from the undisturbed sandur into the ridges. This is supported by the continuity of the turf and near-surface stratigraphy elsewhere in the ridges and sandur. The stratigraphy of the undisturbed sandur is presented in Fig. 6A, that of the ridges immediately up-valley in Fig 6B. The continuity is most clearly seen in the tephra layers (white rhyolitic tephra and black basic tephra), since it is to be expected that the nature of sandur sedimentation would give rise to stratigraphy which is sometimes laterally discontinuous. A bed of dense blue/brown silt marks the lowermost horizon of the undisturbed section (Fig 6A). Above this are 1.2m of sands, silts and sandy silts. Tephra layers occur at 60-65cm, 68-70cm and 102-105cm below the surface. This stratigraphy is traceable continuously into the ridge complex immediately to the south, and is the basis of much of the interpretation carried out below.

Section B (Fig 6) is representative of an internally undisturbed block which has been partly thrust over an identical stratigraphy downvalley, towards section A (Fig 6). As Fig 6B shows, the majority of elements in the stratigraphy are laterally correlated. The lower part of the stratigraphy is extended beyond that visible in Section A, since this element has been tectonically uplifted. This process has exposed an additional 50cm of silt, sand and gravel below the lowermost bed exposed in Section A. As with other parts of this section, a major thrust truncates the stratigraphy through the basal gravels.

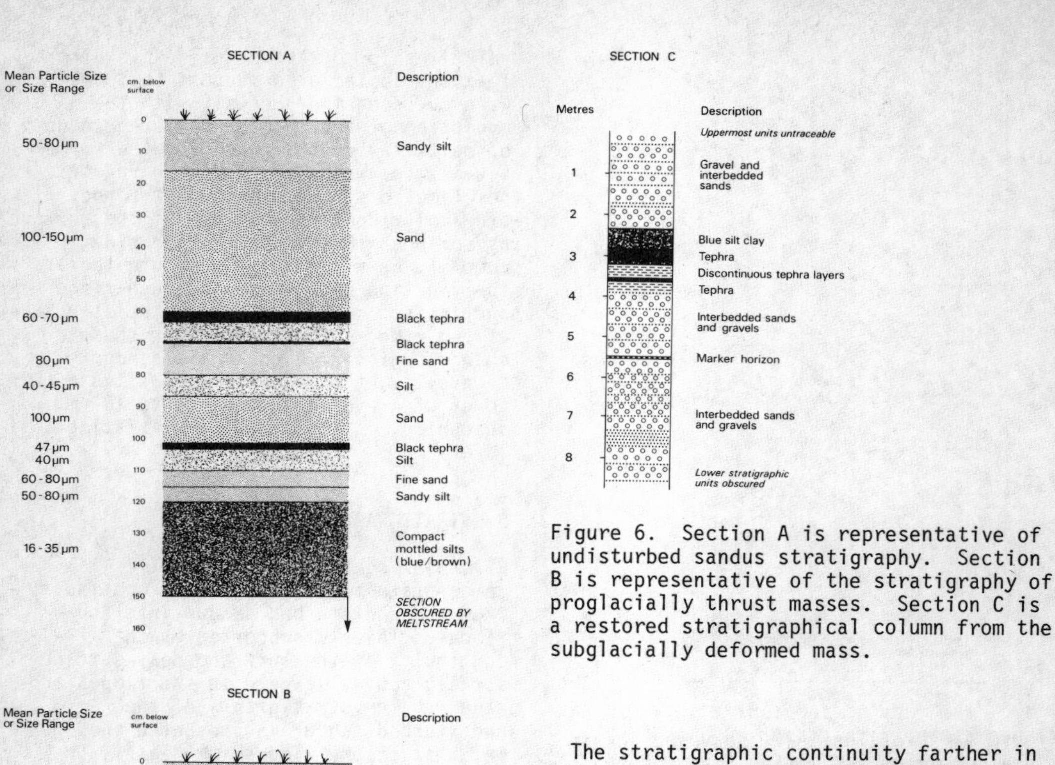

Figure 6. Section A is representative of undisturbed sandus stratigraphy. Section B is representative of the stratigraphy of proglacially thrust masses. Section C is a restored stratigraphical column from the subglacially deformed mass.

The stratigraphic continuity farther in towards and beneath the 1890 ice margin is more difficult to trace. Reconstruction of the original "type stratigraphy" by graphically removing the tectonic movement produced the palinspactic stratigraphic column in Fig 6C (no accurate scale can be produced by this method). The column is substantially different from those of 'A' and 'B' with different sediment types and apparent bed-thicknesses. The lateral hiatus is within the highly disturbed sediments which form the largest ridge at the former ice margin but the uppermost elements of the section are to be found imbricately stacked in this large ridge. If these elements were restored above those of Section C, then the stratigraphy would be very similar to that of A and B.

4 INTERNAL STRUCTURES

Incision through the proglacial ridges has revealed the internal structure of some of them. One such section exposed in the western end of ridge complex 'e' was cleaned by spade and trowel then photographed using a polaroid camera from distances of 1m to 3m. The individual polaroid prints were then used to construct a large-scale photo-mosaic on which section detail was recorded. These

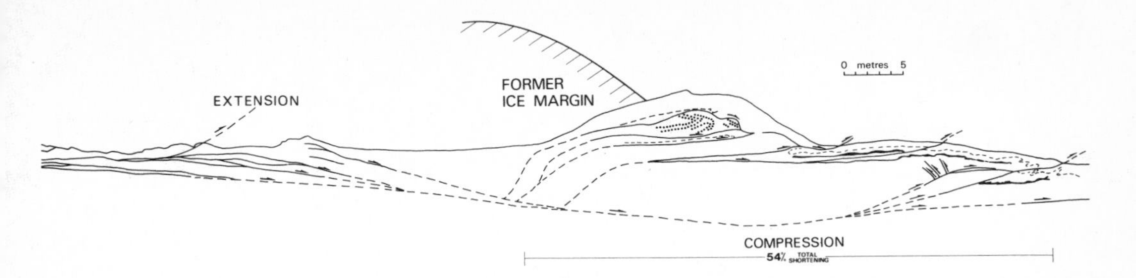

Figure 7. Principle components of the tectonic system comprising ridge complex 'e', divisible into two main zones: proglacial compression and subglacial "extension", separated by a zone of transition.

details were subsequently transferred from the polaroid prints to a reference section compiled from surveying and high resolution photographs taken using a tripod-mounted camera 60m from the face.

For ease of description the section is divided into two elements characterised by contrasting styles of deformation: firstly, the section beneath the former (1890) ice margin (subglacial, Figs 7 and 8) and secondly that portion downvalley from the former ice margin (proglacial, Figs 7 and 9).

The main part of the 'subglacial' section is characterised by a number of low-angle faults dipping in the down-ice direction. A marker horizon of tephra within the sands and gravels which comprise the majority of this section enabled the amount of displacement along each fault to be accurately measured. The downvalley dip of the faults increases in the direction of transport from 4°, to 17°. The direction of slip in this portion is invariably downdip along subhorizontal to low angle fault planes. The amount of slip varies from 0.5m to just less than 6m with no apparent systematic variation. The arrangement of the fault set implies a convergence of faults at a depth of 4-5 metres some 10 metres within the former ice margin. Although the lowest part of the section is obscured, the gravels below the lowest visible fault appear to be compacted, but undisturbed stratigraphically. It is suggested therefore that the lowermost, visible fault is the "floor" of the total movement.

In addition to this set of faults

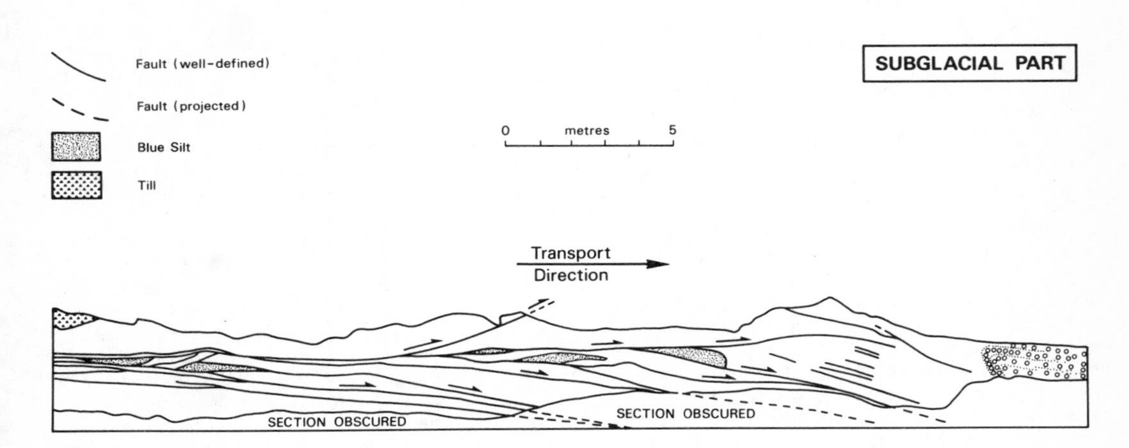

Figure 8. Structure of part of push-ridge 'c' overridden by ice in 1890. (Subglacial).

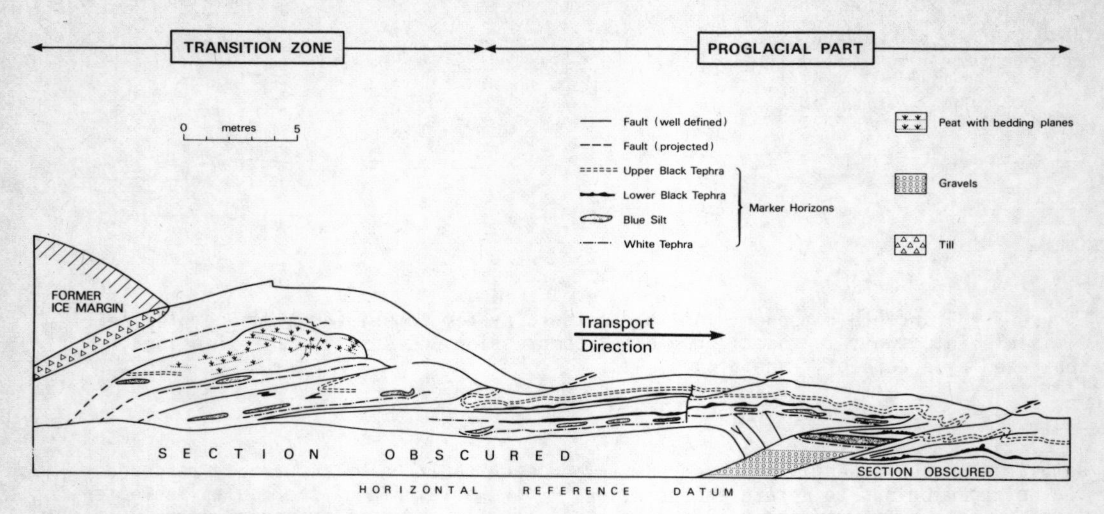

Figure 9. Stratigraphy and Structure of part of push-ridge 'c' which remained proglacial during the rapid advance (surge) of 1890.

dipping downvalley, there is an upper horizontal thrust which has carried horizontally bedded gravels and a mixture of glacial till and gravel downvalley. This overthrust unit is itself bisected by an upward curving thrust which converges with the roof thrust of the extension fault system at a shallow angle.

The downdipping fault system dies out downvalley in the upper part of the section, giving way to a fold structure the axil plane of which dips downvalley at 30-35°. The upper part of the fold has been removed, but the core remains intact.

Between these well defined 'subglacial' structures and those which developed beyond the former ice margin, is a 'transition zone' of substantial deformation which forms the largest ridge of the complex. This ridge apparently formed partly beneath the ice margin and partly in front of the glacier, but structurally the style of deformation is most similar to the to the downvalley section. The ridge comprises at least three sheets of highly disturbed brown sandy silts, compact blue silts and bedded turf layers, bounded top and bottom by well-defined thrusts. The original stratigraphy, as indicated by tephra layers, appears to have been identical to that found immediately downvalley, but the deformation within each thrust sheet has been so extreme that it is impossible to demonstrate it diagrammatically

The majority of the 'proglacial' section (Fig 9) displays quite different characteristics from the 'subglacial' section described above. The whole section is divisible into a number of imbricated thrust sheets, dipping in the up-ice direction. The majority of thrusts themselves are concave in their upper parts, and convex upwards below; a characteristic common to imbricated thrust sheets in major orogenic thrust belts (Boyer and Elliott, 1982). The thrusts are all low angle (between 3° and 20°), the majority cutting out to the surface with associated topographic deformation. The amount of displacement along three of the major thrusts can be assessed using well-defined tephra layers. The thrust which cuts out immediately downvalley from the large topographic hump has ill-defined bedding structures above it, and the amount of displacement attributable to this individual thrust is consequently difficult to ascertain. Down section from this point, however, displacement progressively decreases from 8.4 metres to 2.8 metres and finally to 2.3 metres for the major thrusts which cut the surface.

In addition to the major thrusts which reach the surface, there are a number of blind thrusts which die out before reaching the topographic surface. These are replaced higher in the stratigraphy by fold structures. This is particularly evident in the section farthest downice.

Table 2. Selected mechanical properties of representative proglacial sediments incorporated into push-ridge complex 'e'

Bed Description	Mean particle size (μm)	Porosity %	Coefficient of Permeability (k, cm sec^{-1})	Shear Strength Parameters C(kNm^{-2})	ϕ(kNm^{-2})
COMPACT SILT	37	45	5×10^{-4}	27	23-31
SILT	47	63	4.7×10^{-3}	22	23
SAND	80	54	5×10^{-2}	10	30.5
GRAVEL	10,000	23	1.72		

Within each thrust sheet, post-thrust faulting is evident which is mainly steeply dipping and normal, representing gravity settling after the major thrust movement. Minor folds, developed marginally within each sheet reflect ductile movement of the beds immediately prior to the development of the thrusts.

5 MECHANICAL ANALYSIS OF MATERIALS

Mechanical analysis was carried out in order to asses the properties of the sediments involved in the tectonic structures and their probable behaviour at the time of deformation. Clearly this assumes that the materials have not undergone changes since 1890, or during the process of deformation. Wherever possible, undisturbed samples were taken for analysis, but in most cases this was impossible (eg gravels and sands). Individual samples were taken from each identifiable stratigraphic horizon, but results from similar beds were so close that the data are summarised into broad generic classes (Table 2).

The behaviour of different stratigraphic units during deformation depends on a number of properties: particle size distribution, porosity, permeability, and shear strength being the main criteria. The results given in Table were produced using standard laboratory techniques and show that values of shear strength and permeability vary critically with each type of material in the sedimentary sequence.

Whilst the silts and sands are on the whole more porous, they have low permeability values. The consolidated (blue) silts have exceptionally low permeability. The gravels on the other hand have relatively low porosity, due to the intersitial sands, but are highly permeable. Shear strengths also vary. The unconsolidated sands have little cohesion and relatively low angles of internal friction, and consequently move quite readily by internal shearing. However, the consolidated blue silts are much stronger under direct shear conditions, and consequently would behave as more competent beds than either sands, gravels or uncompacted silts.

A number of workers have clearly shown that frozen earth materials behave in quite different ways from their unfrozen counterparts (Baker 1979, Johnston 1981, Tsytovich 1975, Williams 1982). Johnston states that "the phenomena that control the mechanical behaviour of frozen soils are complex and not fully understood" (p. 80) whilst Tsytovich (1975) argues that shear strength depends on the negative temperature of the soil, the external pressure, and the time of load action in addition to the normal parameters of cohesion and internal friction. Consequently, predicting the behaviour of the frozen sediments in this case, from their unfrozen behaviour, is difficult. However, it is possible to predict the direction in which particular properties would develop with freezing (Williams 1982). The most important properties which must be considered are permeability and shear strength. The permeability of silts and sands, given a reasonable field moisture level on freezing should decrease markedly since pore spaces used in meltwater migration will be occupied by ice. A further significant decrease in permeability could have taken place as pore water is drawn towards the freezing front possibly from a high water table in the gravels below and ice lenses further inhibit water migration (William 1982).

The permeability of frozen gravels will depend to a large extent on the water content at the time of freezing, the availability of freely moving pore water available to migrate to the freezing front, and the pore size. The permeability tests on unfrozen gravels demonstrate their ability to allow free migration of water under pressure and large pore sizes but it is generally accepted that the downward passage of a freezing from through gravels does not generate sufficient suction potential, because of the larger pore space size, to generate upward pore water migration (Williams 1982). The permeability of the gravels in the sedimentary pile would only have been reduced substantially if the water content had been high initially. Freezing of dry gravels, even with water available beneath them, would not necessarily have reduced their permeability significantly.

It would appear from the evidence available that the relative permeabilities of gravel and sands/silts would have remained constant even if the ground were frozen, unless it was totally saturated with water at the time of freezing.

The compressive strength of soils is generally increased by ice-bonding when they freeze (Johnston et al 1981, Williams 1982), although as with unfrozen soils, strength varies with grain size, interparticle contacts and forces, pore water pressure, etc. There are also complex relationships between normal load, temperature, unfrozen water content and time period of excess loading to be considered (Nickling and Bennett 1984). In the analysis considered here, the parameters controlling the temperature and state of the soils/sediments are not known with sufficient accuracy to facilitate prediction of frozen-state behaviour. It is clear from the work on the subject however that shear strength and failure strength generally increase with decreasing temperatures for all sedimentary types, and that although the relative strengths of the gravels, compact silts and normally consolidated silts may have changed this does seem unlikely. The

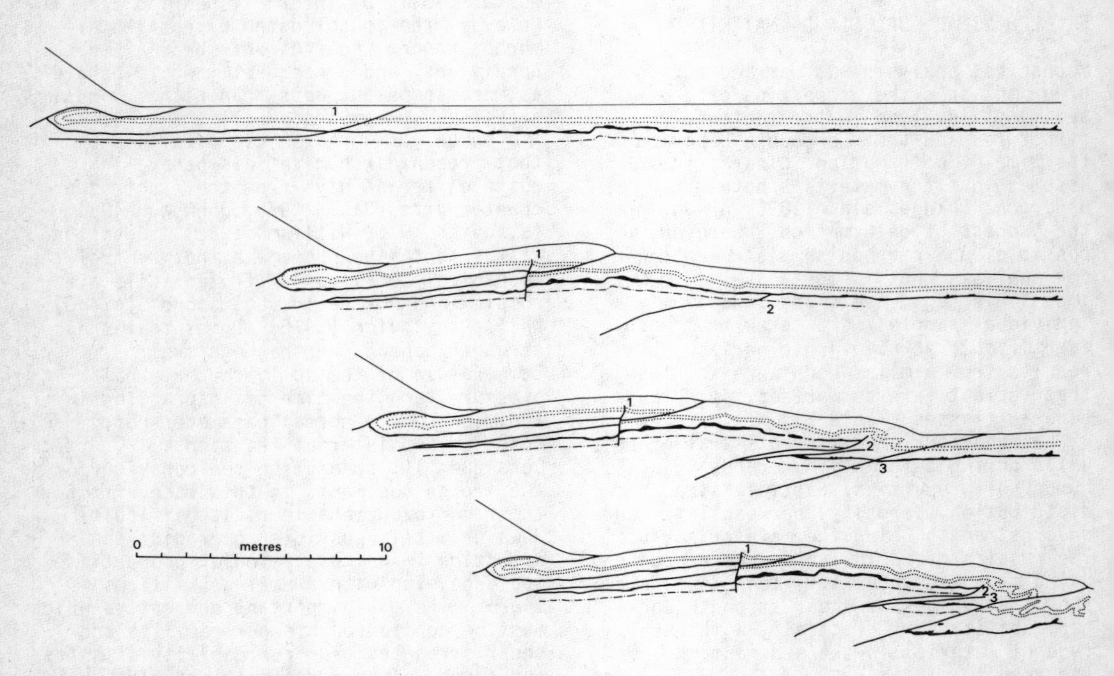

Figure 10. Development of the proglacial part of the push ridge complex 'c', along profile 11 , according to the piggy-back model.

style of deformation structure too indicates that the gravels were the incompetent strata through which faulting and thrusting most readily propagated, and that the compacted silts behaved as a relatively competent strata.

6 RESTORATION AND EVOLUTION OF STRUCTURES

The detailed measurements of the direction, and amount of faulting, thrusting and folding using at least two identifiable marker horizons facilitated the construction of a palinspastic section (Fig 10), using line balancing (Dahlstrom 1969, Hossack 1979).

Restoration of the 'proglacial section' (Figs 10, 11) is relatively straightforward, since the deformation has occurred along well-defined thrusts, without major internal deformation within the imbricate sheets, and with no loss, thickening or thinning, of horizons. Restoring the subglacial section, and the components of the large ridge which marks the transition between the two zones was more problematic. The overall product does however 'balance' according to the rules set out by Dahlstrom (1969), and reviewed by Hossack (1979). Unstraining the total section reveals that the subglacial element has been extended by 385% (following removal of the upper layers of turf, silts and sands), whilst the proglacial section has been shortened by 54% (where shortening = (change in length/restored length) x 100).

A model of the evolution of the landforms and their internal structure, based on the interrelationships of thrusts, faults and fold elements in the overall section is also illustrated in Fig. 10. Prior to deformation the stratigraphy throughout the proglacial area appears to have been very similar to that shown in Figs 3A and 3B. Restoration extends this stratigraphy as a planar surface, some 50-70m farther towards the glacier.

The mechanical attributes of the sediments and the characteristics of surging glaciers are important assumptions in the proposed model. Of particular importance are the relative and absolute permeabilities of the compact silts and underlying gravels, the rate of meltwater production, and the rate of advance of a surging glacier. Whatever the temperature of the ground at the time of the advance, the upper beds of the section comprised relatively impermeable competent strata, overlying openwork, incompetent and permeable gravels.

Field observations at Eyjabakkajokull (Thorarinsson 1939) and elsewhere confirm that surge advances are accompanied by high levels of meltwater discharge (jokulhlaups) which flood the sandur. Rates of advance of a glacier front during a surge have been measured in a variety of locations, including the advance of a neighbouring glacier, Bruarjokull (Thorarinsson 1969, Meier and Post 1969) and it is clear from these observations that the rate of meltwater production during a surge exceeds the permeability of all the proglacial strata, by several orders of magnitude. Consequently, if water penetrated to the gravel horizon not only would it become trapped by the overlying silts and sands, but the rate of glacier advance would create an advancing overburden pressure at a greater rate than that at which it could be dissipated, causing a build up of pore-fluid pressure in the gravel beds.

As Eyjabakkajokull surged to within 50-70m of the maximum limit achieved in 1890 in the vicinity of the lake-filled depressions up-ice from the ridge complexes, it appears to have begun to detach the overlying beds of turf, silt and compact silt in sheets, from the gravels beneath. Such detachment would have been facilitated by high pore pressures generated within the gravels, providing a natural plane of decollement either within the gravels or at the base of the overlying beds. The rate of pressure release by lateral meltwater within the gravels would have been exceeded by the rate of advance of the glacier snout.

These sheets of detached turf and silt became imbricated as the glacier reached its maximum extent and may have played a role in preventing the advance continuing; a similar conclusion was reached by Aber (1981). Meanwhile, the lower gravel beds were subject to high pore water pressures, and combined lateral and downward stresses such hat the upper beds were dragged along and forced downwards by the weight and forward movement of ice. This forward/downward pressure appears to have bent the imbricated thrusts down in their rear sections and caused extension of the original bedding.

The development of the proglacial thrusts appears to have been associated with the development of the normal fault structures in the subglacial section, and was probably synchronous. The two systems appear to be physically linked by a decollement which is concave upwards, and

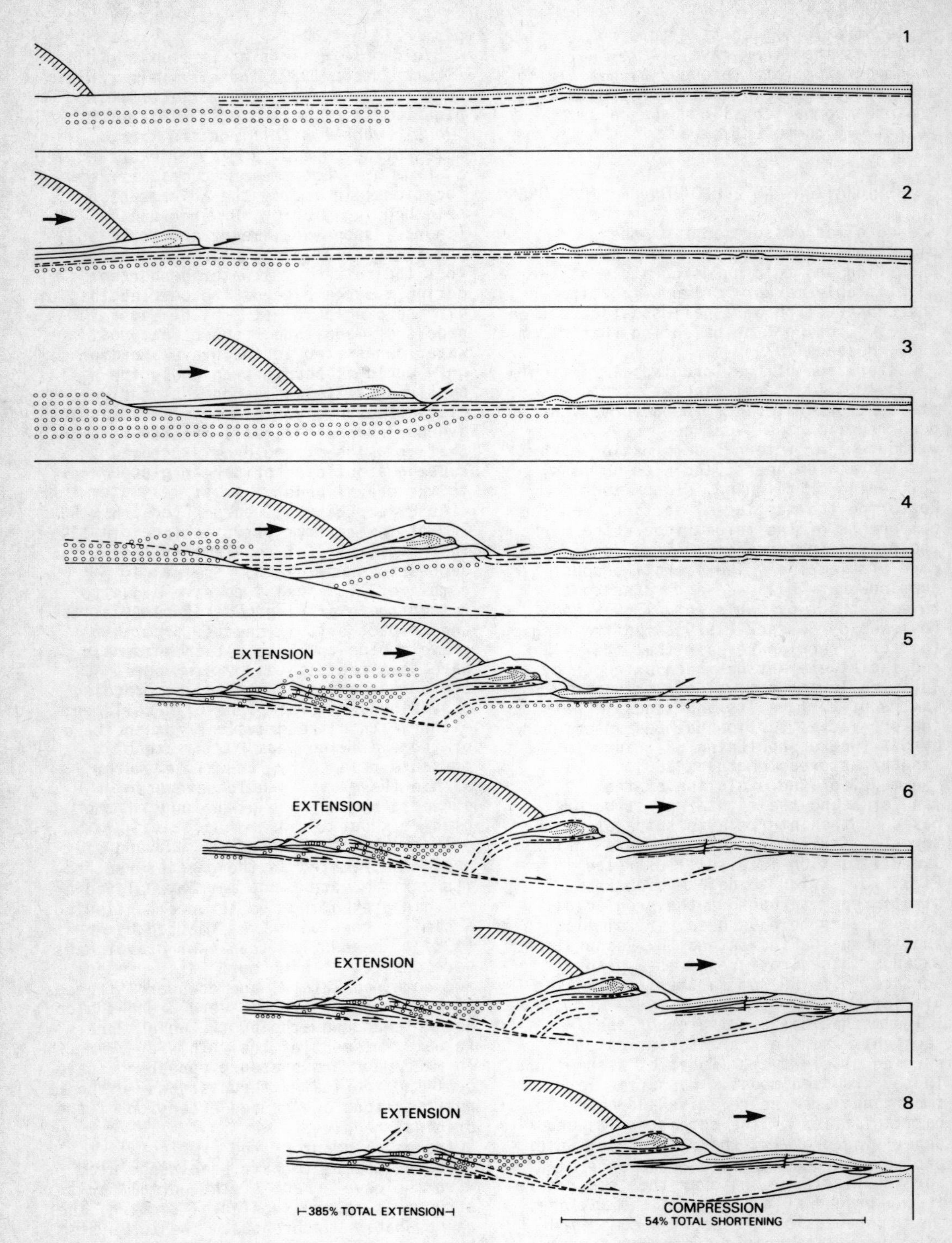

Figure 11. Sequential development of push-ridge 'c' as interpreted from structural evidence contained in the present-day section (8 on the figure).

along which development seems to have propagated. Although only small elements of this floor thrust appear in the section, the projection of all thrust planes in both the subglacial and proglacial sections indicates its presence. The concept of such a floor thrust of plane of decollement is well-established in regard to large scale tectonic features (Price and McClay 1981) and implicit in most models of glaciotectonics, although it is usually expressed as a basal plane of decollement, and commonly assumed to be the depth to the permafrost table.

The interrelationships of thrusts in the proglacial section suggest that they developed in the sequence shown in Fig 10. The upper beds, particularly the compact blue silt, appear to have behaved as competent strata, whilst the thrusts have propagated through the gravels and cut up-section through the rest of the more competent units, a process typical of large-scale tectonic movements. As stresses increased, strain was firstly accommodated by folding, followed by the initial thrust propagated close to the advancing glacier snout.

The relic water courses which originate at many of the thrust traces in the ridges suggests that the thrusts were water-lubricated during movement,and provided natural routeways for the release of encapsulated meltwater or groundwater under high pressure. As the stresses in each thrust were dissipated, such outlets became unnecessary. As the thrust movement waned, the next thrust took up the strain, initially by ductile folding, but subsequently by failure along the gravel; cutting up-section through the silts and turf. The final movements involved sequentially lower levels of stress, and led to blind thrusts and eventually to simple fold structures at the toe of the system. The overall development is analagous to 'piggyback' development of duplex thrust systems in orogenic belts (Boyer and Elliott 1982, Butler 1982).

7 SUMMARY

Recent research on glaciotectonism has resulted in a number of models, the majority of which emphasise the roles of an overriding ice-mass and permafrost (Banham 1975, Bluemle and Clayton 1984, Thomas 1984). Relatively few authors account for glaciotectonic deformation by ice which does not override the area, or in which permafrost does not play a major role in developing not only confined aquifer conditions, but also in providing a natural plane of decollement along which movement propagates (Gripp 1975, Kalin 1971, Oldale and O'Hara 1984, Rabassa et al, 1979).

The landforms and structures described in this paper are partly a product of overriding (causing extension of the original bedding) but are primarily a product of proglacial tectonic activity which caused imbrication and shortening of the original stratigraphic bed lengths. Whilst permafrost may have played a role in the development of these features, it has been shown that they may have evolved without it, but that such evolution will depend on the relative speeds of (i) pore water pressure dissipation, (ii) meltwater production and (iii) glacier front advance. In the case examined here, the rapid advance of the surging glacier and the associated high rate of meltwater production probably contributed significantly to the tectonic deformation.

The overall form of the push-ridges described above is most similar to the push-moraine of the Thompson Glacier, described by Kalin (1971) and the ridges in Spitsbergen described by Gripp (1929, 1938, 1975), but comparisons of internal structure and evolution cannot be made as neither author described their internal fabric.

Of the five types of push-ridges morphology identified at the snout of Eyjabakkajokull the internal structure and evolution of only one type have been described. Further work along similar lines on the sections available in the other ridges will help fill the gaps in our knowledge of the structure and evolution of neoglacial tectonic features.

ACKNOWLEDGEMENTS

The author is grateful to M. Garton (formerly of City Polytechnic, London) for assistance both in the field and in interpretation; to M Coard, K Solman of Plymouth Polytechnic, Department of Geographical Sciences, for assistance in the laboratory, to B Rogers for the cartographic work, to Mrs D Petherick for typing the manuscript and to Plymouth Polytechnic for financial support towards the cost of fieldwork. The Icelandic National Research Council gave permission for the work to be carried out, and many Icelandic friends and colleagues gave us logistical assistance.

REFERENCES

Aber, J.S. 1981: Two-stage model for glaciotectonism. Geological Society of America Abstracts with Programs, 13, 393-

Aber, J., Croot, D.G., Fenton, M.M. 1988. Glaciotectonic landforms and structures, Reidel, (in press)

Baker, T.H.W. 1979: Strain rate effect on the compressive strength of frozen sand. Eng'g. Geology. 13(1-4), 223-231.

Banham, P.H. 1975: Glaciotectonic structures. In Wright, A.E. & Mosely, F.(eds): Ice Ages: Ancient and Modern, 69- 86. Seel House Press, Liverpool.

Banham, P.H. 1977: Gleciotectonites in till stratigraphy. Boreas 6, 101-105.

Berthelsen, A. 1973. Weichselian ice advance and drift successions in Denmark. Bull. Geol. Inst. Uppsala New Series 5, 21-29.

Berthelsen, A. 1978. The methodology of kineto-stratigraphy as applied to glacial geology. Bull. Geol. Soc. Denmark 27, 25-38.

Berthelsen, A., Konradi, P. and Petersen, K.S. 1977. Kvartaere lagfolger og strukturer; Vestmons klinter, Dansk. geol. Foren, Arsskrift for 1976, 93-99.

Bluemle, J.P. and Clayton, L. 1984. Large-scale glacial thrusting and related processes in North Dakota, Boreas 13, 279-299.

Boyer, S.E. and Elliott, D. 1982. Thrust Systems. Bull. Ass. Am. Petrol. Geol. 66(9), 1196-1230.

Butler, R.W.H. 1982. The terminology of structures in thrust belts. Jnl. Struct. Geol. 4(3), 239-245.

Cooper, M.A., Garton, M.R. and Hossack, J.R. 1983. The origin of the Basse-Nomandie duplex, Boulonnais, France. Jnl. Struct. Geol. 5(2), 139-152.

Dahlstrom, C.D.A. 1969. Balanced cross-sections, Can. J. Earth Sci. 6, 743-757.

Elliott, D. 1976. The energy balance and deformation mechanisms of thrust sheets, Phil. Trans. R. Soc. A283, 289-312.

Gripp, K. 1929. Glaziologgische und geologische Ergebrisse der Hambergischen Spitzbergen Expedition. Abh. Naturw. Verein. Hamburg 22, 147-247.

Gripp, K. 1938. Endmoranen, Comptes Rendus Congr. Intern. Geogr. Amsterdam T.II Sect. IIa. Geographie Physique, 215-228.

Gripp, K. 1973. Grundmorane und Geschiebepflaster Meyniana 23, 49-52.

Gripp, K. 1975. Jahre Untersuchungen uber das Geschehen am Rande des Nordeuropaischen Inlandeises. Eiszeitalter u. Gegenwart, 26, 31- 73.

Hossack, J.R. 1979. The use of balanced cross-sections in the calculation of orgenic contraction: a review. J. Geol. Soc. Lond. 136, 705-711.

Johnston, G.H. (ed) 1981. Permafrost: Engineering design and construction. Wiley. Toronto, 540 pp.

Kalin, M. 1971. The active push moraine of the Thompson Glacier. Montreal McGill University, Axel Heiberg Island Research Reports Glaciology no. 4, 68pp.

McClay, K.R. and Price, N.J. 1981. Thrust and Nappe Tectonics. Geol. Soc. London Spec. Publ. 9, 539pp.

Meier, M.F. and Post, A. 1969. What are glacier surges? Can. J. Earth Sci. 6, 807-817.

Nickling, W.G. and Bennett, L. 1984. The shear strength characteristics of frozen coarse angular debris. J. Glaciol. 30, 348-358.

Oldale, R.N. and O'Hara, C.J. 1984. Glaciotectonic origin of the Massachusetts coastal end moraines and a fluctuating late Wisconsinan ice margin. Geol. Soc. Amer. Bull. 45, 61-74.

Petersen, K.S. 1978. Applications of glaciotectonic analyses in geological mapping of Denmark. Danm. geol. Unders. Arbog. 1977.

Price, N.J. and McClay, K.R. 1981. Introduction in McClay, K.R. and Price, N.J. (eds) Thrust and Nappe Tectonics, Geol. Soc. London Spec. Publ. 9, 1-5.

Rabassa, J., Rubulis, S. and Suarez, J. 1979. Rate of formation and sedimentology of (1976-1978) push-moraines, Frias Glacier, Mount Tronador, Argentina. In Schlucter, C. (ed); Moraines and Varves, 65-79.

Rotnicki, K. 1974. Ogolne podstawy teoretyczne powstawania deformacji glacitektonicznych Symp.

Rotnicki, K. 1976a. Theoretical basis for and a model of the origin of glaciotectonic deformations. Quaestiones Geographicae 3, 103-139.

Rotnicki, K. 1976b. Glacitektoniczna struktura poziomogo nasuniecia lusek (Glaciotectonic imbricate structure of Horizontal Overthrust). Badania Fizjograficzne nad Polska Zachodnig 29A.

Ruszczynska-Szcnajch, H. 1984. Relation of large-scale glaciotectonic features to substratum and bedrock conditions in Central and Eastern Poland. Symposium on the Relationship between glacial terrain and glacial sediment facies. INQUA. Alberta Canada (Abstract), p.10.

Sharp, M. 1985. Sedimentation and stratigraphy at Eyjabakkajokull - An

Icelandic surging glacier. Quat.Res. 24, 268-284.
Sjorring, S. 1978. Glazialtektonik und Glazialstratigraphie Einige Beispiele aus Danemark. Eiszeitalter und Gegenwart 28, 119-125.
Thomas, G.S.P. 1984. The origin of the glacio-dynamic structure of the Bride-Moraine, Isle of Man. Boreas 13, 355-364.
Thomas, G.S.P. and Summers, A.J. 1984. Glacio-dynamic structures from the Blackwater Formation, Co. Wexford, Ireland. Boreas, Vol. 13, pp. 5-12.
Thorarinsson, S. 1938. Uber anolmale Gletscherschwankungen mit besondere Berucksichtigung des Vatnajokullgebietes. Geol. Foren. Stockholm Bd 60(1), 412.
Thorarinsson, S. 1943. Oscillations of the Icelandic glaciers in the last 250 years. Geogr. Annlr. 25, 1-54.
Thorarinsson, S. 1951. Notes on Patterned Ground in Iceland with particular reference to Icelandic "Flas". Geogr. Annlr. 33, 144-156.
Thorarinsson, S. 1964. Sudden advance of the Vatnajokull outlet glaciers 1930-64. Jokull Ar 14(3).
Thorarinsson, S. 1969. Glacier surges in Iceland with special reference to the surges of Bruarjokull. Can. J. Earth Sci. 6(4) II, 875-82.
Tsytovich, N.A. 1975. The Mechanics of Frozen Ground, McGraw Hill, New York, 426 pp.
Van der Wateren, D.F.M 1985. A model of glaciotectonics, applied to the ice-pushed ridges in the Central Netherlands, Bull geol soc Denmark, 34, 55-74.
Williams, R.S. Jr. 1976. Vatnajokull ice-cap Iceland. In: Williams, R.S. Jr. et al (eds): ERTS-1 a new window on our planet. U.S.G.S. Prof. Paper 929, 188-193.
Williams, P.J. 1982. The Surface of the Earth, Longman, London 212 pp.

Glaciotectonics: Forms and Processes, Croot (ed.), © 1988 Balkema, Rotterdam. ISBN 90 6191 848 0

Glaciotectonics and surging glaciers: A correlation based on Vestspitsbergen, Svalbard, Norway

D.G.Croot
Department of Geographical Sciences, Plymouth Polytechnic, Plymouth, Devon, UK

ABSTRACT: 29 composite ridges (commonly called push-moraines) have been identified at glacier snout locations in Vestspitsbergen, Svalbard, Norway. Each composite ridge system comprises a series of individual ridge forms, arcuate in plan, marking all or part of the glacier margin position and the limit reached during a recent (surge) advance. The landforms are proglacial and composed of a variety of lithological units (silts, sands, clays). No ridge complex in Spitsbergen is associated with a non-surging glacier. Not all surging glaciers develop large push-ridge complexes. At present, 98 glaciers have been identified to have surged in their known or 'visible' history. Of these, 20 terminate in deep water fjords or open water, and any push-ridge forms are yet to be found; 13 terminate in steeply sloping valleys, (which inhibits push-ridge complex development); and a further 36 glacier termini have yet to be investigated.

1. INTRODUCTION

The reconstruction of Weichselian/Wisconsinan glacial conditions in Europe and North America has been based either partly or wholly on the interpretation of geological and geomorphological evidence. Such interpretation relies on commonly accepted or preferably known relationships linking glacial processes and geological/geomorphological products. The degree of precision which can be achieved can never of course be known, but must be improved by the study of modern analogs enabling researchers to establish strong relationships between form, processes, and environment. This paper is concerned with the probability of a relationship between glaciotectonic features (composite ridges) and glacier surges in a modern glacial environment: Vestspitsbergen, Svalbard Norway.

Composite ridges are distinctive glaciotectonic landforms developed at the snouts of a large number of Arctic, sub-Arctic and temperate glaciers. Aber et al (1988) describe composite ridges as 'ice shoved ridges ... describing an arc in plan ... which is concave up ice'. Such ridges usually comprise a set of imbricately stacked scales or slices of proglacial sediments which may be of bedrock or unconsolidated sediments. The leading edge of each imbricate slice is usually expressed at the surface as a ridge. Some composite ridges comprise buckled or folded sediments rather than stacked thrust sheets. Aber et al subdivide the features into large (>100m relief) and small (<100m relief) forms. Large forms are common around the margins of former Quaternary ice sheets (Aber 1988, Banham 1975, Bluemle and Clayton 1984, Clayton et al 1980, Christiansen and Whittaker 1976, Gijssel 1987, Meyer 1987, Oldale and O'Hara 1984, Ruegg 1981, amongst others) but rarely occur at the margins of smaller modern glaciers. Small composite-ridges are however common features of modern ice-marginal areas, (Croot 1987, Gripp 1929, Hagen 1987, Kalin 1971) and are readily identifiable on air photographs. (Fig 1).

Recent work on other glacial bed landforms in North America, most notably by Clayton and Moran (1974), Mickelson et al (1983) and Clayton et al (1985), has

Figure 1. Air photograph of composite ridges at the snout of Rabotsbreen, (I.D. Number 4W4/300). Norsk Polarinstitutt 1969.

led to suggestions that the late-Wisconsinan Laurentide ice sheet was subdivided into individual lobes which surged repeatedly. Glaciotectonic forms, including large composite ridges are an integral part of the landscapes created by these surging or rapidly advancing ice lobes.

Workers seeking to explain the origin of Wisconsinan/Weichselian composite ridges in North America and Northern Europe have been led to consider the conditions most favourable to glaciotectonic deformation of lightly-consolidated and unconsolidated strata. Van der Wateren (1985), Aber et al (1988), and Banham (this volume) separately conclude that glaciotectonic deformation is most favoured when either most or all of the following conditions apply:

- Lateral pressure gradient
- Elevated groundwater pressure
- Permafrozen proglacial area
- Ice advance up-slope
- Contrasting lithologies in proglacial area
- Compressive flow with basal drag
- High volumes of meltwater

It is evident from brief consideration of these conditions that they are most easily, (and perhaps exclusively) achieved by a glacier, or ice lobe, surging into a topographically low area, in which drainage is restricted and which comprises a suitable sedimentary package.

Earlier work by Croot (1979, 1981) demonstrates that composite ridges are common features at the margins of some of the surging glaciers in Iceland and Vestspitsbergen. This paper demonstrates that there is a correlation between composite ridges and surging glaciers in Vestspitsbergen, Norway.

2. DATA SOURCES

A range of sources were used in the compilation of the database for this work, varying from accounts of early and more recent expeditions to Vestspitsbergen (Gripp 1929, Karczewski 1984, Lamplugh 1911, Garwood and Gregory 1898, Pillewizer 1939), (Liestol 1969), air photographs (Norsk Polarinstitutt 1945-1974), recently compiled work (Liestol, unpublished), satellite imagery (Landsat) and the UNESCO World Glacier Monitoring Service database. Even compiling from all these sources leaves significant gaps in some elements of the database. The database is subdivided into two main elements: surging glaciers and composite ridges.

3. SURGING GLACIERS

Surging glaciers are recorded as a special category of glacier in the World Glacier Monitoring Service. These recorded surges form the basis of Table 1 (listed OBS under DATA SOURCE). Some of them are also specifically identified and dated on the Map of Svalbard: Glacial geology and Geomorphology (Kristiansen and Sollid, 1986). Some glaciers were observed to advance very rapidly by early visitors to Vestspitsbergen (Gripp 1929, Garwood and Gregory 1898, Lampugh 1911). I take these 'phenomenal rates of advance' to be early records of surges, since the glaciers in question have since shown no sign of advance, and have all downwasted in their snout regions: behaviour typical of cyclic surging. A number of glaciers have escaped both methods of direct observation. An excellent example of this category is Battyebreen, at the head of Dicksonfjord (Fig 2). There is no doubt that this glacier has surged repeatedly in the Holocene, but such events have never been observed, probably because the

Table 1. Details of surging glaciers in Vestspitsbergen 1860-1988.

SURGING GLACIERS IN VESTSPITSBERGEN

BASIN	GLACIER ID NO. (1)	GLACIER NAME	DATA SOURCE (2)	DATE(S) OF SURGE(S)	NATURE OF GLACIER SNOUT AREA (3)
4W1					
	100	Negribreen	OBS	1935/6	Tidewater
	203	Usherbreen	OBS	1980	CR
	300	Elfenbeinbreen	OBS	1903	CR
	307	Skruisbreen	OBS	1920	CR
	308	Sveigbreen	OBS	1960	CR
	400	Inglefieldbreen	OBS	1952	CR
	504	Jemelianovbreen	OBS	1972	Tidewater
4W2					
	100	Hambergbreen	OBS	1890 1960	Tidewater
	102	Markhambreen	OBS	1935	Tidewater
	103	Staupbreen	OBS	1960	Tidewater
	404	Korberbreen	OBS	1938	Tidewater
4W3					
	100	Recherchebreen	OBS	1838 1945	Tidewater
	111	Scottbreen	OBS	1880	CR
	112	Renardbreen	MOR		CR
	114	Crammerbreen S.	MOR		Tidewater
	200	Nathorstbreen	MOR		CR
	201	Hessbreen	OBS	1974	CR
	202	Finsterwalderbr.	OBS	1900	CR
	205	Penckbreen	MOR	1800's	CR
	207	Siegerbreen	OBS	1940	
	214	Liestolbreen	MOR		Tidewater
	215	Doktorbreen	MOR		Tidewater

Table 1. (Cont'd) Details of surging glaciers in Vestspitsbergen 1860-1988

BASIN	GLACIER ID NO. (1)	GLACIER NAME	DATA SOURCE (2)	DATE(S) OF SURGE(S)	NATURE OF GLACIER SNOUT AREA (3)
	218	Steenstrupbreen	MOR		
	219	Sysselmanbreen	MOR		
	223	Martinbreen	MOR		
	224	Charpentierbreen	OBS	C.1890	
	400	Paulabreen	MOR		Tidewater
	408	Vallakrabreen	MOR		
	500	Edvadbreen	MOR		
	505	Hyllingebreen	OBS	1970	Steep
	506	Kroppbreen	MOR		
	511	Skutbreen	OBS	1930	Steep
	512	U.Storknausen, E.	OBS	1960	Steep
	513	Slottsmoya	OBS	1960	Steep
	600	Slakkbreen	MOR		Steep
	604	Greinbreane	MOR		Steep
	607	Uvbreen	MOR		Steep
	608	Ankerbreen	MOR		Steep
	612	Marthabreen	OBS		Steep
	614	Lunckebreen	OBS	1930	
	700	Fridtjovbreen	OBS	1861	CR
4W4					
	100	Gronfjordbreen	MOR		CR
	200	Dronbreen	OBS	1900	Steep
	211	Scott Turnerbreen	OBS	1930	
	216	Glottfjellbreen (Moysalbreen)	OBS	1925	Free draining
	300	Rabotsbreen	MOR		CR
	312	Vendombreen	OBS	1935	

Table 1. (Cont'd) Details of surging glaciers in Vestspitsbergen 1860-1988

BASIN	GLACIER ID NO. (1)	GLACIER NAME	DATA SOURCE (2)	DATE(S) OF SURGE(S)	NATURE OF GLACIER SNOUT AREA (3)
	318	Marmorbreen	OBS	1965/70	
	400	Tunabreen	OBS	1930, 1970	Tidewater
	401	Wandbreen	OBS	1980	
	402	Von Postbreen	OBS	1870	Tidewater
	510	Horbyebreen	MOR		CR
	522	Skandalsbreen	OBS	1930	
	600	Battyebreen	MOR		CR
	617	Fyrisbreen	OBS	1960	Steep
	621	U/Breena NW	OBS	1937	Steep
	700	Holmstrombreen	MOR	1890	CR
	713	Seftstrombreen	OBS	1890	CR
	804	Wahlenbergbreen	OBS	1908	Tidewater
	901	Nansenbreen	OBS	1947	Tidewater
4W5					
	300	Osbournebreen	OBS	1987	Tidewater
	400	Avaatmarksbreen	MOR		CR
	411	Comfortlessbreen	MOR		Tidewater
	412	Uversbreen	MOR		
	500	Kronebreen	OBS		Tidewater
	504	Au Broggerbreen	OBS	1890	
	506	Mid-Lovenbreen	OBS	1890	
	510	Kongsvegen	OBS	1948	Tidewater
	514	Blomstrandbreen	OBS	1960	Tidewater

Table 1. (Cont'd) Details of surging glaciers in Vestspitsbergen 1860-1988

BASIN	GLACIER ID NO. (1)	GLACIER NAME	DATA SOURCE (2)	DATE(S) OF SURGE(S)	NATURE OF GLACIER SNOUT AREA (3)
4W6					
	300	Karlsbreen	MOR		CR
	307	Adolfbreen	MOR		
	308	Nygaardbreen	MOR		
	309	Schelderupbreen	MOR		
	400	Vonbreen	MOR		CR
	410	Elnabreen	OBS	1930	CR
	416	Abrahamsenbreen	OBS		CR
	425	U/Svelgfj. S.	OBS	1969	
	700	Lisbetbreen	MOR		CR
	703	Anne-Mariebreen	MOR		
	717	Uggebreen	MOR		Steep
	722	Belshornbreen	MOR		Steep
	724	Bukkebreen	MOR		Steep
	725	Yggbreen	MOR		
	730	Universitetsbreen	MOR		CR
	821	Stubendorfbreen	MOR		Tidewater
	823	Smutsbreen	MOR		
	826	Sandersbreen	MOR		
	828	Tryggvebreen	MOR		
	832	Planckbreen	MOR		
	834	Reinbukkbreen	MOR		
	901	Cookbreen	MOR		
	908	Asgardfonna NW (Longstaffbreen)	OBS	1960	

Table 1. (Cont'd) Details of surging glaciers in Vestspitsbergen 1860-1988

BASIN	GLACIER ID NO. (1)	GLACIER NAME	DATA SOURCE (2)	DATE(S) OF SURGE(S)	NATURE OF GLACIER SNOUT AREA (3)
4W7					
	200	Veteranenbreen	MOR		CR
	201	Rimfaksbreen	MOR		CR
	202	Gullfaksbreen	MOR		CR
	203	Skinfaksbreen	MOR		CR
	300	Hinlopenbreen	OBS	1969/1972	Tidewater
	305	Kosterbreen	OBS	1930	Tidewater

Notes:

(1) UNESCO World Glacier Monitoring Service Identification Numbers.

(2) DATA SOURCE: OBS: Surges recorded by visiting scientists or observed from air photographs in more recent years, (Liestol 1988 pers. comm). MOR: surge behaviour inferred from moraine patterns (Gripp 1929, Meier and Post 1969).

(3) NATURE OF GLACIER SNOUT AREA. CR: Composite Ridges identifiable from air photographs or recorded in field observations and reported by various authors. TIDEWATER: Glacier terminates in open water, and little is known of terminal landforms. STEEP: Air-photo analysis shows proglacial area to slope steeply away from glacier snout. No record: no data available.

glacier is so remote. The evidence for cyclic surging is clear on air photographs, which show the characteristic looped medial moraine patterns which can only be explained by repeated surging of the main valley glacier which causes instability and subsequent surges in tributary glaciers (Fig 3). Such patterns were originally accounted for by Gripp (1929), and were subsequently highlighted by Meier and Post (1969) as a key characteristic in identifying surging valley glaciers. Glaciers exhibiting well-developed looped moraine patterns are also added to Table 1. (MOR under DATA SOURCE).

4. COMPOSITE RIDGES

The excellent quality of Norsk Polarinstitutt air photographs facilitates the identification of composite ridges in the majority of cases tabulated (Table 2). In some cases detailed ground observations, supported by photographs have been recorded in the literature, and provide additional confirmation. (Gripp 1929, Drozdowski 1987, Karczewski 1984, Lamplugh 1911).

An attempt was made to use satellite imagery (Landsat IV) to identify composite ridges. The level of resolution (30m) in

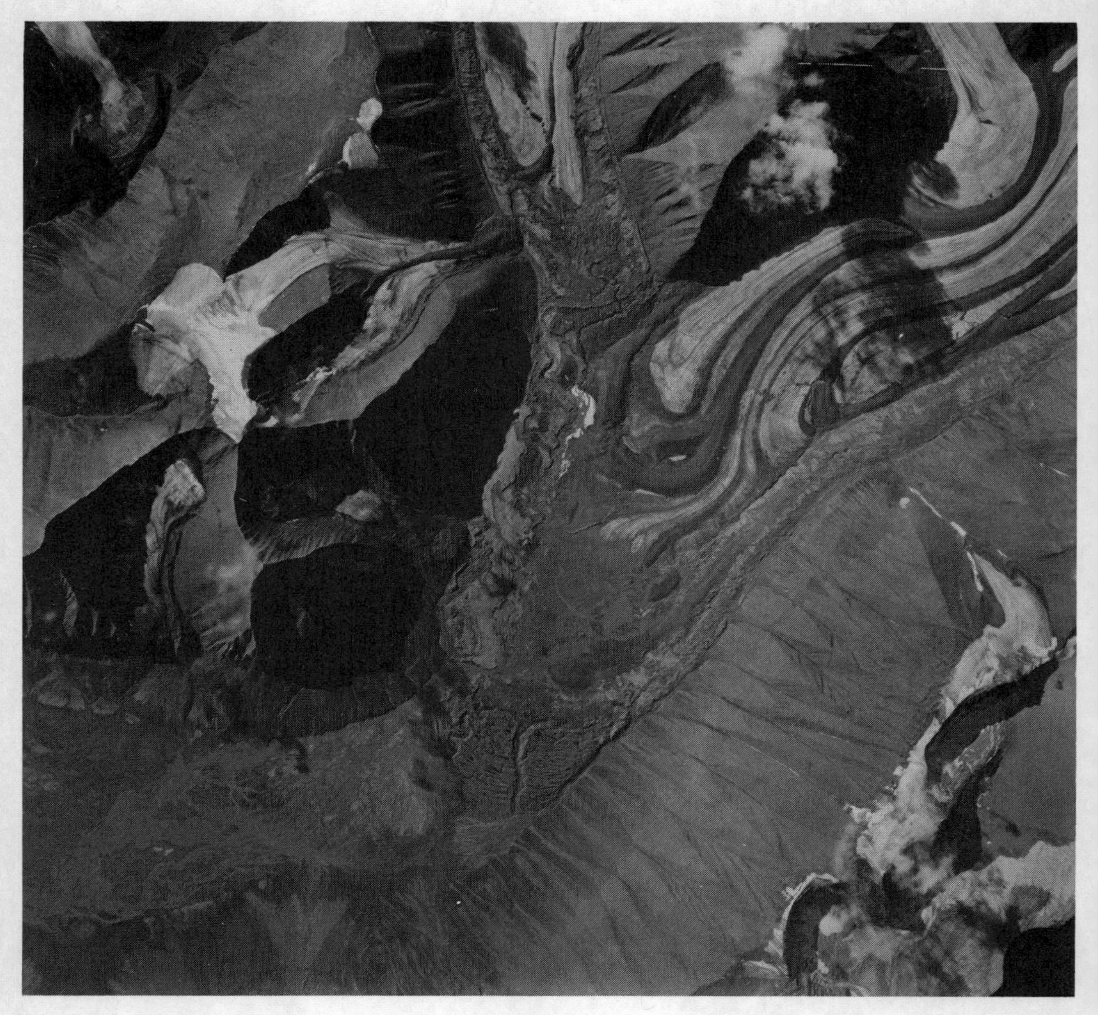

Figure 2. Air photograph of Battyebreen (I.D. Number 4W4/600) Norsk Polarinstitutt 1969. Looped moraines are clear evidence of a previous history of repeated surging. The glacier terminus is dominated by a fine push-ridge complex.

these images is insufficient to differentiate composite ridges from other ice marginal/proglacial forms.

5. RESULTS

The results of the investigations are presented in Tables 1 and 2, and Fig 4. Twenty-eight composite ridge systems have been identified from the data sources listed. As suggested above, air-photo coverage is not complete, and the authors feel that several more composite ridges systems may well be present. Not all composite ridge systems are as well-developed as the one at the snout of Battyebreen (Fig 2). Many do occupy the full extent of the ice margin, but equally composite ridges may only occur in a restricted part of the ice marginal/proglacial zone. (Fig 5).

As Table 2 shows all the composite ridge systems listed are associated with

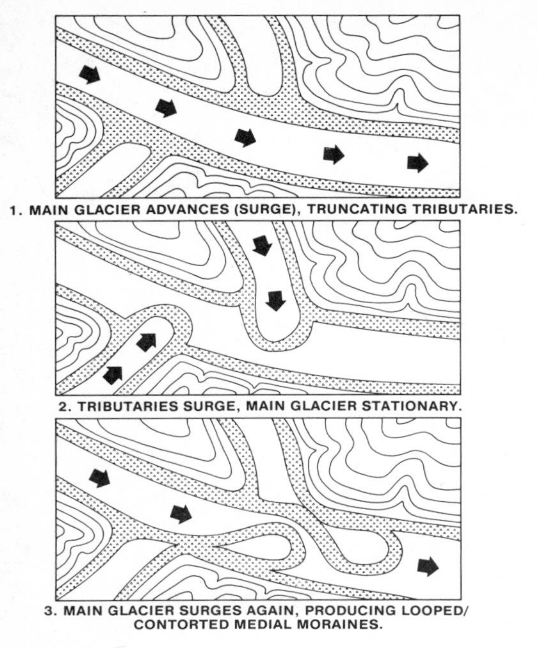

Figure 3. Gripp (1929) recognised the significance of looped moraine patterns, and demonstrated the only possible mechanism by which they could form.

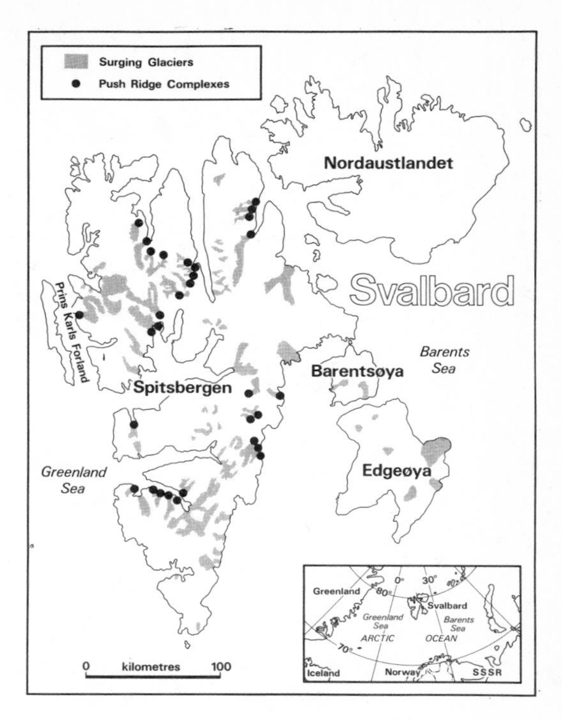

Figure 4. The distribution of surging glaciers and push-ridge complexes (composite ridge systems) in Vestspitsbergen.

glaciers which have surged in the recent past. Of these 29 systems, 12 are associated with glaciers where recent surges have actually been observed, the rest, like Battyebreen, are associated with glaciers which have well-developed looped-moraine patterns.

Table 1 shows that there is clear evidence of at least a further 60-70 glaciers with a history of surge behaviour. In 22 cases, the glacier currently terminates in a fjord, where the bottom topography is not sufficiently well known to identify composite ridges. In 13 cases the glacier surge terminated in a valley which slopes rather steeply away from the ice margin, or comprises coarse gravels which are not susceptible to glaciotectonic deformation since they transmit meltwater generated by surges too readily. Good examples of such termini are Dronbreen and Moysalbreen in Longyeardalen (ID numbers 4W001/4 42200, 42116). A further 36 surging glacier snout areas have yet to be examined in detail for evidence of composite ridge development.

6. DISCUSSION

The data analysis shows quite conclusively that small composite ridge systems in Vestspitsbergen are only developed in front of glaciers with a history of repeated surging. This idea is not new: in a recent paper on 'surge moraines' Drozdowski (1987) suggested that the best modern analog for Quaternary composite ridges in North Poland are the ridges in front of Avaatmarksbreen (4W001/5 54/500) and Seftstrombreen (4W001/4 47/713). Indeed, it is implicit in all models of glaciotectonic deformation, and particularly those used to account for composite ridge systems, that a rapidly advancing glacier is most likely to fulfil the criteria for composite ridge development. Ridges will not develop however, if the proglacial conditions are wrong (particularly where the proglacial area slopes steeply away from the glacier, facilitating free drainage of meltwater

Table 2. Details of composite ridge systems in Vestspitsbergen from various sources (see text for details).

COMPOSITE RIDGES IN VESTSPITSBERGEN

BASIN	GLACIER I.D. NUMBER (1)	GLACIER NAME	AIR PHOTO	SURGE (2)
4W1	203	Usherbreen	69/1940 69/1941	1980
	300	Elfenbeinbreen	69/2062	1903
	307	Skruisbreen	69/2062	1920
	308	Sveigbreen		1960
	400	Inglefieldbreen	69/2068	1952
4W3	111	Scottbreen		
	112	Renardbreen		Moraine Pattern
	200	Nathorstbreen		Moraine Pattern
	201	Hessbreen		1974
	202	Finsterwalderbreen		1900
	205	Penckbreen		Moraine Pattern
	700	Fridtjovbreen		1981
4W4	100	Gronfjordbreane	69/2534	Moraine Pattern
	300	Rabotsbreen	69/3158	Moraine Pattern
	510	Horbyebreen	Needed	Moraine Pattern
	600	Battyebreen	69/1468	Moraine Pattern
	700	Holstrombreen	69/1501	1900
	713	Seftstrombreen	69/1501	1890
4W5	400	Avaatmarksbreen	No	(Tidewater) Drozdowski (Tidewater)
4W6	300	Karlsbreen	No	Moraine Pattern
	400	Vonbreen	66/4123 64/400 66/4408	Moraine Pattern
	410	Elnabreen	66/4406	1930
	416	Abrahamsbreen	66/4121	1978
	700	Lisbetbreen	66/4031	Moraine Pattern
	730	Universitetsbreen	66/4032	
4W7	200	Veteranenbreen	66/4725	Moraine Pattern
	201	Rimfaksbreen	66/4725	Moraine Pattern
	202	Gullfaksbreen	66/4725	Moraine Pattern
	203	Skinfaksbreen	66/4725	Moraine Pattern

Notes:

(1) UNESCO World Glacier Monitoring Service Identification Number.

(2) Information in this column relates to known or inferred history of surge behaviour. Dates are known dates of recent surges. 'Moraine Pattern': no recorded history of surging, but previous history of surging is evident from lobate supraglacial moraines (Gripp 1929, Meier and Post, 1969).

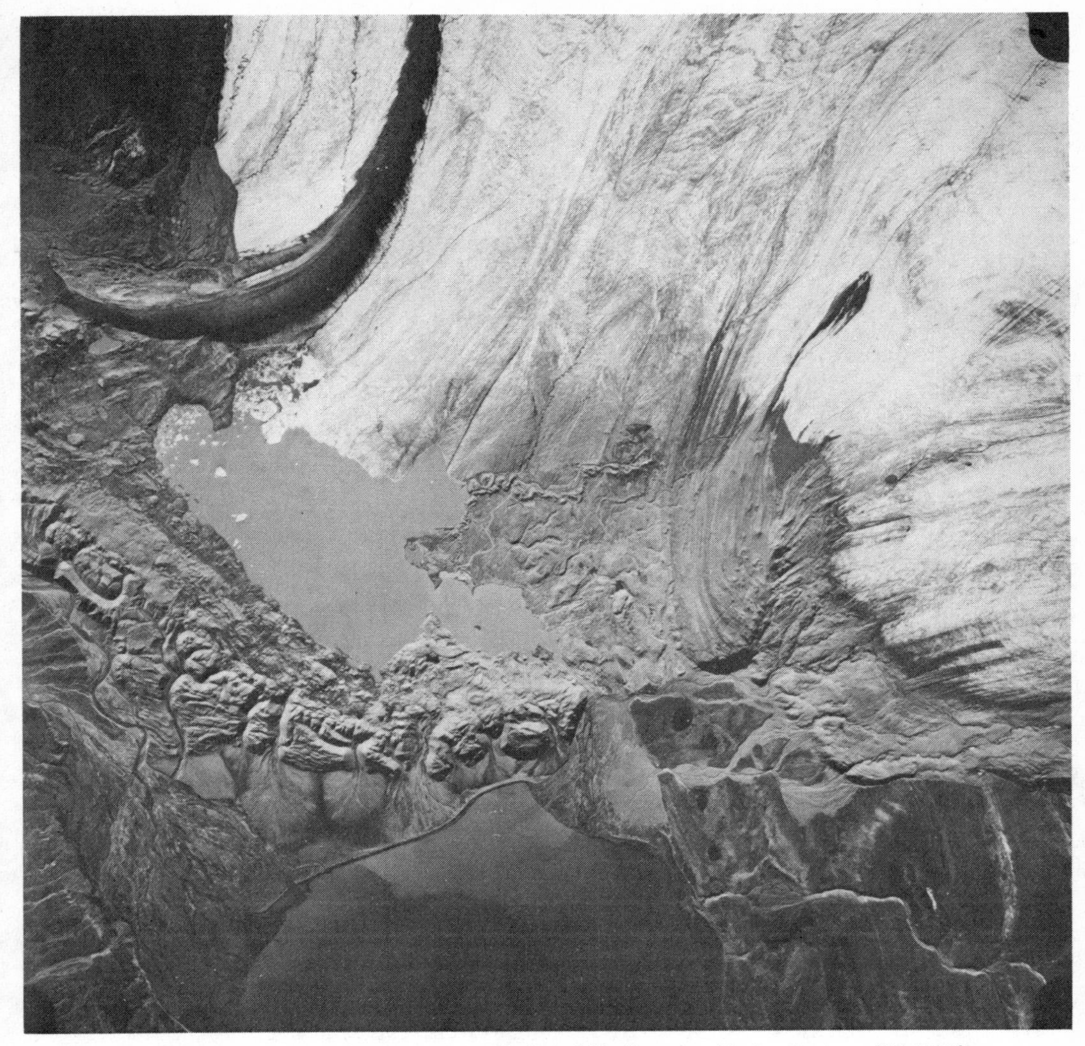

Figure 5. Air photograph of the snout of Gronfjordbreane (I.D. Number 4W4/100), showing incomplete development of composite ridges. (Norsk Polarinstituttt 1969).

produced during the surge). Recent work in North America and Scandinavia has highlighted the conditions in which surging is commonplace. Clayton et al (1985) and Lagerlund (1987) examine the hypothetical surge behaviour of the Laurentide and Scandinavian ice sheets respectively. Both papers are drawn to the conclusion that surge behaviour is intimately associated with glacier bed drainage conditions, and that surging is most probably a natural outcome of water-soaked basal conditions in flat lying areas of sediments with low or very low permeability. The same conditions are listed by Aber et al (1988), and Banham (this volume) as pre-conditions for composite ridge development. Such observations would seem to suggest that composite ridges and surging glaciers are intimately associated.

7. CONCLUSION

Holocene composite ridge systems in Vestspitsbergen are exclusively developed in front of glaciers with a history of surging. Not all surging glaciers develop composite ridge systems because other preconditions are not satisfied. This work now needs to be extended by firstly examining the causal links between surging and composite ridge development and secondly increasing the scope of the present survey to incorporate composite ridges in other parts of the Arctic. Should a strong correlation be established for a wide geographic area the existence of Quaternary composite ridge systems may be used as a key indicator of surging ice lobes/glaciers of the same age.

ACKNOWLEDGEMENTS

The author gratefully acknowledges the following institutions and individuals; Plymouth Polytechnic for financial support in acquiring air photographs and satellite imagery; Norsk Polarinstitutt for data on Svalbard glaciers; World Glacier Monitoring Service for glacier data. B. Rogers kindly drafted the diagrams and Mrs D Petherick typed the original manuscipt. J.O. Hagen (Norsk Polarinstitutt) very kindly reviewed the databases and made critical comments on the draft.

REFERENCES

Aber, J. 1988 'Geomorphic and structural genesis of the Dirt Hills and Cactus Hills, Saskatchewan' Alberta Research Council (in press).

Aber, J., Croot, D.G., Fenton, M.M. 1988 'Glaciotectonic landforms and structures', Reidel, Netherlands, 350 pp (in press).

Banham, P. 1975 'Glaciotectonic structures: a general discussion with particular reference to the contorted drift of Norfolk'. in Wright A.E. and Moseley, F. (eds) 'Ice Ages : Ancient and Modern', Geol. Jnl. Spec. Issue 6 : 69-94.

Banham, P. 1988 'A thin-skinned glaciotectonic model incorporating extensional structures' (this volume).

Bluemle, J.P. and Clayton, L. 1984 'Large scale glacial thrusting and related processes in North Dakota', Boreas 13 : 279-299.

Christianssen, E.A, & Whittaker, S.H. 1976 'Glacial thrusting of drift and, bedrock', in Legget, R.F. (ed) Glacial till', Roy. Soc. Canada Spec. Pub. 12 : 121-30.

Clayton, L. & Moran, S.R. 1974 'A glacial process-form model', in Coates, D.R. (ed) Glacial Geomorphology, SUNY-Binghampton Publications in Geomorphology New York : 89-119.

Clayton, L., Moran, S.R., Bluemle, J.P. 1980 'Explanatory text to accompany the Geologic Map of North Dakota'. North Dakota Geological Survey Report Investigation No. 69.

Clayton, L., Teller, J.T. & Attig, J.W. 1985 'Surging of the Southwestern part of the Laurentide Ice Sheet', Boreas 14: 235-241.

Croot, D.G. 1979 'Depositional landforms and sediments associated with surging glaciers', Unpubl. PhD Thesis, Univer -sity of Aberdeen.

Croot, D.G. 1981 'Depositional landforms at the snouts of five Spitsbergen glaciers', Annals of Glaciology II, (Abstract only).

Croot, D.G. 1987 'Glaciotectonic structures : a meso-scale model of thin-skinned thrust sheets? J. Structural Geol. 9(7) 797-808.

Drozdowski, E. 1987 'Surge moraines' in Gardiner, V. (ed) International Geomorphology, Wiley and Sons, Chichester, and New York, : 675-692.

Garwood, E.J. & Gregory, J.W. 1898 'Contributions to the glacial geology of Spitsbergen' Q.Jnl.Geol.Soc. 54: 197-225.

Gijssel, K. van 1987 'A lithostratigraphic and glaciotectonic reconstruction of the Lamstedt Moraine Lower Saxony, FRG' in Meer, J.M. van der (ed) 'Tills and Glaciotectonics' A.A. Blakema, Rotterdam : 145-155.

Gripp, K. 1929 'Glaciologische and geologische Ergebnisse der Hamburgischen Spitzbergen Expedition 1927' Abh. Naturwiss. Ver. Hamburg. XX11 = 145-248.

Hagen, J.O. 1987, Glacier Surge at Usherbreen, Svalbard, Polar Research 5, n.s. 239-252.

Kalin, M. 1971 'The active push moraine of the Thompson Glacier'. Axel Heiberg Island Res. Report. Glaciology No. 4.

Karczewski, A. 1984 'Geomorphology of the Hornsund Fjord Area Spitsbergen; commentary to the Map' Inst. Geophys. Polish Acad. Sciences 26pp.

Kristiansen, K and Sollid, J.L. 1986 Svalbard: Glacial geology and geomorphology Map 1:1,000,000. Nasjonalatlas for Norge. Dept of Geography, University of Oslo.

Lagerlund, E. 1987 'An alternative Weichselian glaciation model with special reference to the glaciation of Skane, South Sweden', Boreas 16:433-459.

Lamplugh, G.W. 1911 'On the shelly moraine of the Seftstrom Glacier and other Spitsbergen phenomena illustrative of British glacial conditions'. Proc. Yorks geol.Soc 17:216-241.

Liestol, O. 1969 'Glacier surges in West Spitsbergen' Can. J. Earth Sci. 6:895-897.

Meyer, K.D. 1987 'Ground and end-moraines in Lower Saxony', in Meer J.J. van der (ed) 'Tills and glaciotectonics', A A Balkema Rotterdam: 197-204.

Mei er, M.F. & Post, A.S. 1969 'What are glacier surges?' Can. J. Earth Sci. 6(4) :807-818.

Michelson, D.M., Clayton, L., Fullerton, D.S. Borns, H.W. Jr. 1983 'The Late Wisconsinan glacial record of the Laurentide ice sheet in the United States', in Wright, H.E., and Porter, S.C. (eds) Late Quaternary environments of the United States: Vol 1 The Late Pleistocene: 3-37 Univ. Minnesota Press.

Oldale, R.N. & O'Hara, C.J. 1984 Glaciotectonic origin of the Massachusets coastal end moraines and a fluctuating Wisconsinan ice margin: Geol Soc Amer. Bull 95:61-74.

Ruegg, G.J.H. 1981 'Ice pushed Lower and Middle Pleistocene deposits near Rhene: sedimentary-structural and lithological granulomentrical investigations' in Ruegg, G.J.H. and Zandstra J.G. (eds) 'Geology and archaeology of Pleistocene deposits in the ice-pushed ridge near Rhenen and Van en daal. Med. Rijks. Geol. Dienst. 35-2/7:165-177.

Wateren, D.F.M. Vander 1985 'A model of glaciotectonics, applied to the ice-pushed ridges in the Central Netherlands' Geol.Soc Denmark Bull. 34:55-74.

Glaciotectonics: Forms and Processes, Croot (ed.), © 1988 Balkema, Rotterdam. ISBN 90 6191 848 0

On the mechanics of glaciotectonic contortion of clays

V.Feeser
Geologisch-Paläontologisches Institut, Universität Kiel, FR Germany

ABSTRACT: Using the symmetry concept in connection with findings on failure mechanisms, glaciotectonic contortion processes in a Pleistocene stiff, fissured clay are derived from actual macro (joint size) and micro (particle size) fabrics. The results contradict conventional understanding of shear-induced failure genesis. Joint development is shown to be caused by extension within compressive stress fields. Assuming basal sliding velocities as measured recently, the results demonstrate that the ground was frozen at the time of glaciotectonic deformation.

1 INTRODUCTION

There is currently a lack of information on the detailed mechanisms by which Pleistocene clays become glaciotectonically contorted, which this paper aims to redress. Stiff, fissured clays of Pleistocene age, which have remained widely undisturbed since their genesis provide an opportunity to study joint patterns[1] and fabrics produced by glacial tectonism. The approach used in this work is based on the concept of petrofabric symmetry, which leads to a reconstruction of the contortion mechanisms related to stress history of the clays on the actual data of clay fabrics.

Relatively few researchers have studied fracture genesis without a comprehensive mechanical background. Most past investigations into stress history assume that failure has been a phenomenon of shear. Consequently, little attempt has been made to actually describe the failure surface or plane. This is an important omission, since the character of the fracture surface varies according to the failure mechanism. Failure surfaces induced by shear must be smooth or slickensided, but research has shown that glaciotectonically induced failure produces mainly rough surfaces, with frequent plumose markings.

The main purpose of this paper is to advance our current understanding of the mechanics of glaciotectonism as it affects clays. A detailed analysis of fracture genesis, considering both the mechanism and timing of failure initiation will achieve this. Consequently, it is necessary to briefly rewiew the current state of research in this field.

[1] The terms joint, fissure, discontinuity, fracture-, failure- and rupture plane or surface have been applied without any genetic implications.

2 BACKGROUND

2.1 Stiff fissured clays

Up to the last decade a variety of non-tectonic causes have been put foreward in structural geology to explain discontinuities in soft clayey sediments: syneresis, desiccation, chemical activity. For this reason research on joint systematics was initiated by civil engineers (Gregory 1844, Terzaghi 1936) because such discontinuities affect the operational strength of the clay. Subsequent work on the origin of these features has been taken up by engineering geologists. Fookes (1965) for example showed that preferred orientation patterns of joints in Siwalik Clay of Tertiary age were related to underlying bedrock fault, shear and fold patterns. Fookes and Parrish (1969) came to similar conclusions in respect of Tertiary London Clay. Skempton et al (1969), however, suggested that stress release and weathering play the important role in joint and fissure genesis. D'Elia (1973) combined both schools of thought, and proposed that permafrost processes had been responsible for the differential dilatation of sediment layers of different grain sizes.

Little work has been carried out on glaciotectonically contorted clays of Pleistocene age. Banham (1975) argued that joint patterns in such sediments are irregular. Kazi and Knill (1983) distinguished between planar, striated and undulating fissures in their study of glacial lake clays. Assuming that each type of fissure may have a different origin, they suggested that fissuring may in general be attributed to strike-slip stress conditions associated with glacial overriding and subsequent erosion. Recently, Möbus and Peterss (1983) submitted a three-dimensional determination of principle stress fields in tills of the East German Baltic coast related to glacial loading. Most of these studies are based on the assumption that conjugate joint patterns are induced by shear; an assumption that will be shown to be erroreous in this paper.

2.2 Failure mechanics

Ductility and brittleness are key concepts in earth material mechanics. The terms relate to the deformation character of a sample. A material is said to be ductile if it can sustain permanent deformation without losing its ability to resist load (work-hardening, Fig. 1), whereas the ability of brittle materials to resist load decreases with increasing deformation (work-softening, Fig. 2). In addition Fig. 1 and 2 show that brittle-ductile characterization is also associated with a typical stress-volumetric strain relationship. Brittle behaviour corresponds with an increase in volume with increasing stress, termed dilatancy. Ductile material is characterized by a continuous decrease in volume with increasing stress, known as contractancy.

However, ductility and brittleness are affected by environmental conditions around and within the unit of material. There is a marked tendency from brittleness to ductility by increasing the confining pressure and temperature as well as decreasing the rate of deformation (Jaeger and Cook 1976).

Failure surfaces are generally observed in brittle states by splitting the sample parallel to the direction of maximum principle stress, and perpendicular to the direction of minimum principle stress (Fig. 3a). Fractures are typically rough with frequent plumose markings. Within the state of brittle-ductile transition, splitting is replaced by faulting in

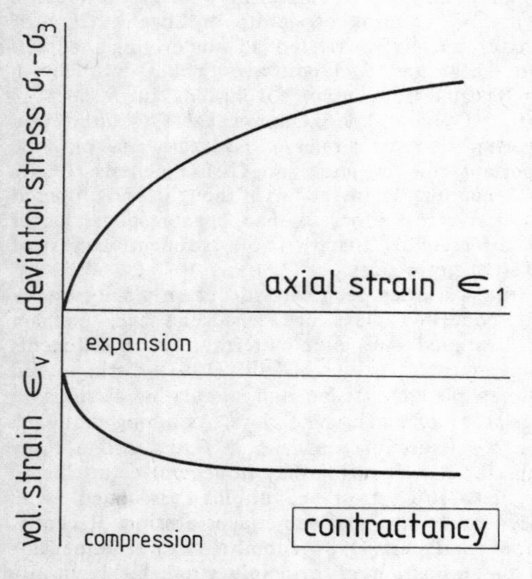

Fig. 1. Typical stress - axial strain - volumetric strain behaviour of ductile material.

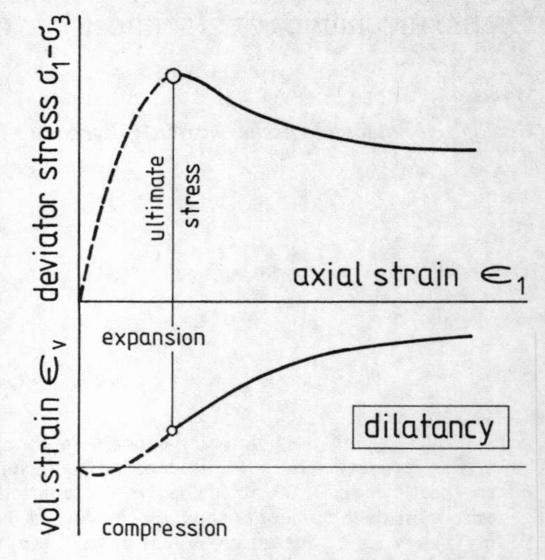

Fig. 2. Typical stress - axial strain - volumetric strain behaviour of brittle material.

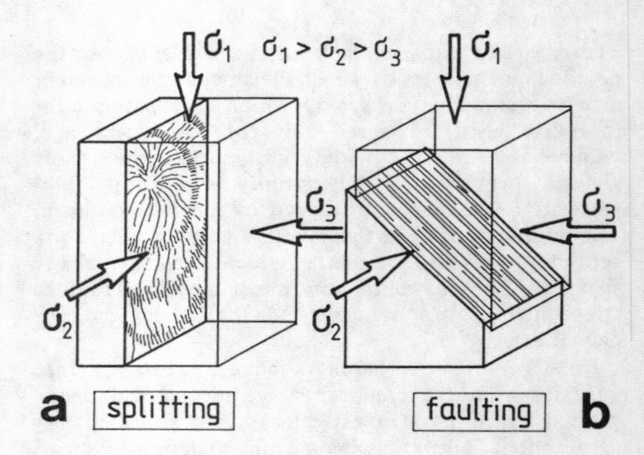

Fig. 3. Basic modes of fracture. a: Longitudinal splitting with plumose surface markings. b: Inclined faulting with smooth and/or slickensided surface.

which a single plane of fracture, inclined at less than 45° to the direction of maximum principle stress, develops (Fig. 3b). If the material becomes fully ductile a network of conjugate faults appears. Fractures developed in ductile and brittle-ductile transition states are characterized by smooth surfaces, frequently polished or slickensided.

Brittle failure associated with dilatancy is usually attributed to the formation and extension of microcracks. The most satisfactory explanation for the initiation of these microcracks, known as the Griffith Theory (Griffith 1924), has been subsequently modified and extended (e.g. Brace and Bombolakis 1963, Hoek and Bieniawski 1965). All theories are based on the assumption that in a compressional stress field brittle fracture starts when tensile stresses, generated around an inherent zone of weakness,reach the cohesive strength of the material (Fig. 4). Subsequently the fracture propagates in a direction along which the strain energy density is at a minimum, i.e. perpendicular to the direction of minimum stress.

A single fracture only leads to failure in simple theoretical models. In reality, propagating fracture surfaces shift from one pre-existing zone of weakness to another, cutting across solid 'bridges' of material, whereas no distinct refracted crossing of the individual failure planes occurrs. Kranz (1979) observed that fracture linking is a result of relatively large tensile stress concentrations between fracture tips. The final fracture image is a result of the coalescene of an array of fracture branches. Even in brittle-ductile transition, where material macroscopically failed in shear (Fig. 5), failure initation may be attributed to pronounced microcracking parallel the direction of maximum principle stress which finally tends to coalesce along a fault (Hallbauer et al 1973).

The foregoing comments apply principally to laboratory investigations, but there is a general consensus that the same processes also apply in field situations (Lajtai 1975, Hoek 1983).

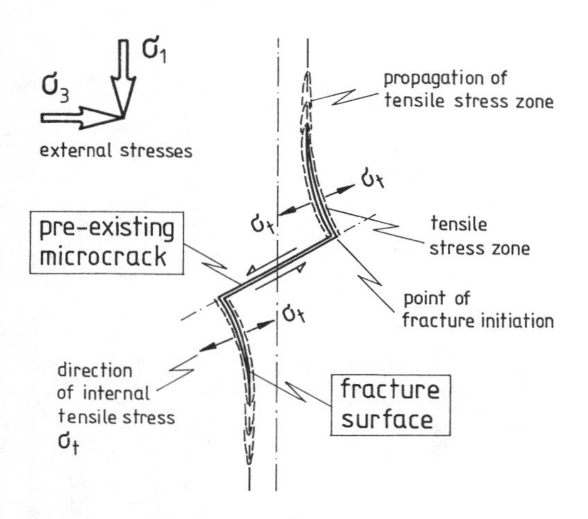

Fig. 4. Model of brittle failure under compression. Fracture development caused by local tensile stress, as a result of sliding on a pre-existing microcrack.

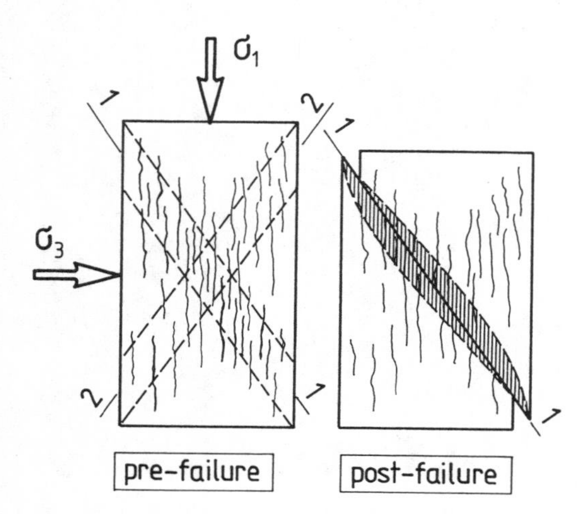

Fig. 5. Schematic sketches of failure by secondary shear. Primary brittle fracturing has caused zones of weakness. Ultimate failure involves complicated fracturing within a shear zone.

3 FIELD WORK

3.1 Investigation sites

Two fresh artificial exposures were chosen, since it is clearly important to avoid sites where joint patterns may have been influenced by weathering or stress relief caused by natural slope instability: first a large open clay pit at Querenstede 50 km west of Bremen, North Germany, and second, a smaller pit at Scharmbeck, close to Hamburg (Fig. 6).

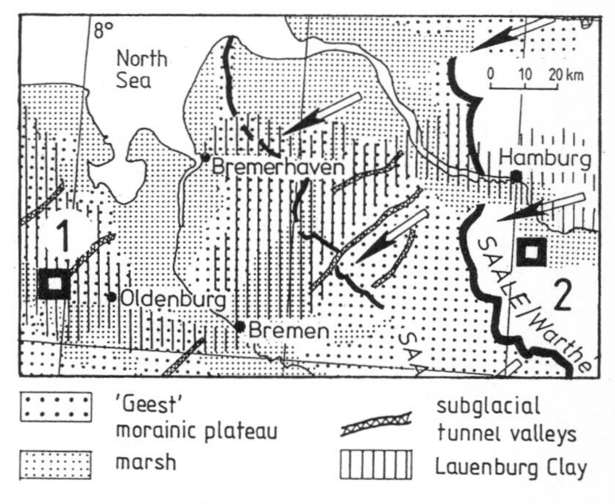

Fig. 6. Location map of investigation sites with approximate limits of Saaleian ice sheets and ice flows. 1: Querenstede site, 2: Scharmbeck site.

The Querenstede pit is excavated about 15 m in depth into Lauenburg Clay, a sequence of up to 150 m of fine-grained clays of late Elsterian age, which occur over a large part of Northern Germany. The Querenstede exposure comprises two elements: a homogenous black clay rarely exhibiting partings or silty layers in excess of 1 mm thickness (maximum), overlain by 10 m of laminated silts and clays. The black clays have a network of discontinuities within them. The Querenstede site was overriden by ice during the Drenthe/Rehburg advance, an event marked at the site by large scale thrusts, imbricate slabs and folds in the upper laminated clays.

The Scharmbeck site is excavated to a depth of 6 m in Drentherian lake clays. As at Querenstede site, the sequence comprises laminated silty clays overlying grey-blue fissured clay. The total extent of clays in this area is not known. However, it is known that only one glacial advance took place across the site, during the Warthe stage. Larger scale deformation produced by this ice advance is restricted to local folding and small scale thrusts within the upper laminated silty clays.

3.2 Observational procedures

Field investigations were undertaken as soon as possible following exposure of vertical faces in the bottom of the pits. Sections studied (typically 2 m long and 1.5 m high) were carefully selected to avoid orientational bias. A layer of clay was carefully removed from one side of recognisable discontinuities to their full height to remove material which may have been disturbed by mechanical excavation. Similary, excavation was made into the face to ascertain the extent of a discontinuity. In order to prevent artificial fractures these procedures were carried out using trowels, putty knives and hair pencils.

A separation scheme, modified from Bock (1974), was used to classify discontinuities to avoid the overvaluation of predominant small scale fissures in stereographic projection. The scheme is based on the geometric interrelationships of failure planes as discussed in chapter 2.2, and classifies the discontinuities according to their relative age (Fig. 7):

i) Category I are discontinuities which are not or only rarely bonded/deformed by interaction with other discontinuities.

ii) Category II are discontinuities which terminate at category I planes or are dislocated by category I planes.

iii) Category III are either truncated by both category I and II planes, or offset by coalescing with them.

There is naturally some overlap between categories, but not of sufficient magnitude to affect the outcome of the research.

Each and every discontinuity was measured for dip, strike, length, and surface topography. Staining and markings were also recorded. Planes and discontinuities less than 5 cm long could not be included because of errors in measurement. In total 800 discontinuities were measured from 6 sections at the Querenstede site and 260 from 2 sections at the Scharmbeck site.

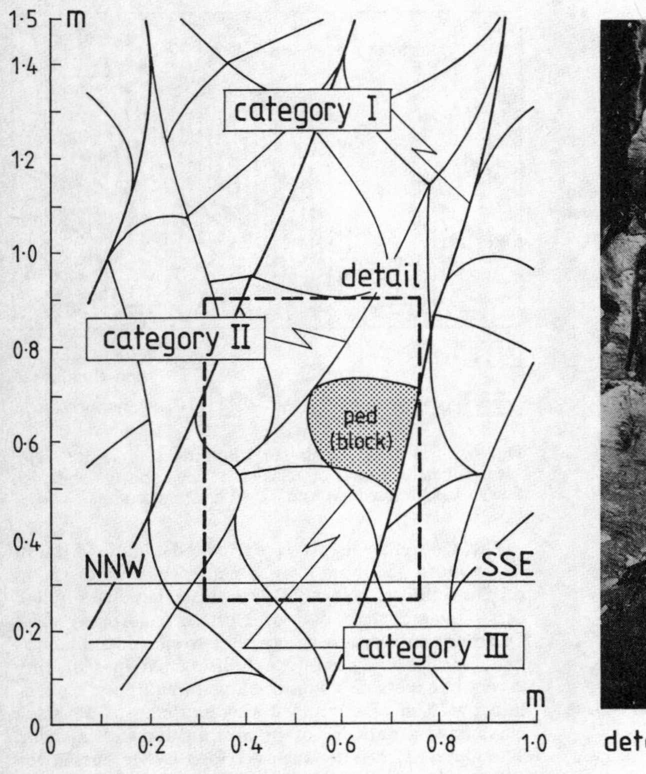

Fig. 7. Actual pattern of discontinuities in Lauenburg Clay. Sketch and detailed photograph of a vertical plane from Querenstede pit, 15 m below ground level.

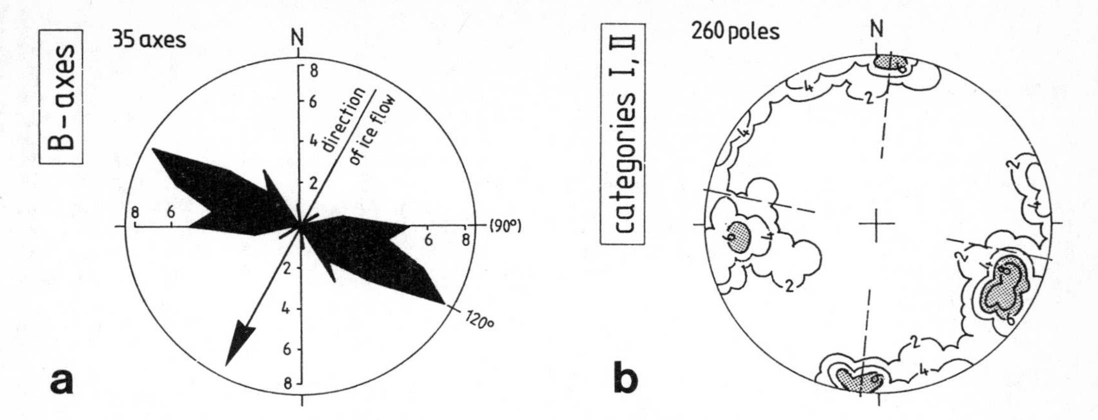

Fig. 8. Glaciotectonic structures in Drentherian lake clay at Scharmbeck. a: Rose diagram of frequency distribution of the strikes of B-axes of microfaults and folds in laminated clay overlying the fissured clay (frame 180°, intervals 10°). b: Composite equal-area lower hemisphere projection of large scale joints (category I and II) in fissured clay. Normalized projection due to rotation of bedding into horizontal plane (density of poles in % per 1% area).

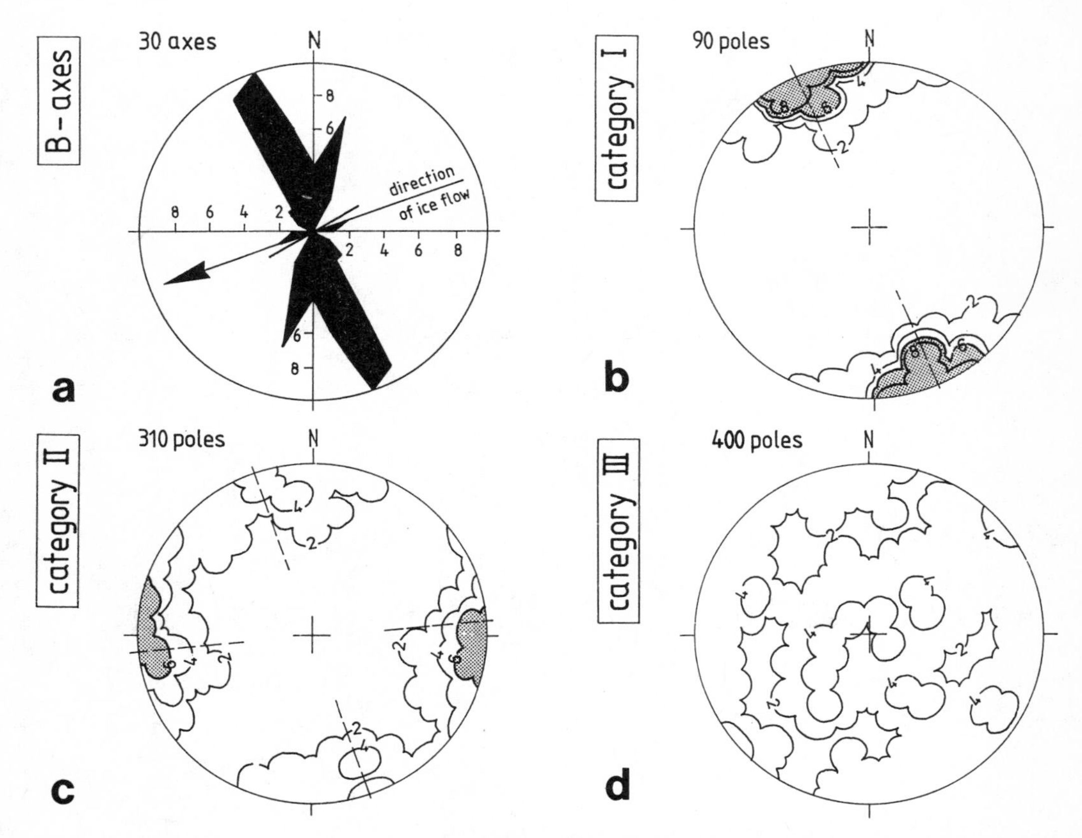

Fig. 9. Glaciotectonic structures in Elsterian basin sediments (Lauenburg Clay) at Querenstede. a: Rose diagram of frequency distribution of B-axes of faults and folds measured in silty clay overlying the fissured clay (frame 180°, intervals 10°). Seperate stereographic projections (lower hemisphere) of discontinuities in fissured clay b: category I, c: category II, d: category III. Joint fabric diagrams show normalized data due to rotation of bedding into the horizontal plane (density of poles in % per 1% area).

3.3 Results

Readings of joint attitudes are plotted as poles normal to the joints on equal-area lower hemisphere projections. To facilitate comparsions, the projections are normalized by rotating them so that the bedding appears in the horizontal plane.

Structural mapping produced evidence of a conjugate set of discontinuities falling into categories I and II, which are nearly perpendicular in strike (Fig. 8b). Apart from this clear geometric relationship, the discontinuities dip very steeply, giving an average dip normal to the bedding planes. As Fig. 8a shows, the B-axes of faults and folds in the overlying laminated clays are perpendicular to the major joint set.

As a separate population, category I discontinuities exhibit a single orientation maximum (Fig. 9b). This orientation is normal to the B-axes of faults and folds in the overlying sediments (Fig. 9a). The pattern is similar to that described by Ehlers and Stephan (1983) and Prange (1987) for tills on the West German Baltic coast. Consequently, it must follow that a dominant set of joints striking in the ac-direction (i.e. in the general direction of tectonism) ought to be a characteristic feature of glaciotectonically stressed clay rich sediments. Ac-orientated joint patterns are also common features of folded, competent lithified sedimentary rocks (Muecke and Charlesworth 1966).

In contrast, category II joints have two maxima (Fig. 9c). The minor set is parallel to planes in category I, whilst the dominant set corresponds with the orientation of B-axes of fault and fold structures. There can be no doubt therefore that the large scale joints falling into categories I and II must be related to the former direction of ice movement.

The small scale discontinuities of category III show no preferred orientation pattern in the normalized projection (Fig. 9d). However, in a non-normalized plot a slight tendency to an orientation parallel to the ground level is revealed. This feature is essentially identical to those observed in other

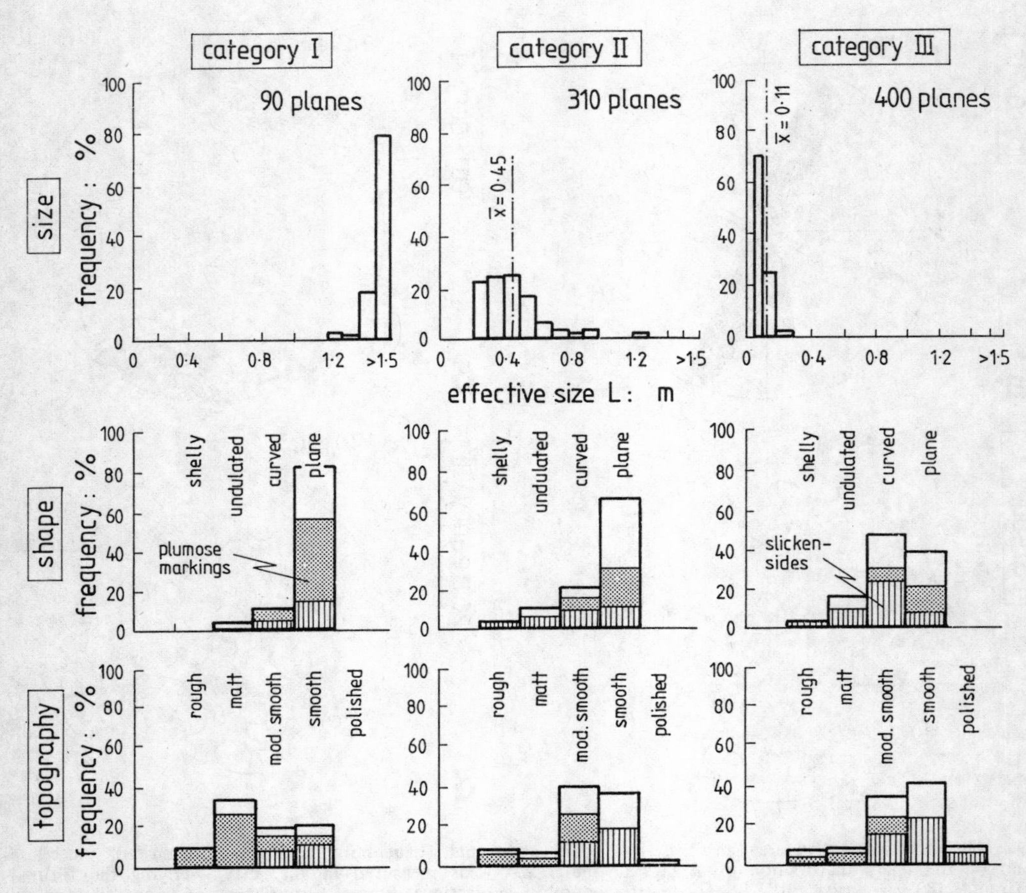

Fig. 10. Joint surface features in Lauenburg Clay at Querenstede. Frequency distribution of size, shape and topography.

Fig. 11. Clay micro fabrics of joint surfaces. Scanning electron micrographs of smooth, polished (a) and rough, matt surface (b).

Fig. 12 Extent of marginal clay particle orientation of smooth joint surfaces. a: Scanning electron micrograph. b: Thin section with isolines of birefringence ratio (ß), (x130, + nicols).

overconsolidated clayey sediments of varying ages (for example Wallrauch 1969).

The histograms of the frequency distribution of joint size (measured as effective size $L = \sqrt{(ab)}$, where a and b are the maximum and minimum dimensions of planes measured) are given in Fig. 10. Inevitably some error occurs in the measurement of category I and III planes since it was rarely possible to observe the full length of category I planes, and joints smaller than 5 cm in length were excluded from category III measurements.

Since the observation procedure was geometrically based, the frequency distributions show strong demarcations between the three categories. Whilst category I joints are generally greater than 120 cm in length, category III joints are tightly clustered around a mean value of 10 cm, and few exceed 20 cm in length. Planes in category II range between 20 cm and 60 cm in size.

Additionally, each size group can be sub-classified according to shape and topography. Large scale category I joints are typically planar and rough or matt. Only a small proportion have curved surfaces and/or smooth textures. The meso-scale category II joints are similar to category I, but there are slightly more smoother surfaces. Category III joints are dominantly curved, or undulating, however, with a high proportion of planar surfaces. Small scale joints are characteristically smooth. Slickensided joints with polished surfaces occur in all categories, but are mainly concentrated in category III. Feather markings are dominant in category I.

4 MICROSCOPIC STUDIES

Scanning electron micrographs (S.E.M.), X-ray texture goniometry and thin section analyses were used to gain an insight into the influence of the failure process on the degree of preferred orientation within the clay matrix, and to underpin the work on the genetic significance of failure markings. Undisturbed block samples, bounded by natural failure planes were collected for these purposes.

4.1 Observational procedures

In the laboratory the blocks were sliced into S.E.M. specimens of 1 cm diameter (maximum), and X-ray specimens of 2 cm diameter (maximum). Since the samples were taken in a state close to the plastic limit, they were prepared by air drying which does not disturb fabric (Barden and Sides 1971, Tovey and Yan 1973). Large specimens (2x4 cm) were cut perpendicular to the margins of block samples for thin sectioning. Standard impregnation techniques using Carbowax and polyester resin as binding agents were used prior to grinding and polishing (Catt and Robinson 1961, Altemüller 1974).

Orientation data are derived from measurements of clay mineral crystallographic basal planes made in X-ray texture goniometry. These data are plotted on lower hemisphere equal-area stereographic projections, and an index of anisotropy (Q̃) derived, which quantifies the degree of preferred fabric orientation. Q̃ provides values ranging from zero (totally preferred orientation) to unity (totally random fabric).

Supplementary analyses of orientation within the clay matrix were carried out on data derived from birefringence measurements on thin sections (Morgenstern and Tchalenko 1967). This technique enables measurements to be made at a more general level within the sample. For a perfectly orientated clay matrix the light intensity transmitted through crossed nicols in a polarising light microscope would exhibit a minimum and maximum value. The birefringence ratio (ß) is the ratio of these two extreme values, and indicates the degree of preferred orientation (ß = 1: random; ß = 0: perfectly parallel orientation pattern).

4.2 Results

Scanning electron micrographs, thin section and X-ray texture analyses confirm the hypothesis that smooth or slickensided surfaces in all three size classes correspond with a high degree of clay particle orientation (Fig. 11, 12 and 13). Both the anisotropy index (Q̃) and the birefringence ration (ß) have values varying between 0.15 and 0.20 (Fig. 12b and 13a).

As thin section examination and S.E.-micrographs show (Fig. 12), the re-orientation of clay particles is restricted to an immediately marginal zone. The most frequent values range between 5 and 15 µm, but in several cases zones of 50 µm were measured. As Fig. 12 shows, the transition from re-orientated particles to the undisturbed core of the clay ped is always abrupt.

The interior of clay peds have anistropy index values (Q̃) of 0.6 to 0.85 (Fig. 13a), with slightly

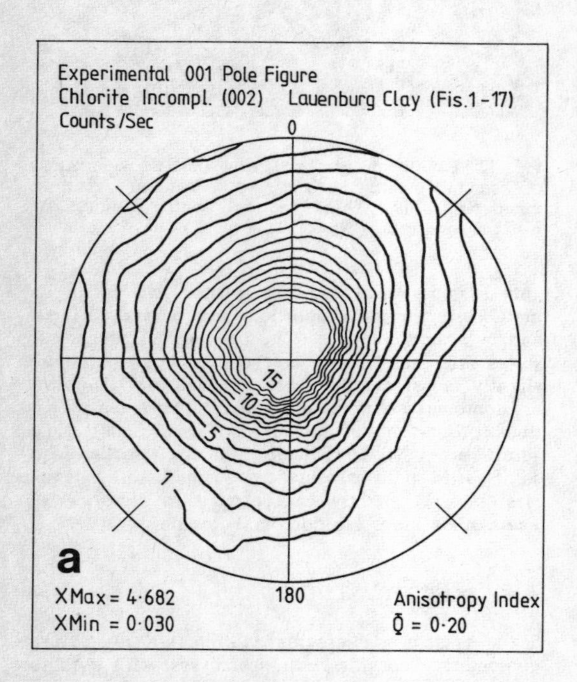

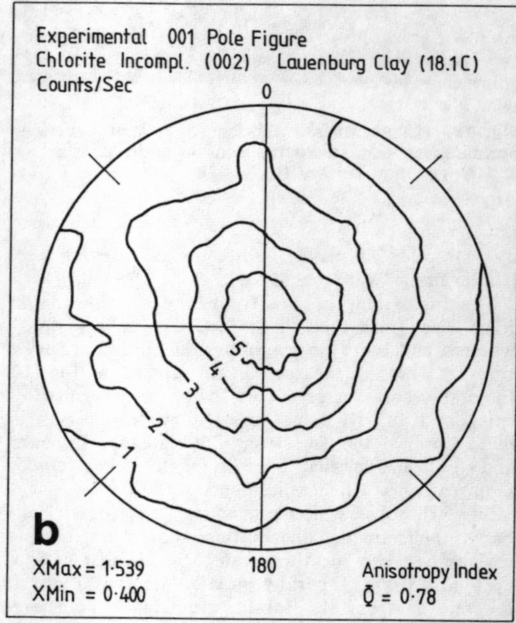

Fig. 13. Clay micro fabrics. X-ray texture analyses. Equal-area lower hemisphere projection of poles of Chlorite (002). Smooth, polished surface (a), rough, matt surface and clay matrix (b). Texture goniometer conditions: CuK radiation, 40 kV, 40 mA.

lower birefringence ratios (ß) of 0.4 to 0.7 (Fig. 12b). Both parameters indicate a moderately strong orientation consistent with normal clay bedding.

Where matt or rough joint surfaces occur, the orientation characteristics of the interior of the clay ped extend right up to the joint margin, without any discernible change. Even in joints with plumose markings the same pattern occurs.

5 STRUCTURAL CONCLUSIONS

A number of conclusions may derived from the site fieldwork and microscopic studies:

i) there are three distinct joint sets, sharply differentiated in size, orientation and shape. In the two groups of large-scale joints, there are conjugate orientation axes. The axes are related to the former direction of ice movement. In the third size category, there is no significant pattern of the joint orientation eaqual to the first and second ones. The great majority of all joints have matt or rough surface textures, with no change in clay microfabric as the faulted margin of the clay ped is approached. Approximately 30% of all discontinuities are slickensided with a high degree of clay particle reorientation indicating shear movement along the joint surfaces.

ii) The fact that slickensided surfaces are mainly concentrated to the small scale joints, whilst the large scale discontinuities, clearly related to a glaciotectonic process, have a rough, un-polished surface suggests that the contortion process responsible for joint development was not a ductile shearing mechanism. In addition, the limited extent of joint margin disturbance in clay particle orientation, even where the margins show both tensile and shear markings, excludes the possibility of a ductile state during joint genesis.

iii) In the cases studied here, glaciotectonic processes must have induced brittle failure. Consequently, joint genesis was extensional, with successive development of joint categories I, II, III. This is a quite different process from that which causes shear failure, and the slickensides observed are a secondary feature.

6 LABORATORY TESTING

A series of unconsolidated-undrained (UU) triaxial tests were carried out on Lauenburg Clay to estimate the conditions leading to the genesis of the discontinuities observed in the field.

Two main factors were controlled: strain was applied at two rates 1% and 0.01% per minute; and confining pressure varied between 0 and 1 MPa. Although it is recognised that temperature is an important factor, the equipment used could not be controlled with sufficient accuracy; further tests are planned. All test specimens (5 cm diameter, 5 cm high) were trimmed from undisturbed blocks. The axial loads are provided by a stiff soil test press with polished and lubricated end plates (which eliminates external constraints on the fracture process).

The trends of the stress-axial strain curves are characteristic of work-softening under uniaxial conditions of zero confining pressure, irrespective of varying stress rates (Fig. 14). Stress-volumetric strain curves examined with the help of a newly developed method for measuring localized deformations of test specimens (Feeser 1984) show a clear dilatancy relationship, typical of brittle behaviour.

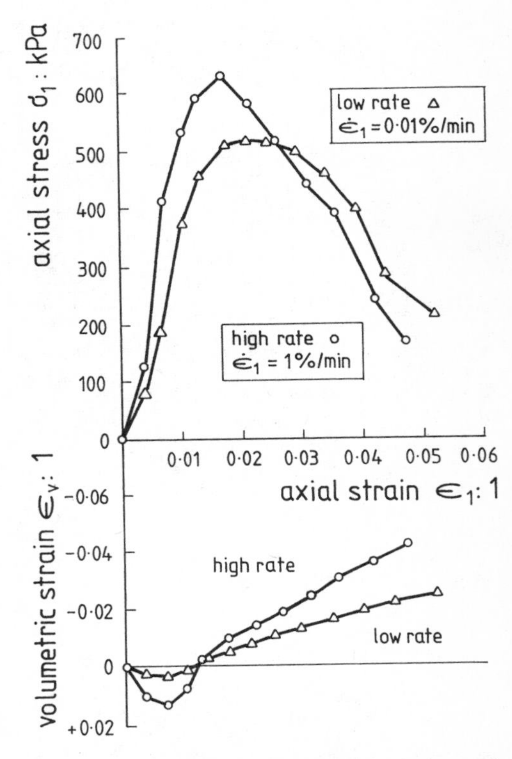

Fig. 14. Stress - axial strain - volumetric strain behaviour of Lauenburg Clay. Uniaxial compression, deformation rates 0.01% and 1% per minute (test no. 82/08-6, 82/08-11).

Additionally, in the post-failure area, specimens macroscopically collapsed by splitting parallel to the direction of maximum principle stress (Fig. 15a). Coalescence of longitudinal fractures along a fault in the central portion of the specimen was also observed (Fig. 15c).

Macro and microscopic studies of laboratory induced failure planes revealed rough surfaces with plumose markings, without parallel marginal orientation of clays. Fracture planes originate at larger particles (such as quartz grains) or inherent discontinuities, and are initially listric in shape (Fig. 15d). These

Fig. 15. Mode of fracture planes in Lauenburg Clay under uniaxial compression. Longitudinal splitting a: front view, b: plan view of test specimen (test no. 82/6-1). Primary longitudinal splitting and secondary inclined faulting c: front view of test specimen (test no. 82/12-3). d: Detail of fracture plane showing plumose markings (test no. 82/12-3).

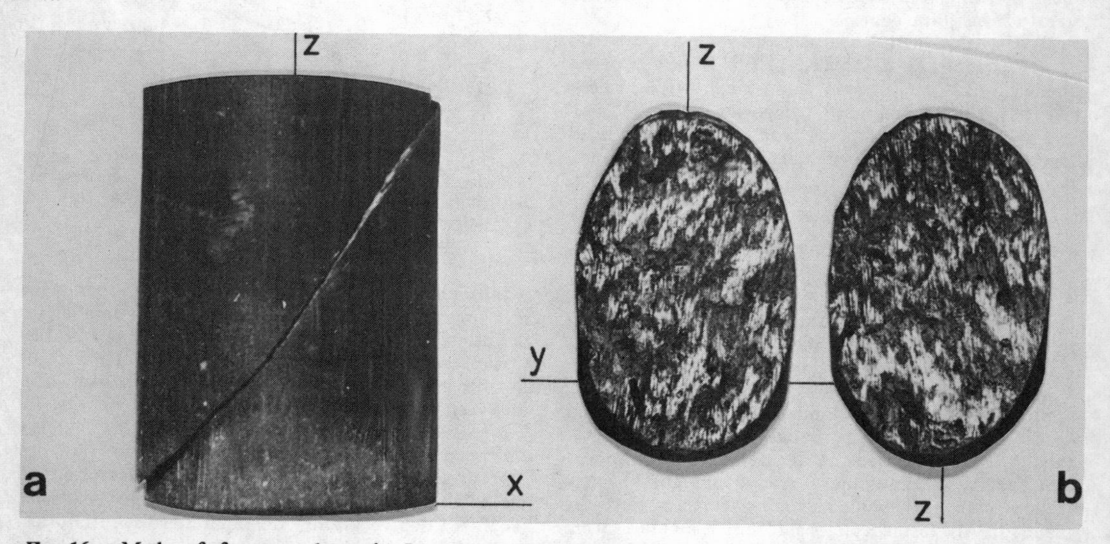

Fig. 16. Mode of fracture planes in Lauenburg Clay under triaxial compression. Inclined faulting a: front view of test specimen, b: view onto the slickensided fracture plane (test no. 82/T-0012).

results show that there is no significant difference between the brittle-state behaviour of stiff clay and that of rock materials.

Under increased levels of confining pressure, behaviour is still brittle, but there is a tendency towards work-hardening. Triaxial conditions does not provide a good medium for the study of micro dilatancy, but void ratio measurements can be used as a surrogate. Measurements before and after loading are compared to yield a value (E) which is the difference between the initial and final, divided by the initial void ratio. Dilatancy is indicated by negative values (E), and contractancy by positive values (E).

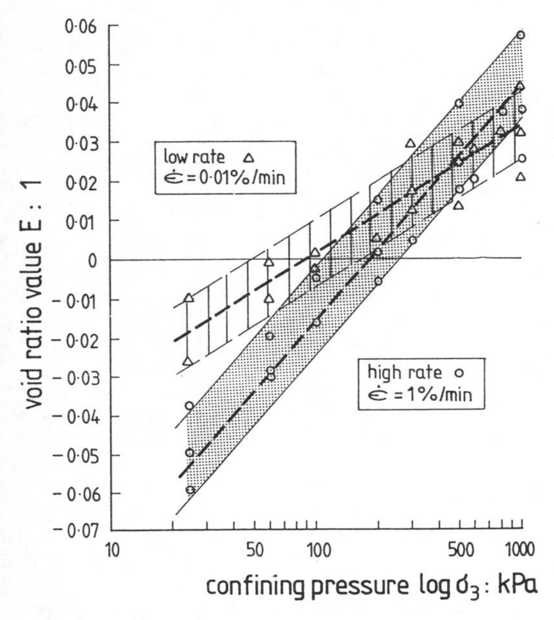

Fig. 17. Dilatancy-contractancy transition of Lauenburg Clay under triaxial compression. Void ratio value E vs. confining pressure; deformation rates 0.01% and 1% per minute.

Fig. 17 (E vs. confining pressure) shows a transition from dilatancy to contractancy between 50 and 250 kPa confining pressure depending on the deformation rate. Tests under confining pressures up to 100 kPa (for low deformation rate) and 200 kPa (for high deformation rate) would be expected to produce subaxial splitting with rough failure planes if failure was brittle. However, all triaxial testing produced slickensided fault planes, without exception (Fig. 16).

In its present field state, but subjected to low confining pressures, moderate to high deformation rates, and laboratory testing temperatures, the behaviour of Lauenburg Clay falls within the brittle/ ductile transition. In order to produce fully brittle triaxial behaviour, as indicated by tensile fractures, either the deformation rates have to be substantially increased, or temperatures substantially decreased.

7 MECHANICAL CONCLUSIONS

Laboratory testing confirms the hypothesis generated from field observations, that certain features of natural joints and fractures, particularly surface geometry and topography, characterise tensile fracturing. Similarly, laboratory evidence confirms the brittle state of the clay at the time of fracturing and the sequential development of joint categories I, II and III. The environmental conditions in which the joint patterns developed may be discussed in more detail.

7.1 Environmental conditions

It is reasonable to assume that the deformation took place at up to 20 m below the glacial ground surface level, with consequent maximum confining pressures of c. 300 kPa. Naturally occuring deformation rates in excess of those used in testing (1% per minute) are inconceivable. Given that Pleistocene glacier bed sliding velocities did not exceed 1000 mm per day, deformation rates in the substratum could not exceed 0.001% per min. Laboratory tests show that at today's normal temperatures such confining pressures and deformation rates do not induce brittle failure as has been observed in the field. Hence, we must con-

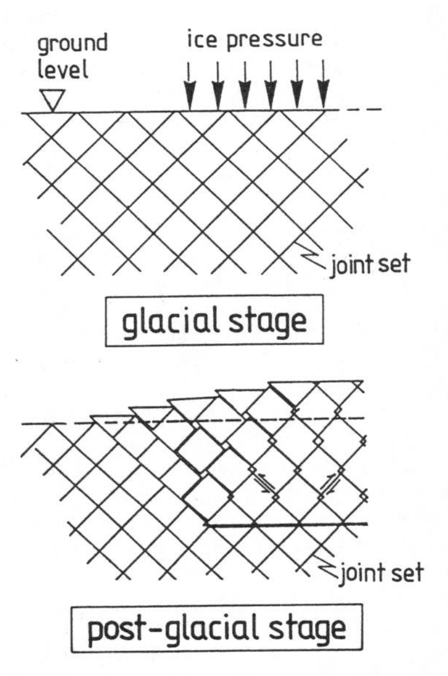

Fig. 18. Model of development of slickensides along pre-existing rough joints caused by differential swelling of clay blocks in post-glacial region due to unloading.

clude that brittle failure occured under low temperatures. Additionally, in its post-deformation state, Lauenburg Clay is much more solid than it would have been before deformation. It is therefore even less likely that brittle failure could have occurred without freezing to substantial depths.

Some caution must be expressed about the relationship between tensile induced failure and slickenslide surfaces observed on 30% of cases. In the past many authors have assumed slickensided surfaces to be exclusive to shear induced failure.

Given the fact that totally preferred clay particle orientation occurs in the surface layers of a shear zone following very short displacements in the range of a fractional part of a millimetre (Rizkallah 1977), differential swelling of clays may well be sufficient to produce a surface which appears slickensided. In this study slickensides occur mainly in category III joints, which are accounted for by unloading stresses following glacial retreat. As Fig. 18 shows, unloading stresses are accommodated by differential movement of individual clay blocks along joint sets developed during overriding. Such small differential movements probably account for the superimposition of slickensides on predominantly rough fracture surfaces, and not primary shear-induced failure.

7.2 Stress history

Given the field and laboratory data, we can now reconstruct the glacially induced stress history of the Lauenburg Clay. Firstly, a chronological sequence of principle stress states can be developed which follows from stress fields associated with the development sequence of category I, II and III planes . The stresses are modelled on an infinitessimally small clay element at a moderate depth below the preglacial surface. The x-axis is in the direction of strike of category I joints, and the z-axis gravitationally downwards (Fig. 19).

Point 3 indicates element position just before and, point 5 just after glacial advance. The clay is normally consolidated prior to glacial loading (between points 1 and 2), and during the phase of total ice overriding (between point 4 and 5). Over-consolidation prevails from point 2 to 4, within an immediately proglacial environment, and during unloading from the time of ice retreat to the present-day (point 6). Limit states in the collapse process of brittle failure are denoted by the points I, II and III. When the seperate points are connected, the path describes the successive stresses due to glaciotectonism.

The second stage in the reconstruction involves coupling this stress path with particular glaciotectonic events. Assuming the overriding glacier transmits lateral forces through the tip of the glacier tongue, and that the structural co-ordinate system a-b-c corresponds with the clay element system x-y-z, then the plane of the fabric forming direction (ac-axis) corresponds with the x-z plane of the stress system.

The pattern of category I joints shows that maximum stresses were in the longitudinal (a-axis), medium stresses in the vertical (c-axis) and minimum stresses in the lateral (b-axis) directions. This limit stress distribution is consistent with a stress field on the ice foreground (Fig. 20), corresponding to point I in Fig. 19. Category I joints are therefore developed proglacially along with the major thrust and fault systems in the overlying beds.

As the glacier approaches, the axes of the principle stresses are rotated, as vertical ice-overburden forces increase and longitudinal forces become less important. The major stress axis, previously horizontal and parallel to the direction of ice movement, becomes increasingly steep (still parallel to the direction of ice movement, Fig. 20). Concurrent with this axial shift, the positions of the medium and minor axes also change. Minor stresses initially perpendicular to the ice flow (a-axis) rotate and turn into the direction of ice movement (b-axis) as the glacier approaches (Fig. 20). This change in axial symmetry of the stress system is correlated

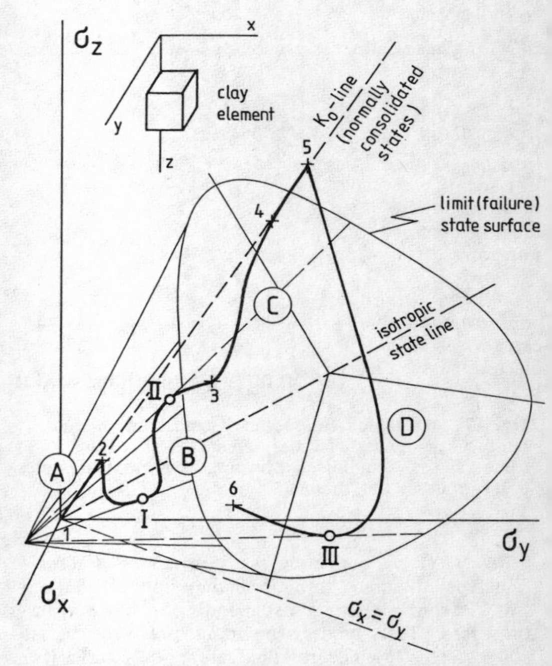

Fig. 19. Reconstruction of stress history of glaciotectonically contorted clays. Stress path in three-dimensional principle stress diagram.

with the transition to extensional loading which is necessary to generate category II joints (point II, Fig. 19). Following dynamic glaciation, no longitudinal forces are transmitted, and the element returns to a geostatically normal-consolidated state (path 4-5, Fig. 19).

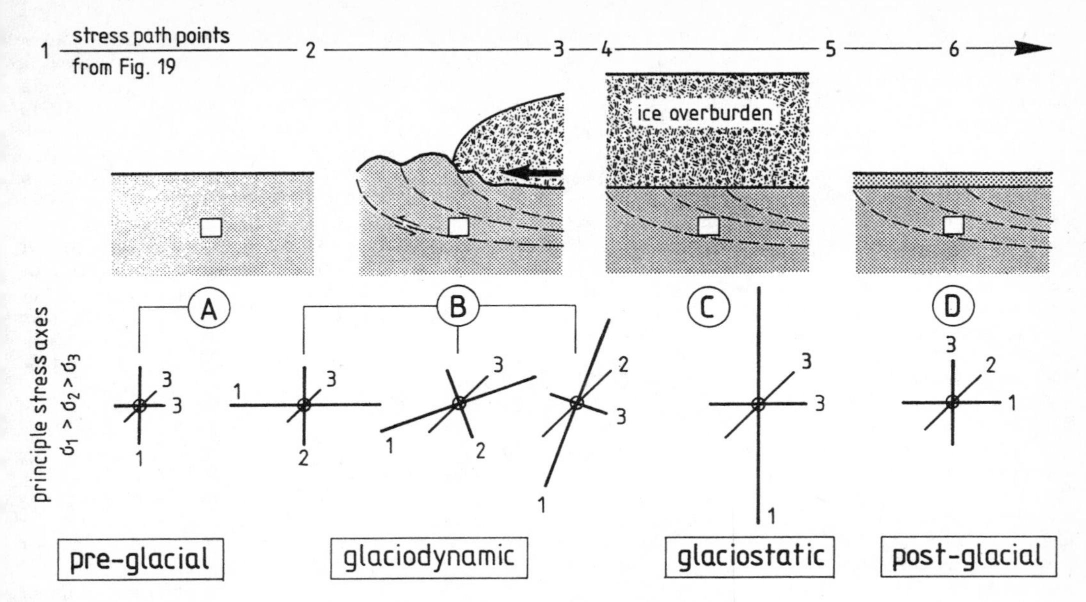

Fig. 20. Classification of glacial contortion mechanisms due to principle stress variation.

Category III joints display no symmetry which can be related to the overriding process, and must therefore be assigned to a second stage of genesis. Since like many other over-consolidated clayey sediments they are slightly parallel to the ground surface, it must be assumed that they were caused by geostatical unloading following glacial retreat (point III, Fig. 19).

Following the suggested stress path glaciotectonic contortion mechanisms in Pleistocene clays can be categorized into four phases: pre-glacial phase Ⓐ, glaciodynamic phase Ⓑ, glaciostatic phase Ⓒ and post-glacial phase Ⓓ (Fig. 19 and 20).

ACKNOWLEDGEMENTS

The work reported comprises part of an investigation carried out at the Technical University of Braunschweig and was supported by the German Research Community DFG within the special programme "Engineering geological problems on the border between soft and hard rocks".

The author wishes to record his thanks to Prof H.J. Altemüller (Braunschweig), Prof H. Siemes (Aachen), Prof B. Mattiat (Hannover) for their help in microscopic studies; all students and colleagues who assisted in field work; the Klinkerwerke Röben (Zetel) for permitting access to their clay pit and supporting the site investigations and especially to Dr D.G. Croot (Plymouth) for critically reading the draft.

REFERENCES

Altemüller, H.-J. 1974. Mikroskopie der Böden mit Hilfe von Dünnschliffen. In H. Freud (ed.), Handbuch der Mikroskopie in der Technik IV, 2: 309-367. Frankfurt/Main, Umschau.

Banham, P.H. 1975. Glacitectonic structures: a general discussion with particular reference to the consorted drift of Norfolk. In A.E. Wright and F. Moseley (eds.), Ice ages - ancient and modern. Geol.J.Spec.Issue 6: 69-94.

Barden, L. and Sides, G.R. (1971). Sample disturbance in the investigations of clay structures. Géotechnique 21: 211-222.

Bock, H. 1974. Ein Beitrag zur statistischen Kluftmessung. In E. Fecker, H.P. Götz, G. Sauer and G. Spaun (eds.), Festschrift Leopold Müller-Salzburg zum 65. Geburtstag, p.99-111. Karlsruhe.

Brace, W.F. and Bombolakis, E.G. 1963. A note on brittle crack growth in compression. J.Geophys.Res. 68: 2709-2713.

Catt, J.A. and Robinson, P.C. 1961. The preparation of thin sections of clay. Geol.Mag. 98: 511-514.

D'Elia, B. 1973. Metodi di osservazione impiegati per lo studio di discontinuitá strutturali nell'argilla di Londra. L'Ingenere 48: 218-233.

Ehlers, J. and Stephan, H.-J. 1983. Till fabric and ice movement. In J. Ehlers (ed.) Glacial deposits in North-West Europe, p. 267-274. Rotterdam, Balkema.

Feeser, V. 1984. Optographic trace recording: a new method of strain measurement in geotechnical testing. Géotechnique 34: 277-281.

Fookes, P.G. 1965. Orientation of fissures in stiff overconsolidated clay of the Siwalik system. Géotechnique 15: 195-206.

Fookes, P.G. and Parrish, D.G. 1969. Observations on small-scale structural discontinuities in the London Clay and their relationship to regional geology. Quart.J.Eng.Geol. 1: 217-240.

Gregory, C.H. 1844. On railway cuttings and embankments; with an account of some slips in the London Clay on the line of the London and Croydon Railway. Min.Proc.Instn Civ.Engrs 3: 135-145.

Griffith, A.A. 1924. The theory of rupture. Proc. 1st Int.Con.Appl.Mech. (Delft): 55-63.

Hallbauer, D.K., Wagner, H. and Cook, N.G.W. 1973. Some observations concerning the microscopic and mechanical behaviour of quartzite specimens in stiff, triaxial compression tests. Int.J.Rock Mech.Min.Sci. & Geomech.Abstr. 10: 713-726.

Hoek, E. 1983. Strength of jointed rock masses. Géotechnique 33: 187-223.

Hoek, E. and Bieniawski, Z.T. 1965. Brittle fracture propagation in rock under compression. Int.J.Frac. Mech. 1: 137-155.

Jaeger,J.C. and Cook,N.G.W. 1976. Fundamentals of rock mechanics (2nd ed.). London: Chapman.

Kazi, A. and Knill, J.L. 1973. Fissuring in glacial lake clays and tills on the Norfolk coast, United Kingdom. Eng.Geol. 7: 35-48.

Kranz, R.L. 1979. Crack-crack and crack-pore interactions in stressed granite. Int.J.Rock Mech.Min. Sci. & Geomech.Abstr. 16: 37-47.

Lajtai, E.Z. 1975. Failure along planes of weakness. Can.Geotech.J. 12: 118-125.

Möbus, G. and Peterss, K. 1983. Kluftstatistische Spannungsanalyse im Geschiebemergel. Analysis of glacitectonical structures - IV Glacitec.Symp. Wyzsza Szkola Insynierska (Zielona Gora): 139-153.

Morgenstern, N.R. and Tchalenko, J.S. 1967. The optical determination of preferred orientation in clays and its application to the study of microstructure in consolidated Kaolin I. Proc.Royal Soc. A.(1463) 300, p.218-234. London.

Muecke, G.K. and Charlesworth 1966. Jointing in folded Cardium Sandstone along the Bow River. Can. J.Earth Sci. 3: 579-596.

Prange, W. 1987. Gefügekundliche Untersuchungen der weichseleiszeitlichen Ablagerungen an den Steilufern des Dänischen Wohlds, Schleswig-Holstein. Meyniana 39: 85-110.

Rizkallah, V. 1977. Stress-strain behaviour of fissured stiff clays. Proc. 9th Int.Con.Soil Mech. Found.Eng. (Tokyo) 1: 267-270.

Skempton, A.W., Schuster, R.L. and Petley, D.J. 1969. Joints and fissures in the London Clay at Wrasbury and Edgware. Géotechnique 19: 205-217.

Terzaghi, K. 1936. Stability of slopes of natural clay. Proc. 1st Int.Conf.Soil Mech. 1: 161-165.

Tovey, N.K. and Yan, W.K. 1973. The preparation of soils and other geological materials for the S.E.M.. Proc.Int.Symp.Soil Struc. (Gothenburg): 60-68.

Wallrauch, E. 1969. Verwitterung und Entspannung bei überkonsolidierten tonig-schluffigen Gesteinen Südwestdeutschlands. Thesis Univ. Tübingen.

Glaciotectonics: Forms and Processes, Croot (ed.), © 1988 Balkema, Rotterdam. ISBN 90 6191 848 0

The Halland Coastal Moraines: Are they end moraines or glaciotectonic ridges?

Joanne M.R.Fernlund
Department of Quaternary Geology, Uppsala University, Uppsala, Sweden

ABSTRACT: The "Halland Coastal Moraines" refer to the numerous NW–SE trending ice-transverse ridges in the Varberg–Falkenberg area along the west coast of Sweden. Previously all of the ridges have been interpreted as end moraines, formed during the Late Weichselian deglaciation, circa. 13,000 to 13,500 ^{14}C years B.P. (Fig. 1). Stratigraphical studies have revealed that not all of the ridges are end moraines. Some are glaciotectonic ridges formed by a readvance of the ice after 12,600 ^{14}C years B.P. (Fig. 10). This readvance along the west coast followed an ice-free period of at least 500 years. The readvance is suggested to have originated east of the Göteborg End Moraine and have extended west of the Tofta Ridge and probably severals km off the present swedish west coast. The ridges associated with drumlins are inferred to have formed during this readvance (Fig. 1). In such case the lateral extent of the readvance must at least have been from Falkenberg to Onsala. The advancing glacier was possibly a warm based ice stream or surge which protruded out from the main ice sheet.

1 HISTORY

The group of ridges in the Varberg–Falkenberg area were first described by Svedmark (1893) and De Geer (1893), and are collectively called the "Halland Coastal Moraines", HCM (Berglund 1979). Various lateral correlations of the HCM have been presented using morphological studies suggesting that they are end moraines (Berglund 1979, De Geer 1893, Hillefors 1969, 1979, Mörner 1969,). However, their mode of formation has never been stratigraphically determined.

Svedmark's (1893) interpretation of the ridges was based upon their composition, which he presumed to be till, and their trend, perpendicular to the main ice flow direction which was from NE to SW. He described ridges in two other directions which he interpreted to be remnants of end moraines from earlier glacial periods (inferred from older striae perpendicular to these ridges). Although Svedmark's ridges in directions other than NW–SE have never since been discussed, still other ridges trending NE–SW have been interpreted as lateral moraines to valley glaciers, drumlins or subglacial streamlined forms (Caldenius 1942, Hillefors 1969, 1975, 1979).

De Geer (1893) proposed that the ice-transverse ridges were formed by waves casting debris up against a stationary ice margin. He presented three ice-stationary positions which formed during the glacial retreat along the west coast: 1. the Påarp–Sundrum Line, 2. the Halmstad–Falkenberg–Varberg Line, 3. the Fjärås Line. The first two are included in the Halland Coastal Moraines. The latter is more often referred to as the Göteborg (Gothenburg) End Moraine, GEM, a complex end moraine indicating a glacial readvance circa. 12,800 to 12,600 ^{14}C years B.P. (Fig. 1) (Berglund 1979, Hillefors 1975, 1979).

The HCM are often associated with drumlin-like forms. Those on the Ölmanäs drumlin are suggested to have been formed as push moraines by large calving icebergs (Påsse 1986). The extention of the HCM between the Ölmanäs and Onsala drumlins was attempted by echo sounding without success (Fält 1975).

2 DEFINITIONS OF RIDGES

In the 1900's basically all transverse glacial ridges were classified as end moraines, "a ridge like accumulation of drift built along any part of the margin of an active glacier" (Flint 1971 p. 200) which "indicate that the glacier terminus occupied these positions longer than neighboring positions" (Flint 1971 p. 203). An exception is a push moraine which is formed by the bulldozing of an advancing terminus (Chamberlin 1894, p. 525). Although formed at the ice margin, a push ridge is a type

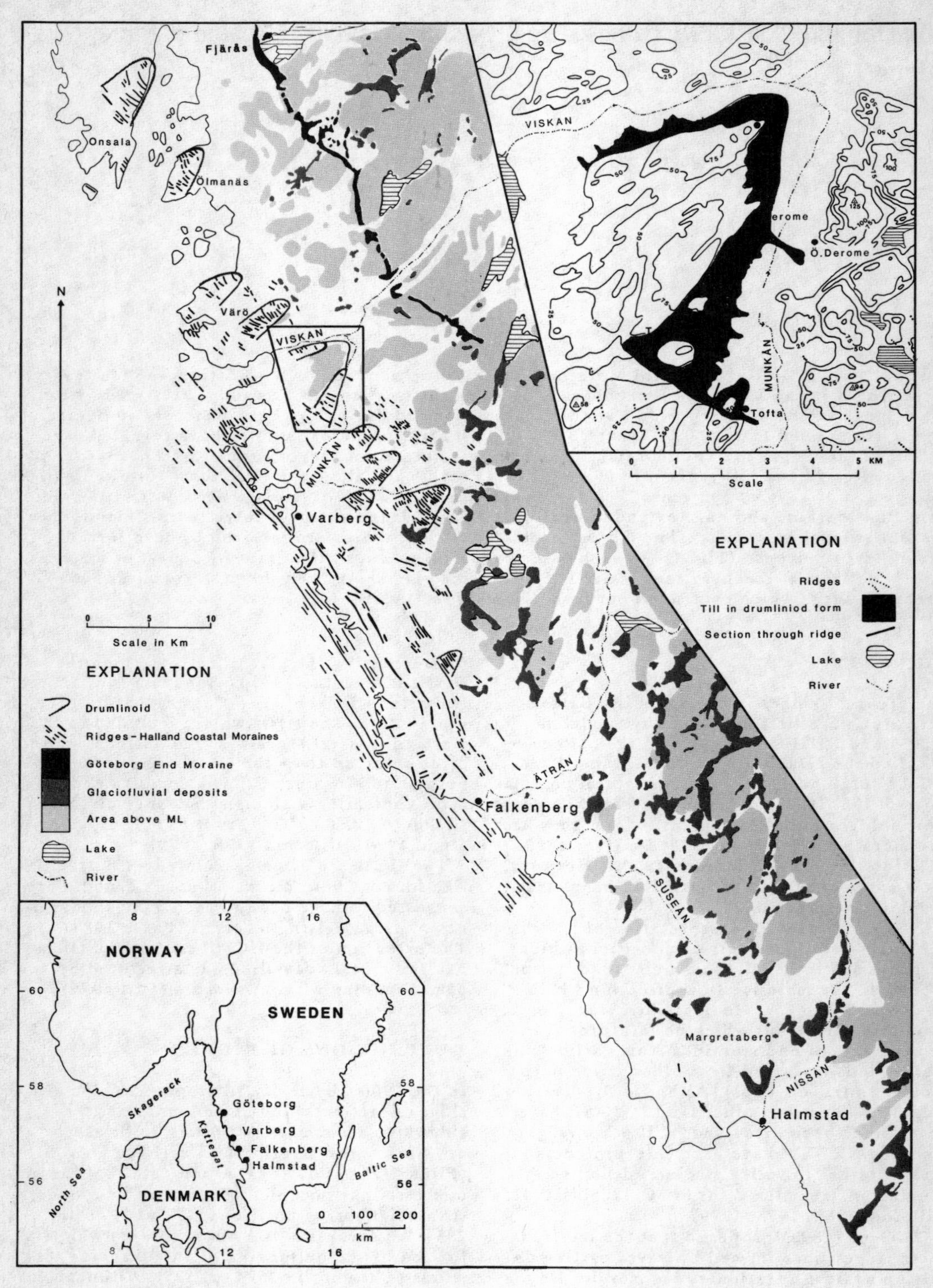

Fig. 1 Location map.

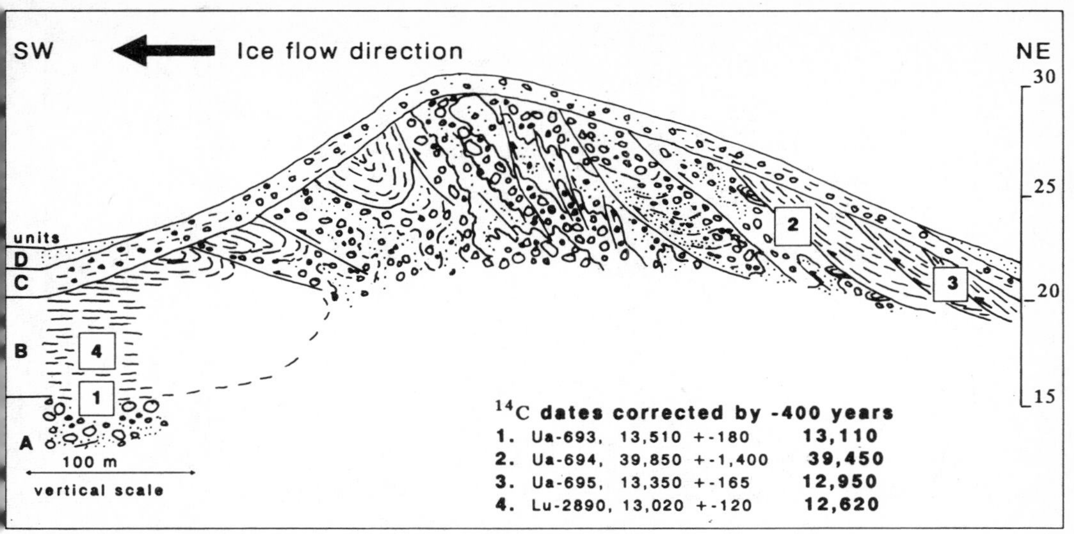

Fig. 2 Schematic cross section of the Tofta Ridge.

of end moraine which marks the maximum position of a glacier advance. Several ice-transverse ridges have been erroneously interpreted as end moraines (Bluemle & Clayton 1984, Möller 1987, Moran 1971). Stratigraphic studies are necessary to distinguish their mode of formation.

Glacial tectonic ridges, GTR, are the product of deformation and dislocation of pre-existing bedrock and/or drift as a direct result of glacier-ice movement or loading (Aber 1985). Push moraines are considered GTR although not all GTR are push moraines. The two main theories for GTR formation are subglacial and proglacial thrusting. Subglacial thrusting is inferred to occur under cold-based glaciers, due to upward thrusting of large blocks of substratum into the ice. Proglacial thrusting involves detatchment of pre-existing stratum along a plane of décollement, either at the base of a permafrost layer or a natural aquaclude. The stratam is then pushed in front of the ice rather than entrained up into the ice. This most often occurs in front of cold-based glaciers but also in front of warm-based glaciers. In both theories, thrusting is thought to be enhanced by high pore-water pressure in an underlying confined aquifer (Aber 1980, 1982, 1985, Berthelsen 1979, Bluemle 1970, Bluemle & Clayton 1984, Clayton & Moran 1974, Croot 1987, Dreimanis 1976, Gripp 1929, Gry 1940, Jessen 1931, Moran 1971, Moran et al. 1980, Nielsen 1967). When a GTR is overridden by a glacier the upper part is often deformed into a melange (Aber 1980, 1982, Banham 1975).

3 GENERAL SETTING

The HCM trend nearly parallel to the present coast line and are most frequent in the Varberg – Falkenberg area (Fig. 1). They are almost always located below the marine limit, ML. In this area the ML is somewhat controversial, about +60 to +70 m, and several transgressions have been described (Caldenius 1942, Gillberg 1956, Hillefors 1969, Mörner 1969, Påsse 1986, 1988, Svedmark 1893, Wedel 1971). According to Påsse (1988) the ML, was formed at the time of deglaciation and was followed by a rapid regression.

Below ML there is a thick sequence of clay. Above ML the landscape changes character around the Falkenberg area. Northwards it is characterized by naked gneiss bedrock, nearly lacking till and glaciofluvial deposits. In contrast, towards the south there is a nearly continuous till cover and numerous glaciofluvial deposits.

3.1 Morphological types of HCM ridges

The HCM ridges vary with regards to morphology and geographical distribution. They have, therfore, been divided into four types which possibly have different genesis and age: Type 1 – long broad flat ridges which occur singularly and trend roughly NW – SE, Type 2 – groups of ridges which trend NW – SE and are often associated with drumlin forms, Type 3 – ridges oriented in directions other than NW – SE and Type 4 – ridges which can be classified into more than one of the above groups.

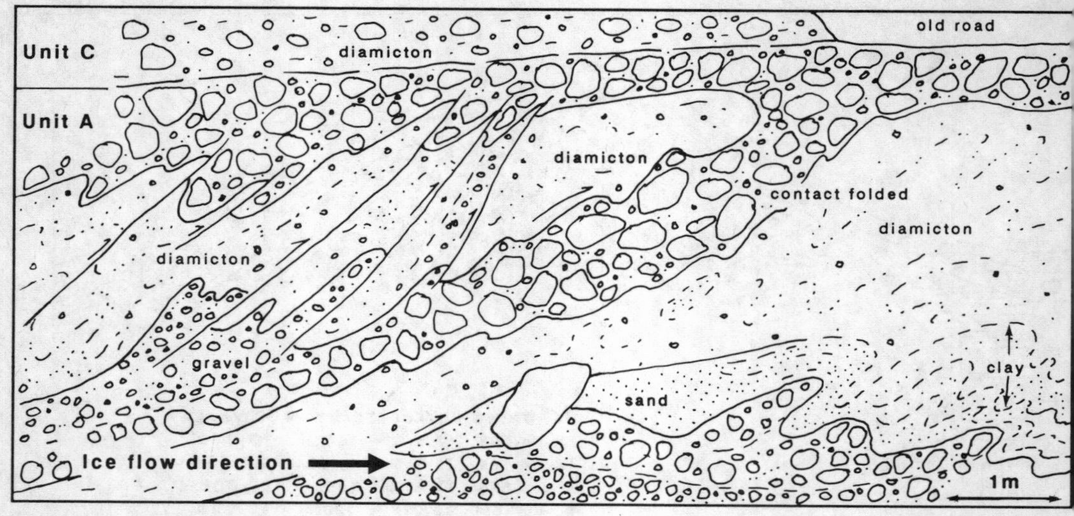

Fig. 3 Section at the ridge crest.

3.2 The Torpa – Tofta Ridges

The Torpa – Tofta Ridges are situated about 10 km NNE of Varberg along the west side of the Munkån River Valley (Fig.1). The valley is between 5 and 10 m above sea level, about 1.5 – 2 km wide and bounded on each side by steep hills of gneiss. Streamlining of the bedrock is pronounced where the foliation is parallell to the ice flow direction (NE and ENE).

The series of NW – SE trending ridges coincides with a drumlin form. This is common along the west coast of Sweden; eg. Hunnestad, Värö, Ölmanäs, Onsala, Dösebacka and Ellesbo (Fält 1975, Hillefors 1983, Påsse 1986). The ridges are usually only evident along the drumlin's side, which gives them a scalloped appearance. In the study area two of the ridges are longer and extend out into the valley while the others stop at the valley side. One of the longer ridges extends from the ice-proximal NE end of the drumlin, between V. Derome and Ö. Derome. The other, which is described in this paper, is at the ice-distal SW end near Tofta.

Based upon the classification of ridges, described earlier, the Torpa – Tofta Ridges as a whole, are clearly Type 2, however the two longer ridges could be considered Type 1 and therefore might be classified as Type 4.

3.3 The Tofta Ridge

The longer ridge near Tofta is a distinct ridge for 1 km and a composite ridge for about 3 km (Fig. 1). The icc proximal, NE side, has a slightly more gentle slope than the distal, SW side. The height of the ridge crest varies be-

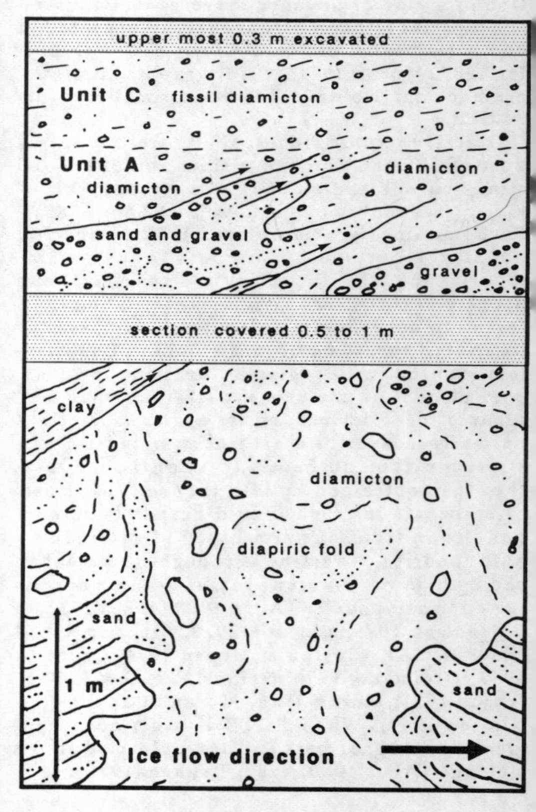

Fig. 4 Section 10 m SW of the ridge crest.

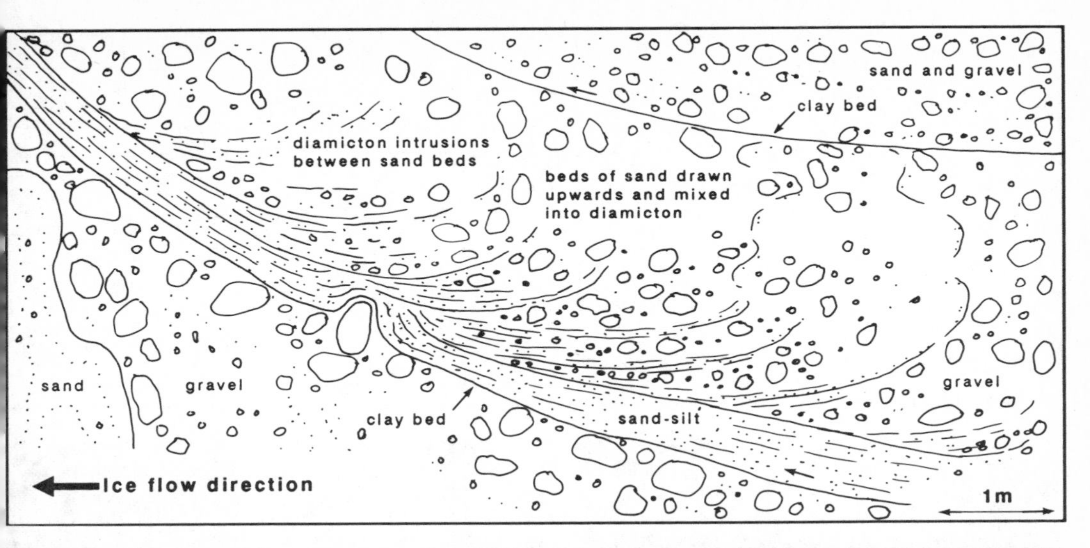

Fig. 5 Section 100 m NE of the ridge crest at +20 m.

tween +25 to +35 m and the ridge width is about 700 m. The section was studied in conection with the construction of a new road, highway 41, which cuts through the Tofta Ridge. The exposure extended from the crest at +30 m down to +20 m, with one exception where it extended down to +16 m. The internal composition and structures have been studied in all parts of the ridge periodically as the excavation progressed.

4 STRATIGRAPHY

The section has been divided into four main units, (Fig. 2) from oldest to youngest: Unit A – glaciomarine (gravel, sand and diamicton), Unit B – still-water marine (rhythmiclly bedded clay and silt), Unit C – glacial diamicton glacially deformed clay, sand and gravel , and Unit D – marine near shore (sand and gravel).

Unit A makes up the central portion of the ridge. The grain size of this unit varies widely from coarse cobble-gravel to sand, but also contains clay and a few large boulders 1 to 2 m in diameter (Fig. 3). The clasts are exclusivly composed of crystalline rocks. Most of Unit A sediments are water sorted, although few primary structures have survived post depositional deformation. About 25 – 35% of Unit A is a diamicton, possibly till or a product of soft sediment deformation (Fig. 3, 4, 5). Coarse cobble beds often maintain their tabular form where as the interbeds of sand and silt are often highly deformed, mixed into irregularly formed diamicton masses. The contacts between beds are often slightly folded and there is often a zone (10 cm) of mixing between adjacent beds (Fig. 3, 8). The deformation structures are described in more detail in section 7.

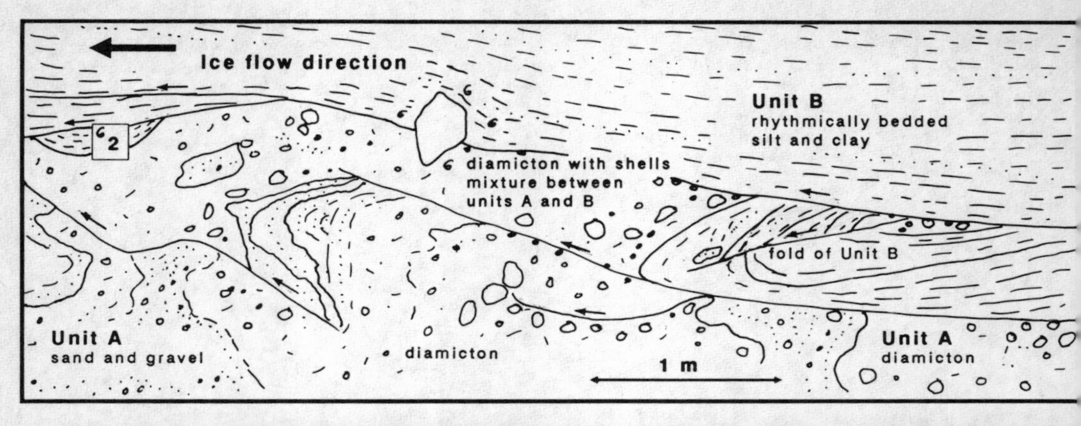

Fig. 6 The tectonic A/B contact on the NE side of the ridge.

Unit B is 4 metres thick on each side of the ridge, however, it wedges out half way up the slopes due to folding and/or faulting (Fig. 2). Its contact to Unit A is depositional on the SW side of the ridge (Fig. 7) and tectonic on the NE side (Fig. 6), where it occurs as imbricate slices. Unit B is composed of rhythmic-bedded clay and silt as well as some interbeds of sand and gravely sand. The beds are in the order of 1 to 2 cm thick (Fig. 7). The unit contains drop stones, up to cobble size, of both crystalline and sedimentary rocks, primarily of flint and limestone.

Shells occur at several places in Unit B. Along the SW side of the ridge there are two main horizons (Fig. 2, 7) and along the NW side there are several occurances, however, the imbrication complicates the vertical stratigraphy (Fig. 2, 6). The fauna assemblage includes: *Astarte borealis, Balanus hameri, Hiatella arctica, Macoma calcarea, Mytilus edulis* and *Neptuna despecta*.

Unit C is composed of both clay rich and sandy-gravelly diamicton. These two facies reflect the grain size of the underlying sediments. Unit C is a clay-rich diamicton on the sides of the ridge where it overlies Unit B, and a sandy-gravely diamicton where it overlies Unit A. Changes in facies occur abou 5 m, in a down-ice direction, after changes in composition of the underlying sediments (Fig. 2). The contact between Unit C and the under lying Units A and B is gradational, downward diminishing degree of deformation (Fig. 7, 8). The clasts are often up to cobble size and are composed of both crystalline and sedimentary rocks. It is uniform in thickness (0.5 – 1 m) over the entire ridge. It is less distinct where it overlies Unit A which has a very similar con position, however, Unit C contains sedimentary-rock clasts. There are numerous vertical to nearly vertical silt-filled partings, with widths on the order of 1 to 3 cm and lengths which seldom exceed 1 m (Fig. 7, 8). These partings extend down from the lower part of the Unit C into the underlying Units A and B, and trend roughly parallel to the ridge crest.

Unit D is predominatly composed of silt, sand and gravelly sand. It only occurs below +22 m and is about 50 cm thick. Primary bedding is nearly lacking possibly due to cryoturbation or solufluction mixing. Its erosional contact to Unit C is generally planer, however it is somewhat diffuse.

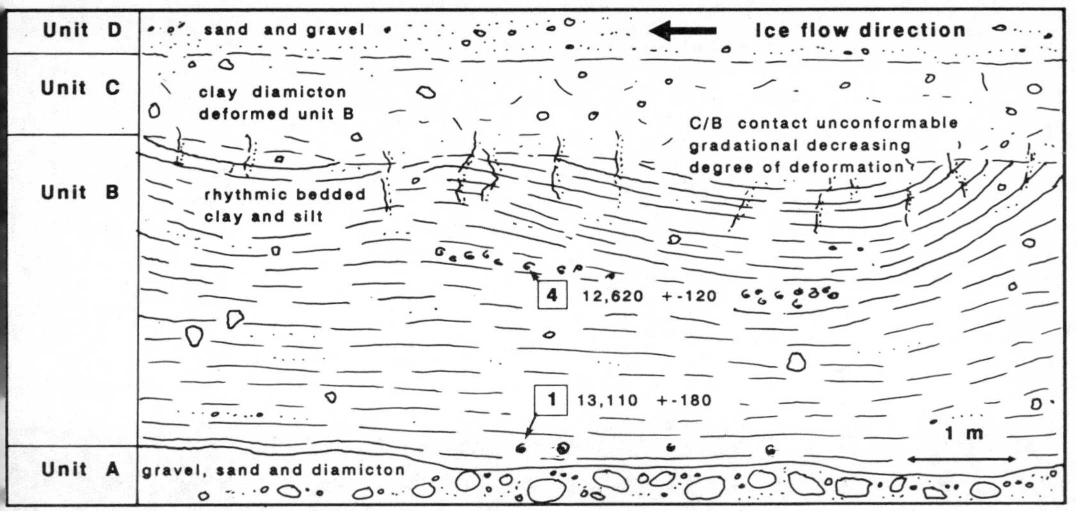

Fig. 7 Section of Unit B 200m SW of the ridge crest.

5 CARBON 14 DATES

Four ^{14}C dates have been determined from the shells in Unit B. Three are based on the accelerator method using the "inner" fractions (carried out by Göran Possnert at The Svedberg Laboratory at Uppsala University). One date was determined by the conventional method (by Göran Skog at the RadioCarbon Dating Laboratory at Lund University). The dates are corrected equivalent to $\delta^{13}C = 0.0$ 0/00 and by minus 400 years for the reservoir effect of sea water (Fig. 2). The ^{14}C date, sample UA-694 39,850 +-1,400 B.P., from the clay on the NE side (Fig. 2, 6) suggests that it could be upthrusted clay of interglacial age which stratigraphically is lower than Unit A.

6 MODE OF DEPOSITION

The environment of depositon is interpreted from the stratigraphy presented above.

Unit A, is interpreted as **glaciomarine sediments**. This is based upon the rapid lateral and vertical change in grain size (cobble gravel to silt and clay), and its position along the west coast where ML is in the order of 60 to 70 m above sea level. Due to the extent of the deformation it is difficult to determine the exact origin of the sediments. It is possibly a highly deformed delta, esker bead, or a sequence of subaquatic flow till intercalated with glaciomarine water-sorted sediments.

Unit B is interpreted to be deposited in a **still-water marine environment**. The difference in the ^{14}C dates as well as the 4 m thick sequence of clay suggest a sedimentation period of at least 500 years. The provenance of the ice-rafted sedimentary-rock clasts is the Baltic Sea Basin, southernmost Sweden and/or Denmark.

Unit C is interpreted as **glacially deformed pre-existing sediments**. The deformation is the product of a glacial readvance in the Varberg–Falkenberg area which caused the formation of the Tofta Ridge and the deformation of the upper parts of Units A and B. It is not interpreted as a "true till" (Driemanis 1982) since there is no evidence that it has ever been incorporated into the glacier.

Unit D is interpreted to be deposited in a **near shore environment**. The timing of the deposition is still unclear. It could have been formed as the sea regressed following the glacial readvance. It could also have been formed during the post glacial transgression, although this is only reported to have reached about +15 m in this area (Påsse 1988).

7 DEFORMATION STRUCTURES

Units A and B are highly deformed with dips that vary from 0° to 90°, folds, faults, and soft sediment mixing (Fig. 2). On the SW slope there is a large recumbent isoclinal fold (Fig. 8) as well as dewatering structures. The more competent Unit B beds usually form continuous bodies, such as long fold limbs, and display less soft sediment deformation than Unit A sediments.

The central part of the ridge (Fig. 3, 4, 5, 6) is characterized by steep dips (40° – 60° NE). There are thrust faults, which occur along clay beds and fold axis, as well as some minor high angle normal faults. Bed boundaries are often slightly folded and mixed into diamicton (10–

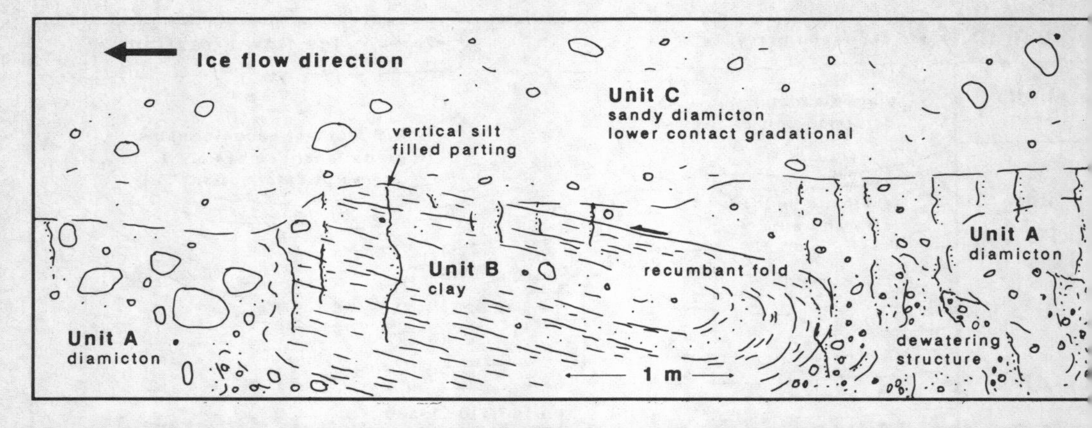

Fig. 8 Section 100 m SW of the ridge crest.

20 cm thick). Existing primary sedimentary structures often can be traced laterally into diamicton bodies (Fig. 3, 4, 5, 6). Most of the diamiction is, therefore, interpreted as deformed sediments caused by mixing, shearing or liquefaction. The bed-boundary folds have spacings of 20 to 30 cm and heights of about 10 cm (Fig. 3). Fold axis and faults generally trend parallel to the ridge crest, N60°W. There are diapiric folds, several of which are ruptured upwards (dewatering structures) (Fig. 4, 6, 8) and clastic intrusions of diamicton sediments into more cohesive, finner grained, sediment bodies (Fig. 5). These structures indicate that Unit A was water saturated and that high pore-water pressure existed during deformation.

The contact between Units A and B along the NE slope is tectonic. There is often a zone of mixing, up to 2 m wide, between the two units. Beds of Unit B are incorporated, as folds, within Unit A and the diamiction near the contact contains shells, and is evidentely the result of shear mixing of the two units (Fig. 6).

Unit B occurs as a sequence of imbricate slices which dip 0–20° NE. The thrust faults are often along clay beds. The bedding within the individual thrust slices is usually only slightly disturbed. However, one of the thrust slices is completely mylonitized.

The mylonite zone is situated about 40 m from the A/B contact. The imbricate slices of Unit B on either side are only slightly deformed. A shell rich layer is situated at the base of the 5 m thick mylonite zone. This zone is characterized by dark grey clay diamicton consisting of brecciated clay cubes, 0.5 cm in diameter. There are numerous shear planes with vertical spacings of 5 to 10 cm and wavy cross sections with wave lengths of 10 to 20 cm. Along the shear planes are fine slickensides indicating movement from the northeast (Fig. 9)

8 GLACIOTECTONIC MODEL

The glaciotectonic model proposed for the formation of the Tofta Ridge is a type of proglacial thrusting, however, it differs slightly from most previous models. It is based both upon the stratigraphy and deformation structurs as well as laboratory models of thin skinned tectonic deformations in orogenic belts (Mulugeta

Fig. 9 The mylonite zone, A-shear planes B-slickensides.

1988 in prep). This laboratory model predicts that with low friction along the décollement plane, such as in the case of high pore-water pressures in a lateraly extensive confined aquifer, the pressure gradient would propagate until it reached a barrier. The deformation would occur at the barrier, first as a clastic-volcanic eruption, followed by imbricate thrusting of the overlying beds. Similar clastic volcanos are reported along subduction zones (Langseth et al. 1988, Westbrook & Smith 1983). Recently the role of water in orogenic belts and the existence of mud volcanos, consisting of melange material, has been reported to be much more important than previously thought (Barber et al. 1986). Clastic dyke and plug intrusions of diamicton have been described for glacial sediments in Denmark (Humlum 1978, Berthelsen 1974) and is probably much more common than previously acknowledged.

The glaciotectonic model for the development of the Tofta Ridge is characterized in Fig. 10. The first stages of development probably occured in front of the advancing ice margin, possibly 1 – 2 km. Loading of the sediments during the advance caused high pore-water pressures to form in Unit A (a confined aquifer). The initial deformation was due to the high pore-water pressures pressing up the aquiclude. A barrier, such as a rise in bedrock or pinching out of the aquifer, caused the pressure to be directed upwards. When the aquaclude ruptured Unit A sediments extruded as a clastic volcano during which the soft sediment deformations and clastic intrusions occured. The next stage of development was imbricate thrusting of Unit B (the aquaclude) onto the ice-proximal side of the ridge. Following this the glacier advanced over the ridge. The upper 0.5 – 1 m of sediments were deformed into diamicton, whereas the sediments between 1 – 2 m were subject to extention and vertical fractures were formed and filled with silt.

9 GLACIAL MODEL

A preliminary three phase model for the glacial history based on the study of the Tofta Ridge has been constructed (Fig. 10).

Phase 1 represents the initial deglaciation of the Swedish west coast which took place around 13,000 to 13,500 ^{14}C years B.P. (Berglund 1979). The ^{14}C dates from this site suggest that deglaciation occurred before 13,110 B.P. (Fig. 2). The ML of about +60 to +70 m was probably formed at this time (Hillefors 1969, Mörner 1969, Påsse 1986, 1988, Wedel 1971). Glacial deposits from this deglaciation phase do not contain sedimentary-rock clasts but are instead totaly composed of crystalline rock types. Some of the HCM ridges were pobably formed as end moraines during this deglaciation phase. However the results of this present study indicate that some of these ridges are instead glacial tectonic ridges formed during a later glacial readvance.

Phase 2 was an ice free period with a still-water environment existing below ML, during which the clay, Unit B, was deposited. The drop stones of sedimentary-rock clasts in this unit were deposited from icebergs originating from the Baltic Basin, southernmost Sweden and/or Denmark. The ^{14}C dates (Fig. 2), of shells occuring in this unit, suggest that this phase lasted at least 500 years.

Phase 3 is a period during which a glacier readvanced from the northeast, and caused the glaciotectonic formation of the Tofta Ridge. The maximum extent of the readvance is not known but must be west of this site. The youngest ^{14}C date, 12,600 B.P., is a maximum age of the glacial readvance. Although the glacier did not erode or deposit extensive amounts of sediments it did incorporate sedimentary-rock clasts from the Phase 2 sediments into its deposits. The sea level both prior and after the readvance is unknown.

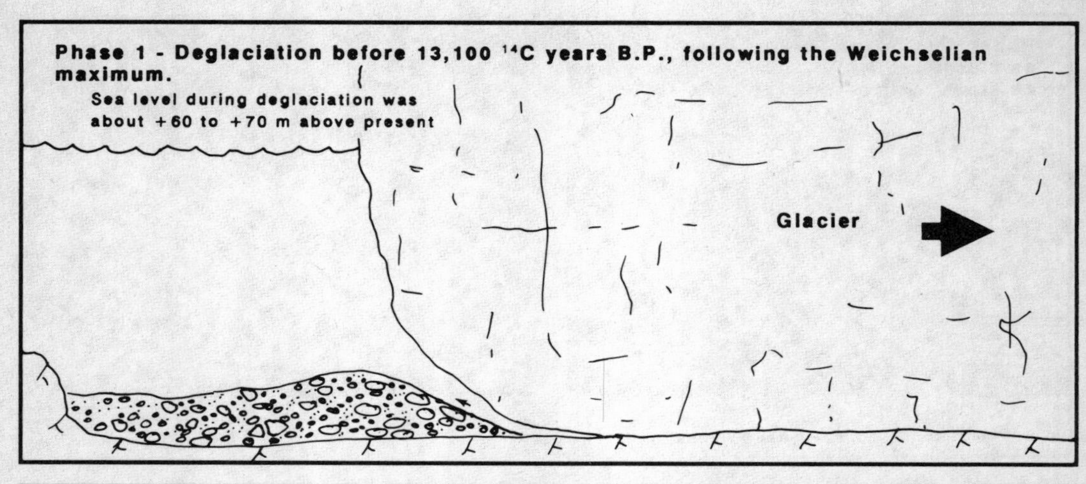

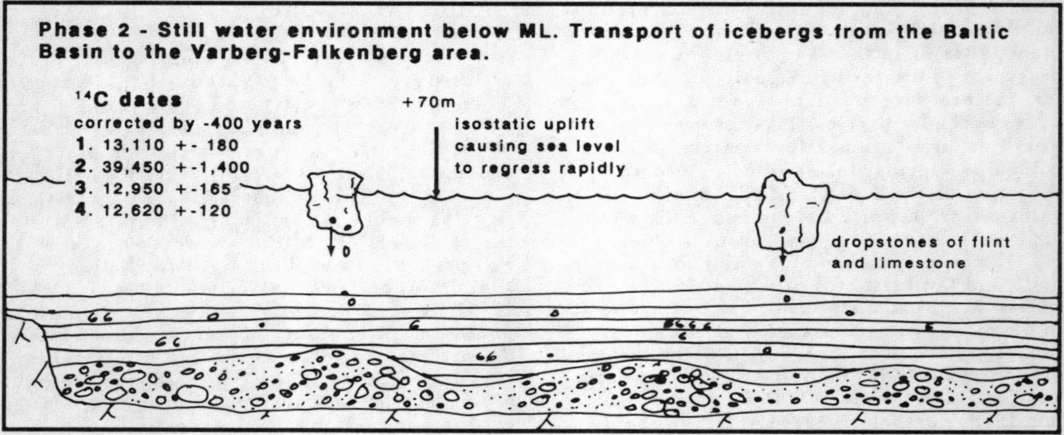

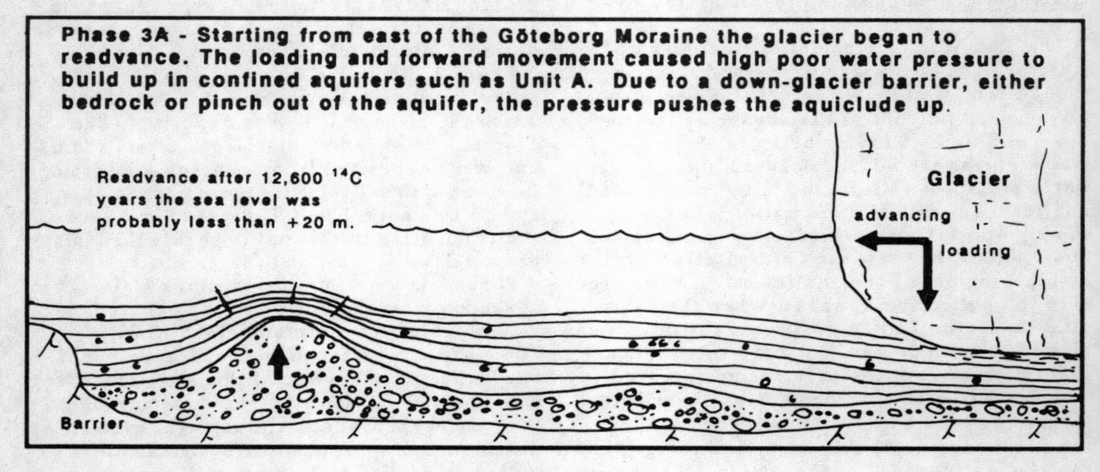

Fig. 10 Cartoon of the glacial development for the Torpa Ridge.

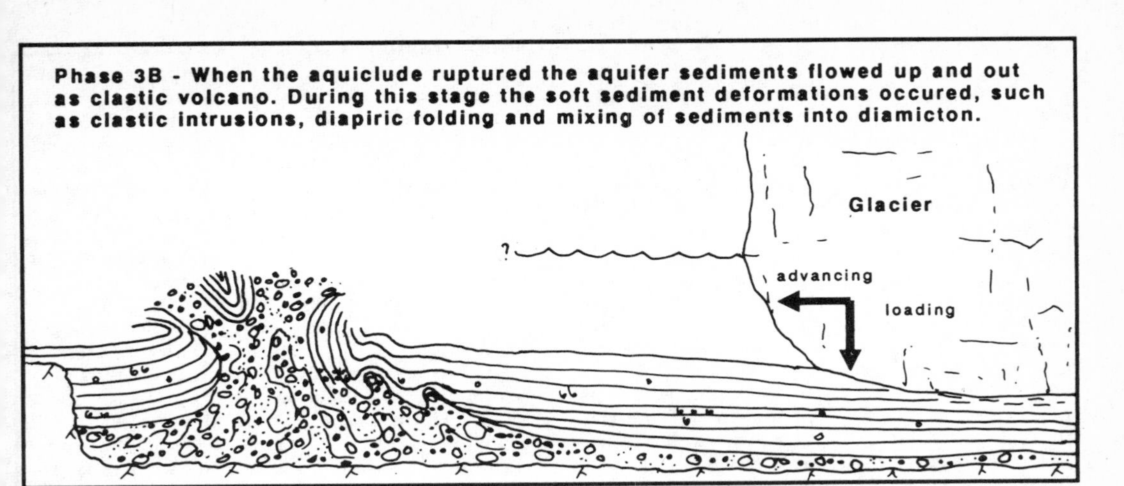

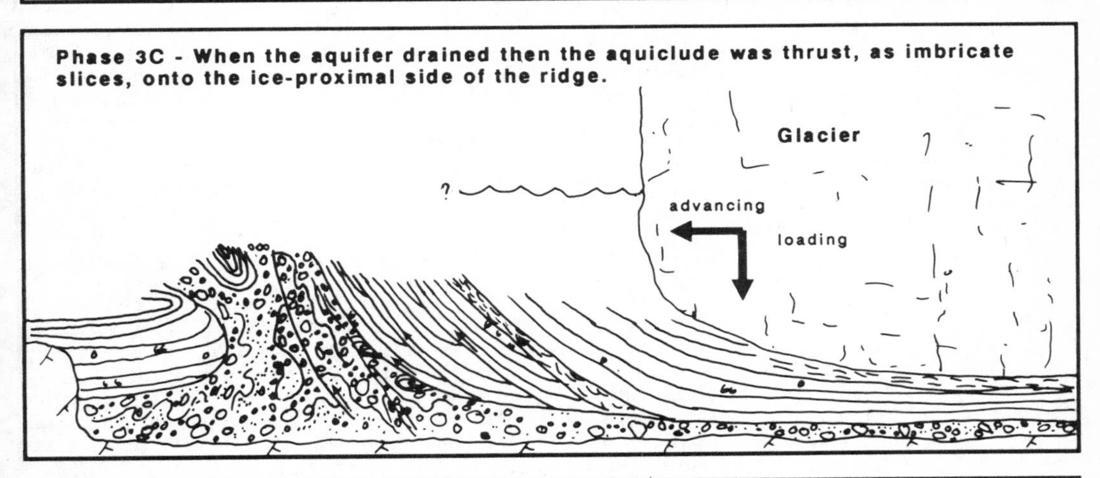

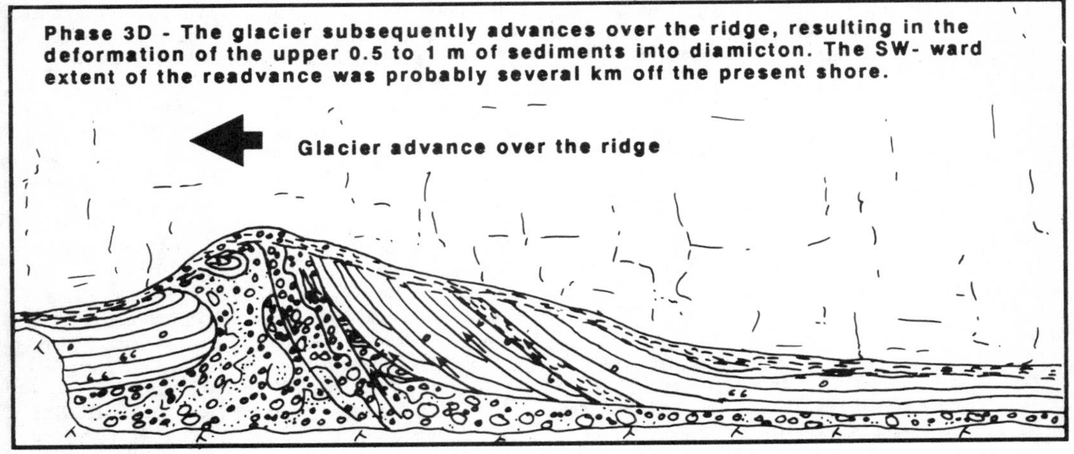

Fig. 10 Continued.

10 DISCUSSION

The HCM ridges as a whole have previously been described as end moraines. It is evident that there has been a readvace along the west coast of Sweden which caused the formation of the Tofta Ridge and probably other ridges. The different morphological types of HCM ridges could in theory be: end moraines from the deglaciation of Phase 1, end moraines from the maximum extent of the Phase 3 readvance, end moraines from the deglaciation following the Phase 3 readvance, glacial tectonic ridges formed by the Phase 3 readvance, or even still of some other genesis. Since at present it is not possible to determine, by means of morphology, which ridges are end moraines from the Phase 1 deglaciation or glaciotectonic ridges from the Phase 3 readvance, the definition of the HCM should possibly be modified, nongeneticly, as all the ice-transverse ridges along the west coast of Sweden in Halland County.

According to the youngest ^{14}C dates of the shells in Unit B, 12,600 B.P., the Phase 3 readvance could have occurred at about the same time as the glacial readvance proposed for the GEM (Berglund 1979, Hillefors 1975). Hillefors (1975) presented a three phase glacial history for the GEM at Fjärås Bräcka: 1. deglaciation, 2. ice-free period with clay containing shells dated to 12,580, 12,650, 12,770, and 12,750 ^{14}C years B.P. (corrected dates), 3. glacial readvance (Fig. 2).

It is possible that the glacial readvance at GEM is the same as the glacial readvance that formed the Tofta Ridge. This would require that GEM was overridden by the glacial advance, however, there is no record of this in the literature. Another explanation for the formation of the Tofta Ridge would be the advance of a local glacier, situated westward of GEM. However, if the Tofta Ridge is representitive of all Type 2 ridges (those associated with drumlins), then the regional extent of the readvance would be from Onsala to Falkenberg. In which case the probability of local glaciers between GEM and HCM is unlikely. Instead a major readvance of the ice sheet would be required. Preliminarily the readvance at GEM is correlated to the readvance that formed the Tofta Ridge.

The formation of the Type 2 ridges possibly is governed by basement topography, the existence of a confined aquifer, weakness in the aquaclude or by selective flow conditions in the glacier. The aerial-photo interpretation of the area revealed that the bedrock is highly foliated and has substantial vertical relief. In most areas it trends parallel to the ridges which suggests that they might initiate ridge formation. However ridges also occur in areas where the bedrock foliation is perpendicular to the ridges, in which case the bedrock morphology would not seem to be a governing factor. Instead changes in the composition of the substratam could have been a factor.

The Type 2 ridges suggest that the drumlin's stratigraphies might be advantageous for proglacial thrusting. The ridges' relationship to the drumlin forms has not been studied. However, several large drumlins associated with ridges are composed of interglacial and interstadial sediments, ex: the Dösebacka-Ellesbo drumlins are composed of older sediments, 24,020, 30,300, 36,000 ^{14}C years B.P. (Hillefors 1983). At Margreteberg (Lagerlund & Fernlund in prep) there is a drumlin form, although lacking ridges, which is composed of sediments from Eem and Brörup (Påsse et al. 1988). The oldest ^{14} C date at Tofta was 39,850 +-1,400 B.P. which suggests that pre-Late Weichselian sediments are also present in this drumlin. It is possible that these large drumlin forms are residual forms, composed predominately of older Quaternary sediments. The formation of ridges on the drumlins could be due to the existence of thicker sequences of sediments associated with isolated aquifers.

The ridges often only occur on the sides of the drumlins and could possibly be a transitional feature between the valleys and drumlins. Possibly the result of differential glacial erosion, with more extensive erosion in the valleys than over the drumlins.

It is somewhat difficult to determine the thermal regime of the glacier. The glacier in general did not erode or deposit extensive debris. Instead the upper part of preexisting sediments are deformed into a diamicton. The soft sediment deformation suggests that the sediments were not frozen at the time of the readvance. A cold based glacier could advance and retreat without eroding or depositing extensive material as well as deform the upper sediments. If the readvance was soon after 12,600 ^{14}C years B.P. then sea level was probably about +10 to +20 m. This would suggest that the readvance was in a subaquatic environment. In which case it would be unlikely that the glacier would be cold based. A study of a warm based ice stream (Boulton & Hindmarsh 1987) showed that the glacier was moved forward by the continual deformation of the underlying sediments. The sediments were under high-pore water presure and dialated and deformed down to about 1 m. Below 1 m the sediments were subject to extention. This is similar to the stratigraphy displayed at Tofta where the upper 0.5 – 1 m of sediments are deformed, where as below 1 m there are numorous extension cracks filled with silt. Therefore the glacier could have been warm based, possibly a surge or ice stream which protruded from the main ice sheet.

11 SUMMARY

A three phase glacial model is presented for the Varberg – Falkenberg area.

Phase 1 – Deglaciation of the Tofta area occurred before 13,100 ^{14}C years B.P., and the marine limit of about +60 to +70 m was probably formed at this time (Påsse 1986, 1988). It is likely that some of the ice-transverse ridges included in the Halland Coastal Moraines were formed as end moraines during this deglaciation. The initial part of Göteborg End Moraine was formed and then the ice retreated even further to the NE.

Phase 2 – Following the initial deglaciation in Phase 1, there was an ice free period which lasted at least 500 years, possibly longer. During this time rhythmic bedded clays were deposited in a still-water environment. These sediments are characterized by the occurance of sedimentary rock clasts which originate from the Baltic Basin, Denmark and/or southernmost Sweden. Shells occuring in these sediments are ^{14}C dated to 13,100, 12,950, and 12,600 B.P..

Phase 3 – There was a glacial readvance from the NE, after 12,600 B.P., which tectonized the Göteborg End Moraine and advanced over it and formed the Torpa – Tofta Ridges and probably the other ice-transverse ridges associated with drumlin forms. It advanced southwest of the Torpa – Tofta Ridges, possibly several km off the present coast. The sea level prior to and following the readvance is not known. In general the glacier did not erode or deposit till but instead deformed the uppermost 0.5 to 1.0 m of pre-existing sediments into a diamicton. It is possible that this was a warm based, surge or ice-stream glacier protruding from the main ice sheet.

The glaciotectonic model for the formation of the ridges is complex (Fig. 10). The ridges probably first formed due to high pore-water pressure deforming the aquiclude upwards. When the aquiclude ruptured the water saturated sediments of the aquifer erupted as a clastic volcano. This was followed by imbricate thrusting of the aquiclude onto the ice-proximal side of the initial ridge. This could have occurred in front of the ice margin, possibly 1 – 2 km. Subsequently the glacier advanced over the ridge and deformed the upper sediments into diamicton.

The Tofta Ridge is not an end moraine, from either Phase 1 or 3, but a glacial tectonic ridge formed and overridden during the Phase 3 glacier advance, sometime after 12,600 B.P.. The Tofta Ridge is one of a series of ridges associated with a drumlin. Therefore, other Halland Coastal Moraine ridges associated with drumlin forms are preliminarily inferred to have been formed during the Phase 3 readvance. If this inference is correct, then the regional extent of the readvance would include all the Halland Coastal Moraines associated with drumlins (Fig. 1), and would have extended at least from Falkenberg to Onsala.

Acknowledgements – This study has been financed by the Swedigh Natural Science Council to whom I owe my sincere thanks.

REFERENCES

Aber, J. S. 1980. Kineto-stratigraphy at Hvideklint, Mön, Denmark and its regional significance. Bull. Geol. Soc. Denmark, Vol. 28, p. 81-93.

Aber, J. S. 1982. Model for glaciotectonism. Bull. Geol. Soc. Denmark, Vol. 30, p. 79-90.

Aber, J. S. 1985. The character of glaciotectonism. Geologie en Mijnbouw, Vol. 64, p. 289-395.

Banham, P. H. 1975. Glacitectonic structures: a general discussion with particular reference to the contorted drift of Norfolk. In: Wright, A. E. & Moseley, F. (eds.) Ice Ages: Ancient and Modern. p. 69-94.

Barber, J. A., Tjokrosapoetro, S. & Charlton, T.T. 1986. Mud volcanoes, shale diapirs, wrench faults, and melanges in accretionary complexes, eastern Indonesia. American Association of Petrolium Geologists Bulletin, Vol. 70, Nr. 11, p. 1729-1741.

Berglund, B. 1979. The deglaciation of southern Sweden 13,500 – 10,000 B. P. Boreas, Vol. 8, p. 89-118.

Berthelsen, A. 1974. Nogle forekonster af intrusivt moraeneler i NÖ-Sjaelland. Dansk. Geol. Foren. Årsskrift for 1973. p. 118-131.

Berthelsen, A. 1979. Recombent fold and boudenage structures formed by subglacial shear. In: van der Linden (ed.) van Bemmelen and his search for harmony. Geol. Mijnbouw, Vol. 58, p. 253-260.

Bluemle, J. P. 1970. Anomalous hills and associated depressions in central North Dakota. Geol. Soc. Am. Abstracts with programs 2, p. 325-326.

Bluemle, J. P. & Clayton, L. 1984. Large-scale glacial thrusting and related processes in North Dakota. Boreas, Vol. 13, p. 279-299.

Boulton, G. S. & Hindmarsh, R. C. 1987. Sediment defornation beneath glaciers: Rheology and geological consequences. Journal of Geophysical Research, Vol. 92, Nr. B9, p. 9059-9082.

Caldenius, C. 1942. Gotiglaciala israndstudier och jökelbäddar i Halland. Förelöpande meddelande. Geol. Fören. Stockh. Förhdl, Vol. 64, p. 163-183.

Chamberlin, T. C. 1894. Proposed genetic classification of Pleistocene glacial formations. Journal of Geology, Vol. 2, p. 517-538.
Clayton, L. & Moran, S. R. 1974. A glacial process-form model. In: Coates, D. R. (ed.) Glacial Geomorphology. p. 89-119.
Croot, D. G. 1987. Glacio-tectonic structures: a mesoscale model of thin-skinned thrust sheets? Jornal of Structural Geology, Vol. 9, p. 797-808.
De Geer, G. 1893. Praktiskt geologiska undersökningar inom Hallands län. Sveriges Geologiska Undersökning, Ser. C, Nr. 131, p. 1-38.
Dreimanis, A. 1976. Tills: their origins and properties. In: Legget, R. (ed.) Glacial Till. p. 11-49.
Driemanis, A. 1982. Work group (1) - Genetic classification of tills and criteria for their differentiation: progress report on activities 1977 – 1982, and definition on glacgeneic terms. In: Schüchter, Ch. (ed.) INQUA commission of genesis and lithology of Quaternary deposits. Report on activities 1977 – 1982, ETH, Zuerich, p. 12-31.
Fält, L. M., 1975. Ändmoräner, kustområdet mellan Kungsbacka och Värö. C-kursarbete, Chalmers Tekniske Högskola – Göteborgs Universitet, Geologiska Instituionen Publication B35, p. 1-43.
Flint, R. F. 1971. Glacial and Quaternary Geology. John Wiley & Sons, Inc. New York. p. 1-892.
Gillberg, G. 1956. Den glaciala utvecklingen inom sydsvenska höglandets västra randzon III. Issjöar och isavsmältning. Geol. Fören Stocklm. Förhandling, Vol 78, p. 357-458.
Gripp, K. 1929. Glaciologische und geologische ergebnisse der Hamburgischen Spitzbergen, expedition. Abh. Verh. Naturw. Ver. Hamburg, Vol. 22, p. 147-247.
Gry, H. 1940. De istektoniske forhold: moleromraadet. Meddr. Dansk Geol. Foren, Vol. 9, p. 586-627.
Hillefors, Å. 1969. Västsveriges glaciala historia och morphologi. Naturgeografiska studier. Medd. Lunds Univ. Geolgr. Inst. Avh. 60, p. 1-319.
Hillefors, Å., 1975. Contribution to the knowledge of the chronology of the deglaciation of western Sweden with special reference to the Gothenberg Moraine. Svensk Geogr. Årsbok 51, p. 70-81.
Hillefors, Å. 1979. Deglaciation models from the swedish west coast. Boreas, Vol. 8, p. 153-169.
Hillefors, Å. 1983. The Dösebacka and Ellesbo drumlins – morphology and stratigraphy. In: Ehlers, J. (ed.) Glacial deposits in north-west Europe. p. 142-250.
Humlum, O. 1978. A large till wedge in Denmark: implications for the subglacial thermal regime. Bull. Geol. Soc. Denmark, Vol. 27, p. 63-71.
Jessen, A. 1931. Lönstrup Klint. Danm. Geol. Unders. V, raekke 2, p. 1-195.
Langseth, M. G., Westbrook, G. K. & Hobart, M. A. 1988. Geophysical survey of a mud volcano seaward of the Barbados Ridge accretionary complex. Journal of Geophysical Research, Vol. 93, Nr. B2, p. 1049-1061.
Möller, P. 1987. Moraine morphology, till genesis, and deglaciation pattern in the Åsnen area, south-central Småland, Sweden. Lundqua Thesis, Vol. 20, p. 1-146.
Moran, S. 1971. Glaciotectonic structures in drift. In: Goldthwait, R. R. (ed.) Till, a symposium. Ohio State University Press. p. 127-148.
Mörner, N. A. 1969. The late Quaternary history of the Kattegatt Sea and the swedish west coast. Sveriges Geologiska Undersökning, Ser C, Nr. 640, p. 1-487.
Moran, S. R., Clayton, L., Hooke, R. LeB., Fenton, M. M. & Andriashek, L. D. 1980. Glacier-bed landforms of the praire region of north america.Journal of Glacialogy, Vol. 25, Nr. 93, p. 457-476.
Mulugeta, G. 1988. In prep. Hans Ramberg Tectonic Laboratory, Institute of Geology, University of Uppsala, Sweden.
Nielsen, A. V. 1967. Landskabets tilblivelse. In: Nörrevang, O. & Meyer, T. J. (eds). Danmarks Natur, Bd 1, Landskabernes Opståen. Politikens Forlag, Copenhagen, p. 251-344.
Svedmark, E. 1893. Beskrifning till kartbladet Varberg. Sveriges Geologiska Undersökning, Ser. Ab, Nr. 13, p. 1-82.
Påsse, T. 1986. Beskrivning till jordartskartan Kungsbacka SO. Sveriges Geologiska Undersökning. Ser. Ae, Nr. 56, p. 1-106.
Påsse, T., 1988. Beskrivning till jordartskartan Varberg SO/Ullared SV. Sveriges Geologiska Undersökning. Ser. Ae, Nr. 86. in press.
Påsse, T., Robertsson, A. M., Klingberg, F. & Miller, U. 1988. A Late Pleistocene sequence at Margreteberg, southwestern Sweden. Boreas, Vol. 17, p. 141-163.
Wedel, P. O. 1971. Kvartärgeologiska studier i norra Halland. Chalmers Tekniska Högskola – Göteborgs Universitet, Geologiska Institutionen Publ. 1971 A:1, p. 1-85.
Westbrook, G. K. & Smith M. J. 1983. Long décollements and mud volcanoes: evidence from the Barbados Ridge complex for the role of high fluid pressure in the development of an accrecionary complex. Geology, Vol. 11, p. 279-283.

Glaciotectonics: Forms and Processes, Croot (ed.), © 1988 Balkema, Rotterdam. ISBN 90 6191 848 0

Glaciotectonic unconformities in Pleistocene stratigraphy as evidence for the behaviour of former Scandinavian icesheets

Michael Houmark-Nielsen
Institute of General Geology, University of Copenhagen, Denmark

ABSTRACT: Glacitectonic structures are the products of glacier-induced tectonic displacement in soft sediments. Often their orientation is related to the direction of glacier flow. In Pleistocene sedimentary sequences structures form glacitectonic unconformities that may appear as marker horizons on a regional scale. In stratigraphic studies glacitectonic and glacidynamic structures and textures are connected with the boundaries and lithology of till formations. Examples are given of how the regional distribution of till formations and related ice flow directions can be used to reconstruct the pattern of glaciation and deglaciation during the Late Weichselian. It is concluded that neither classical glaciation models nor modern models dealing with one central ice dome or alternatively operating with several central and marginal ice domes are developed fully enough to explain the course of Middle and Late Pleistocene glaciations in northwest Europe.

1 INTRODUCTION

Before glacitectonic structures were recognized in Denmark and used as evidence for the flow direction of active glaicers, early models of the evolution of the Scandianvian ice sheet were based on the dispersal of indicator erratics from southern Norway, middle Sweden, Finland and the Baltic Sea. It was recognized that the direction from which ice streams crossing Denmark and Skåne (fig. 1) came, changed progressively from N, to NE, to E, and eventually to SE especially during the last glaciation (cf. Milthers 1909, Holmström 1904).

Studies of tectonic structures and their morphological expressions have a long history in Danish geological investigations (cf. Madsen 1897, 1900, 1916, 1928; Madsen & Nordmann 1940, Berthelsen 1973). Madsen argued that the nature of the so called dislocated cliffs on Ristinge, Møn, Mors and Lønstrup (fig. 1) originated from crustal tectonism, even though Jessen (1918, 1930, 1931) suggested that the dynamics of transgressing glaciers was the active agent responsible for the tectonic displacement. Gry (1940), using methods based on the study of orogenic features initiated detailed structural analyses to describe the nature of the dislocated cliffs and other disturbed strata found in excavations in Denmark. Referring to Gripp (1929), Gry (1940) refuted the glacidynamic displacement of thrust slices as suggested by Slater (1927) and Richter (1929).

Gry (1940) explained folds and thrust faults in the Pre-quaternary as well as Quaternary deposits at Mors and Lønstrup in northern Jylland (fig. 1) as the result of active glacier push (as opposed to static / glacidynamic loading) and estimated the relative

Fig. 1: Key map with names mentioned in the text.

Fig. 2: Schematic presentation of the evolution of glacitectonic structures and deformational mechanisms in relation to other glacigenetic events and time.

age and direction of ice deformation. The intense discussions that followed are summarized by Gry (1941). Rosenkrantz (1944) adopted Gry's approach, explaining the tectonic structures found in the cliff at Ristinge (fig. 1) as products of glacier push rather than crustal uplift, as had been suggested by Madsen (1916)

In Denmark the method of structural analysis of glacitectonic features has systematically been applied during stratigraphic investigations of Pleistocene deposits over the last two decades (Aber 1979; Berthelsen 1971, 1974, 1975; Berthelsen et al. 1977; Frederiksen 1975, 1976; Houmark-Nielsen 1976, 1980, 1981, 1983, 1987; Houmark-Nielsen & Kolstrup 1981; Jacobsen 1976, 1981,1985; Jensen 1977; Jensen & Knudsen 1984; Kronborg 1983; Kronborg & Knudsen 1985; Larsen et al. 1977; P.E. Nielsen 1980, J.B.Nielsen 1987; Petersen 1973, 1978; Rasmussen 1973, 1974, 1975; Rasmussen & Petersen 1980; Sjørring 1974, 1977, 1981; Thamdrup 1970).

2 SOME PRINCIPLES OF GLACITECTONISM

Glacitectonic structures can be defined as glacier induced tectonic structures developed above a surface of décollement in soft sediments in connection with the advance, overriding and decay of ice sheets (fig. 2).

Glacitectonic structures can be distinguished from orogenic structures or features developed by crustal movements and halokinesis primarily because (i) they are restricted in size and (ii) glacitectonic structures die out with depth (this property can not always be directly observed in the field). Glacitectonic structures can also be distinguished from structures produced by landslides and isolated slumping within a sedimentary sequence as well as featues developed by periglaical deformation or sedimentary loading. Glacitectonic structures may have gradational or abrupt transitions to glacidynamic structures.

The development of mobile glacitectonic structures is due to a combination of load pressure and active glacier movement, and takes place both pro- and subglacially (fig. 2, I). Supraglacial tectonic structures may develop under static conditions after glacier movement ceases (fig. 2, II).

The position of the lower boundary of glacitectonic structures (the surface of décollement), may depend on the position of high porosity / low permeability deposits and / or the depth of possible permafrost.

2.1 PROGLACIAL STRUCTURES

Proglacial glacitectonic structures may include overturned folds, reverse listric faults, overthrusts, reverse upthrusts and normal listric faults developed in proglacially deformed beds due to collapse. Proglacial structures usually die out distally into periglacial zones. This progressive decline is often indicated by a decrease in both size and inclination of thrust slices and the development of increasingly open fold styles until the undeformed extramarginal foreland is reached.

Proglacial structures produced by active folding and thrusting of the glacier foreland are excellent indicators of the direction of glacier movement. Proglacial deformation has a constructive effect on landforms and may lead to the creation of push moraines.

2.2 SUBGLACIAL STRUCTURES

Upglacier, subglaical structures develop primarily by superimposition of subglacial shear on preexisting glacitectonic - and/or other structures. Secondly, structures may form by diapirism and other intrusive mechanisns due to porewater mobilization under glacier load.

According to Berthelsen (1979a) subglacial structures may include recumbent shear folds, local rotation and change in layer thickness, internal buckling or boudinage and reorientation of textural properties.

Such structures, formed by subglaical shear are only marginally useful indicators of the direction of glaicer movement. Subglacial deformation has a destructive effect on preexisting landforms and may eventually lead to the formation of smooth till plains. The orientation of structures created by intrusive mechanisms may often have no relation to the direction of movement of glaciers.

2.3 SUPRAGLACIAL STRUCTURES

Glacitectonic structures formed under stagnant ice conditions (fig. 2) after cease of glacier movement may include conjungate normal faults (Selsing 1981) and slump folds may develop in former dead ice supported supraglacial sediments. Supraglacial structures have only little or no relation to previous ice movement.

3 GLACITECTONICS AND PLEISTOCENE STRATIGRAPHY

In open exposures, pro- and subglacial glaci-otectonic structures often form recognizable glacitectonic unconformities (fig. 3) which appear as marker horizons in the stratigraphic record.

S.A. Andersen (1933. 1945. 1950, 1957) contributed to the foundation of modern Danish glacial stratigraphic methodology by developing a multiple set of working methods. He combined stratigraphical methods (including the lithic characters of till units and biostratigraphical determination of interglaical or interstadial beds) with observations on glacially striated boulder pavements and pre-Qquaternary bedrock, and the interpretation of glacier movement directions based on glacitectonic features. Andersen was thus able to set up a polyglacial stratigraphic framework, which was radically different from the established monoglacial model (cf. K. Milthers 1942). According to Andersen, the tills of the last glaciation can be grouped into individual units on the basis of erratic contents and stratigraphic position. They are vertically bound by glacitectonic unconformities and separated by meltwater deposits (supposedly deposited under subaerial, relatively ice-free conditions between stadials). Berthelsen (1973, 1978) took Andersen's work a stage further and formulated the principle of kinetostratigraphy in an attempt to introduce a structural line of thinking into the stratigraphic studies of Pleistocene deposits in Denmark. Berthelsen (1978, 1979b) summarises the Weichselian till stratigraphy developed during the seventies.

In more recent lithostratigraphic studies

Fig. 3: Examples of glacitectonic unconformities. Upper: Rhoden, Vejle Fjord. Late Saalian deposits (Lillebælt Till overlain by outwash material) folded and thrust from the northeast during the Main Weichselian Advance and unconformably overlain by proximal sandur deposits. Lower: Vestborg, Samsø. Late Middle Weichselian deposits (East Jylland Till and associated outwash material) folded by ice deformation from the east and unconformably overlain by the Bælthav Till. Further explanation fig. 4.

of Pleistocene deposits (Houmark-Nielsen 1987) till formations are used as the basic framework to describe the geology of former glaciated lowlands like Denmark, which for the larger part comprises non-fossiliferous, clastic deposits. Till Formations show a reasonably high degree of lithic uniformity, and even though not directly mappable, they can, after simple field investigations and laboratory treatment, be recognized in a large number of exposures.
Till units are described in terms of lithic characteristics with emphasis placed on those features which express the direction of glacier flow together with provenance dependant, compositional elements.

The recognition of boundaries between units is of prime importance in lithostraigraphic classification. Special emphasis has been placed upon the study of glacitectonic unconformities because they give information on the direction of ice transgression as well as adding a structural chronology to a given body of deformed strata (cf. Houmark-Nielsen & Berthelsen 1981).

4 GLACIAL STRATIGRAPHY IN THE CENTRAL PART OF DENMARK

Houmark-Nielsen (1987) recognized till units deposited during the last three glaications i.e. the Elsterian, the Saalian and the Weichselian. A schematic NE-SW'oriented cross section through a c. 300 km wide sector covering Denmark and stretching from the Øresund region and southern Kattegat southwestwards to the Danish / German borderland is presented in fig 4.

The course of the Weichselian, the Saalian and probably also the Elsterian glaciations across Denmark seems to follow a directional pattern in which northerly ice advances, originating in the southern part of Norway were replaced by northeasterly advances crossing southwestern Sweden. The final stage consist of southeasterly advances through the Baltic depression. With modifications, the above-mentioned clockwise movement of ice-flow directions is reflected in the change of indicator erratics of tills and striations on bedrock and boulder pavements and was recognized early in the Danish literature (cf. Milthers 1909).

5 THE WEICHSELIAN GLAICATION

Six Weichselian till Formations (fig. 4, e-j) have been identified in open exposures in the central parts of Denmark. Their compositional features, textural properties and associated glacitectonic boundaries suggest

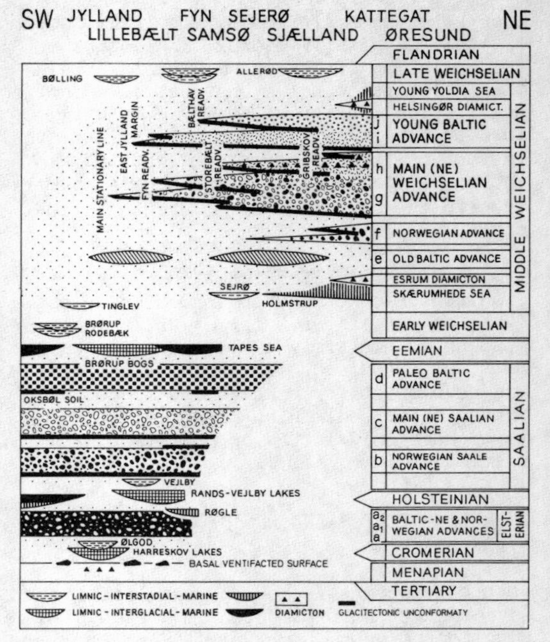

Fig. 4: Event-stratigraphic scheme in a NE-SW oriented transect covering southern and central Denmark. Till formations (a-j) and associated glacitectonic unconformities are related to major Middle - and Late Pleistocene ice advances. d: Lillebælt Till, e: Ristinge Klint Till, f: Kattegat Till, g: Mid Danish Till, h: North Sjælland Till, i: East Jylland Till, j: Bælthav Till. Modified from Houmark-Nielsen (1987).

that they were deposited by four major glacier transgressions separated by intervals of glacioaqueous and periglacial activity during the Late Weichselian (sensu Mangerud & Berglund 1978). The four major ice transgressions are 1) the Old Baltic advance, 2) the Norwegian advance, 3) the Main Weichselian advance and 4) the Young Baltic advance.

5.1 OLD BALTIC AND NORWEGIAN ADVANCES

Glacitectonic activity associated with the ice stream which deposited the Kattegat Till (fig. 4, f) apparently affected the Ristinge Klint till (fig. 4, e) in north Sjælland. It is suggested that the Ristinge Klint Till was deposited by the Old Baltic advance after the Sejerø Interstadial (37 ka BP). Petersen (1984) argued that this first Weichselian ice advance took place arround 60 ka BP. Houmark-Nielsen (1987) opposed this view questioning the validity of the data presented by Petersen. However, the age of this

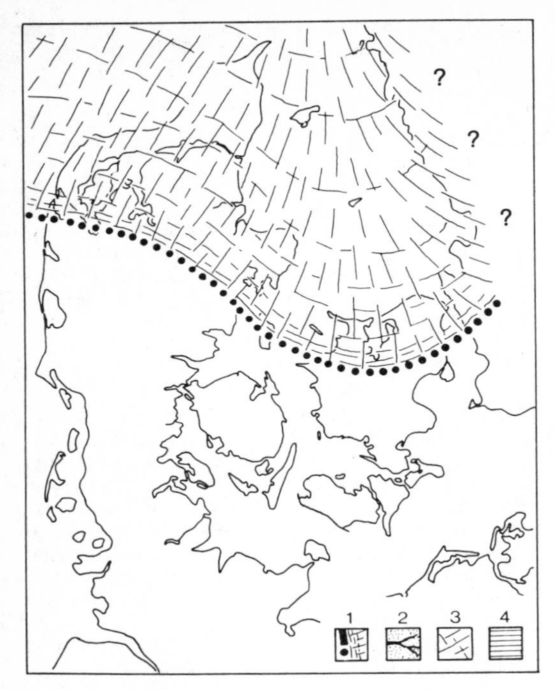

Fig. 5: Norwegian Advance. Possible ice distribution and flow directions. 1: Major icemarginal features and inferred connections 2: Sandur deposits and outwash drainage patterns. 3: Flow lines of glaicer ice. 4: Late Glaical Younger Yoldia Sea.

advance is for the time being regarded as an open question because biostratigraphic control and radiometric age determinations are still very sparse.

Later than 25 Ka BP Norwegian ice advanced from the north, deposited the Kattegat Till and reached the central and northern parts of Denmark and the Swedish west coast (fig. 5).

5.2 MAIN WEICHSELIAN ADVANCE

The Mid Danish Till (fig. 4, g) is found in the whole of the country east of the maximum limit of Weichselian icestreams: Main Stationary Line (fig. 6 M), and is bound by glacitectonic unconformities which indicate ice-deformation from the NE and the E. It was deposited by the Main Weichselian ice advance (also refered to as the NE-ice). This event probably occurred shortly after 20.000 BP, with ice from north-easterly directions. The Mid Danish Till overlies the Kattegat Till, and is overlain by the North Sjælland Till in the Northeastern part of the region studied.

During the decay of this ice sheet, readvances took place across parts of eastern Denmark presumably between 18.000 BP and 15.000 BP. The first recognized of these readvances (Fyn readvance fig. 6.F) transgressed Fyn and the eastern parts of Jylland and dislocated the previously deposited Mid Danish Till from the NE and formed the central dead ice deposits on Fyn. The North Sjælland Till is separated from the Mid Danish Till by waterlain outwash-material and a glacitectonic unconformity, the structures in which indicate ice-deformation from the E in northern Sjælland. The North Sjælland Till was deposited from an easterly direction presumably by a younger readvance (Storebælt Readvance ,fig. 6 S) of the Main Weichselian ice sheet. It is suggested that the Mid Danish Till and the North Sjælland Till merge into one single till unit either in the north-eastern part of Sjælland or the Øresund region. An even younger readvance is assumed to have formed the NW-SE trending so-called Gribskov ice marginal hills in NE-Sjælland (fig. 6 V).

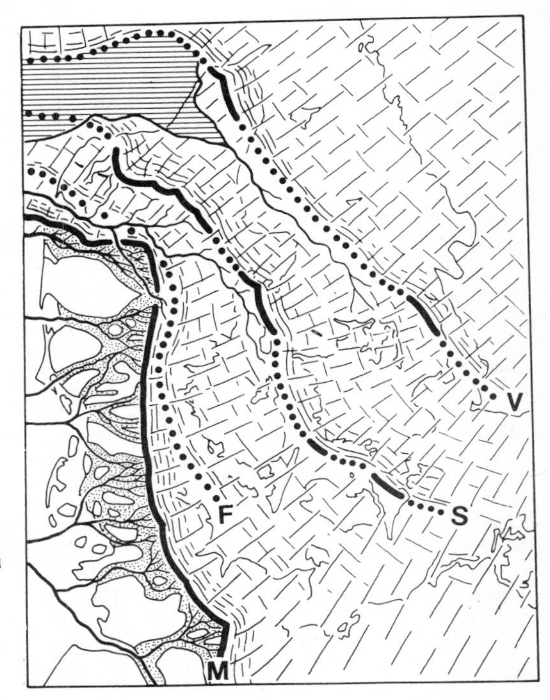

Fig. 6: Main Weichselian Advance. Ice distribution and flow directions. M: Main Stationary Line, F: Fyn readvance, S: Storebælt readvance, V: Vendsyssel-Gribskov readvance. Further explanation see fig. 5. From Houmark-Nielsen (1987).

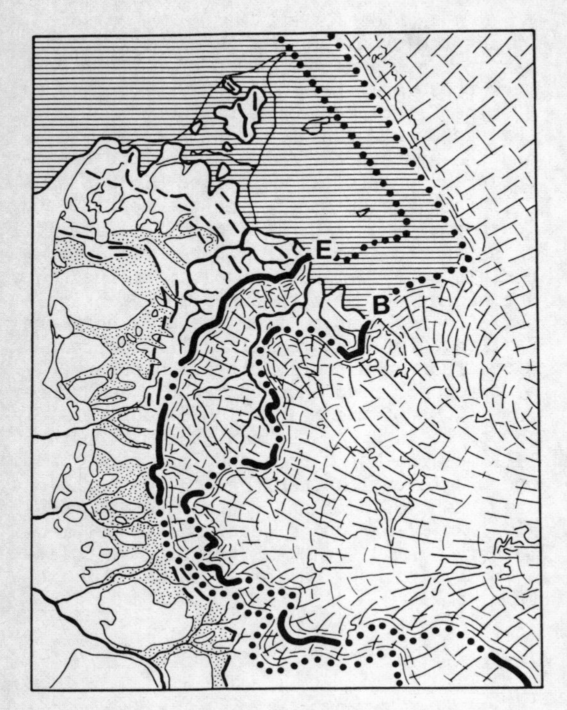

Fig. 7: Young Baltic Advance. Ice distribution and flow directions. E: East Jylland advance, B: Bælthav readvance. Further explanation see fig. 5. From Houmark-Nielsen (1987)

5.3 YOUNG BALTIC ADVANCE

The East Jylland - and the Bælthav Tills (fig. 4, i-j) were deposited after the retreat of the NE-ice by the Young Baltic icesheet which came generally from south-easterly directions (fig. 7) most probably between 15.000 and 14.000 Bp. They overlie the Mid Danish Till towards the west and the North Sjælland Till in the eastern part of the country. The lower boundary of these till units is marked by glacitectonic unconformities indicating ice-deformation from a southeasterly direction. At an early stage of maximum extension along the East Jylland Ice Border (fig. 7, E) the East Jylland Till was deposited in the area west of Sjælland. Shortly afterwards a readvance of the Young Baltic ice sheet (Bælthav readvance fig. 7, B) deposited the Bælthav Till in the eastern parts of the region and succeding readvances formed terminal moraines southward through Storebælt.

In the Øresund region the Bælthav Till is laterally replaced and/or overlain by ice rafted diamict material (Helsingør diamicton, Houmark-Nielsen & Lagerlund 1987) deposited in connection with a Late Weichselian (sensu Mangerud & Berglund 1978) transgression of the Younger Yoldia Sea.

6 MODELLED BEHAVIOUR OF SCANDINAVIAN ICESHEETS

Fundamentally different glaciation models have been proposed to elucidate the apparent cyclic change in directional pattern of Scandinavian sheets in general and to explain or deny the presence of Baltic ice streams in particular.

In accordance with the classical outline as well as more recent publications, Ehlers (1981) and Ehlers et al. (1984) discussed the possibility that the Scandinavian ice-divide migrated in response to an eastward shift of the glaciation center. They argue that this constitutes the primary cause of the clockwise change recorded in the periphery of the ice sheet. Another factor mentioned by these authors is the possiblity that a negative mass balance developed towards the end of each glaciation in the western part of the ice-sheet owing to rapid melting in near coastal regions.

Despite published evidence on Young Baltic glacitectonic activity Overweel (1977) rejected the idea that Baltic glaciers travelled across Denmark and western Skåne during the Weichselian. He argues that the scattering of Baltic indicator erratics supposedly arose primarily from ice-rafting through a proglacial lakebasin linking the Baltic with the Skagerrak and Kattegat.

A different glaciation model, partly founded on glacitectonic evidence is presented by Lagerlund (1980, 1987). One or more marginal domes situated in the periphery of the ice-sheet are thought to develop and decay independently of a large central dome. Thus, for the Weichselian, it is suggested that the existence of Norwegian and Baltic ice streams is brought about by surging of marginal domes situated in the western and southern rim of the Scandinavian ice-sheet that affected Denmark and surrounding regions.

Boulton et al. (1985) present a computerized model for the Scandinavian ice-sheet during the last glaciation, in which a Fennoscandian ice-stream crossed Denmark from northeasterly directions and terminated along the Main Stationary Line. This model clearly depicts the behaviour of the ice-sheet during the Main Weichselian advance. Moreover, by excluding the influence of Lagerlunds marginal domes as beeing the primary cause of Baltic icestreams Boulton et al. find their model unable to reproduce Baltic icestreams that cross western Skåne, Denmark and Schleswig-Holstein.

7 CONCLUSION

The clockwise changes in the directional pattern must reflect a fundamental trend in the evolution of consecutive Scandinavian icesheets. It is suggested that whatever attempts are made to model the evolution of Scandinavian ice sheets, the fundamental and repeated changes of direction of ice advances as recorded by modern glacial-stratigrphic research including glacitectonic studies must be taken into consideration.

It must be concluded that Norwegian and Baltic icestreams apparently only exist when glaciers from central Sweden are not occupying the Kattegat and the western Baltic depressions. Baltic and Norwegian icestreams are supposed to have been brought about by an interaction between the evolution of a main (central) icedome and local factors at the fringes of the ice-sheet.
The evolution of the central dome, which is likely to give rise to the migration of the ice divides, may consequently produce the clockwise change of movement. Changes in bedrock composition may also influence the regional behaviour of icesheets. Local factors in the periphery may strengthen or suppress the course of icestreams. Primarily, local climatic changes may trigger the building of marginal domes and alternations in bedrock/ ice temperature, glacier-induced morphological modifications such as proglacial depressions and other isostatic responses, neotectonic movements and eustatic fluctuations will govern the course of glaciers that flow from the pre-Cambrian, Fennoscandian shield and the Palaeozoic platform out on to the soft, mostly Cainozoic north European lowlands.

References

Aber, J. 1979: Kineto-stratigraphy at Hvideklint, Møn, Denmark and its regional significance. Bull. Geol. Soc. Denmark, vol. 28, 81-93.

Andersen, S.A. 1933: Det danske Landskabs Historie. Danmarks Geologi i almenfatteligt Omrids. Levin & Munksgaard, Copenhagen, 111-124.

Andersen, S.A. 1945: Isstrømmenes Retning over Danmark i den sidste Istid, belyst ved Ledeblokundersøgelser. Kritiske bemærkninger til K. Milthers: Ledeblokke og Landskabsformer i Danmark (DGU II, No. 69 med svar og gensvar). Meddr. dansk geologisk Forening, vol. 10, 594-615.

Andersen, S.A. 1950: Rågeleje Egnens Geologi. Meddr. dansk geologisk Forening, vol. 11, 543-577.

Andersen, S.A. 1957: Lolland i den sidste istid. Meddr. dansk geologisk Forening, vol. 13, 225-235.

Berthelsen, A. 1971: Fotogeologiske og feltgeologiske undersøgelser i NV-Sjælland. Dansk Geol.Foren., Årsskrift for 1970, 64-69.

Berthelsen, A. 1973: Weichselian ice advances and drift successions in Denmark. Bull.Geol.Inst.Univ.Upps., 5, 21-29.

Berthelsen, A. 1974: Nogle forekomster af intrusivt moræneler i NØ-Sjælland. Dansk geol. Forening, Årsskrift for 1973, 118-131.

Berthelsen, A. 1975: Geologi på Røsnæs. Ekskursionsfører No. 3, VARV, 78 p.

Berthelsen, A. 1978: The methodology of kineto-stratigraphy as applied to glacial geology. Bull.geol.Soc.Denmark, vol. 27, Special Issue, 25-38.

Berthelsen, A. 1979a: Recumbent folds and boudinage structures formed by subglacial shear: an example of gravity tectonics. Geol.Mijnbouw, 58, 253-260.

Berthelsen, A. 1979b: Contrasting views on the Weichselian glaciation and deglaciation of Denmark. Boreas, vol. 8, 125-132.

Berthelsen, A., Konradi, P. & Petersen, K.S. 1977: Kvartære lagfølger og strukturer i Vestmøns klinter. Dansk geol. Forening, Årsskrift for 1976, 93-99.

Boulton, G.S., Smith, G.D., Jones, A.S. & Newsome, J. 1985: Glacial geology and glaciology of the last mid-latitude icesheets. J.geol.Soc.London, vol. 142, 447-474.

Ehlers, J. 1981: Different till types in North Germany and their origin. In: Tills and Related Deposits. Proceedings of the INQUA Symposia on the genesis and lithology of Quaternary deposits USA. 1981, Argentina 1982, A.A.Balkema, Rotterdam, 68-80.

Ehlers, J., Meyer, K-D. & Stephan, H-J. 1984:Pre-Weichselian glaciations in Northwest Europe. Quaternary Science Reviews, vol. 3, 1-40.

Frederiksen, J.K. 1975: Glacialtektoniske og -stratigrafiske undersøgelser i udvalgte områder i det sydlige Danmark. Unpubl. prize essay, Univ. Copenhagen.

Frederiksen, J.K. 1976: Hvad de sønderjyske klinter fortæller, VARV, 1976, No. 2, 35-45.

Gripp, K. 1929: Glaciologische und geologische Ergebnisse der Hamburgischen Spitzbergen Expedition 1927. Abh.d.na-

turviss.Vereins, Hamburg. Bd. 22, Heft 2-4.
Gry, H. 1940: De istektoniske Forhold i Moler området. Meddr. dansk geol. Foren., vol. 9, No. 5, 586-627.
Gry, H. 1941: Diskussion om vore dislocerede Klinters Dannelse. Meddr. dansk geol. Foren., Vol. 10, 39-51.
Holmström, L. 1904: Öfversikt av den glaciala afslipningen i Sydskandinavien. Geol. Fören. Stockh. Förh., Vol. 26, 365-432
Houmark-Nielsen, M. 1976: Nordsamsøs Istidsnekrolog. VARV, 1976, No. 3, 89-96.
Houmark-Nielsen, M. 1980: Glacialstratigrafien omkring det nordlige Bælthav. Unpubl. Ph.d thesis, Inst. General Geol., University of Copenhagen, 194 p.
Houmark-Nielsen, M. 1981: Glacialstratigrafi i Danmark øst for Hovedopholdslinien. Dansk geol. Foren., Arsskrift for 1980, 61-76.
Houmark-Nielsen, M. 1983: Glacial stratigraphy and morphology of the northern Bælthav region. In: Ehlers, J. (edit.): Glacial deposits in northwest Europe. A.A.Balkema, Rotterdam, 211-217.
Houmark-Nielsen, M. 1987: Pleistocene Stratigraphy and glacial History of the central part of Denmark. Bull. geol. Soc. Denmark, vol. 36, part 1-2, 189 p.
Houmark-Nielsen, M. & Berthelsen, A. 1981: Kineto-stratigraphic evaluation and presentation of glacial-stratigraphic data, with examples from northern Samsø Denmark.
Houmark-Nielsen, M. & Kolstrup, E. 1981: A radiocarbon dated Weichselian sequence from Sejerø, Denmark. Geol. Fören. Stockh. Förh., vol. 103, Pt. 1, 73-78.
Houmark-Nielsen, M. & Lagerlund, E. 1987: The Helsingør Diamicton. Bull.geol.Soc. Denmark, vol. 36 part 3-4, 237-247
Jacobsen, E.M. 1976: En morænestratigrafisk undersøgelse af klinterne på Omø. Dansk geol. Foren., Arsskrift for 1975, 15-17.
Jacobsen, E.M. 1981: Some glaciotectonic features from the southern part of Sjælland, Denmark. Abstract. INQUA & IGCP 73/1/24. Field meeting in Denmark, may 1981.
Jacobsen, E.M. 1985: En råstofgeologisk kortlægning omkring Roskilde. Dansk geol. Foren., Arsskrift for 1984, 65-78.
Jensen, J.B. & Knudsen, K.L. 1984: Kvartærstratigrafiske undersøgelser ved Gyldendal og Kås Hoved i det vestlige Limfjordsområde. Dansk geol. Foren., Arsskrift for 1983, 35-54.
Jensen, V. 1977: Store Karlsminde Klint, materialer og strukturer. Dansk geol. Foren., Arsskrift for 1976, 47-55.
Jessen, A. 1918: Vendsyssels Geologi. Danm. geol. Unders. V, No. 2, 195 p.
Jessen, A. 1930: Klinten ved Halk Hoved. Danm. geol. Unders. IV, vol. 2, No. 8, 26 p.
Jessen, A. 1931: Lønstrup Klint. Danm. geol. Unders. II, No. 49, 142 p.
Kronborg, C. 1983: Glacialstratigrafien i Øst- og Midtjylland. Unpubl. Ph.d. thesis Geol. Inst., Univ. of Aarhus, 259 p.
Kronborg, C. & Knudsen, K.L. 1985: Om kvartæret i Rugård: En foreløbig undersøgelse. Dansk Geol.Foren., Arsskrift for 1984, 37-48.
Lagerlund, E. 1980: Lithostratigrafisk indeling av Västskånes Pleistocen och en ny glaciationsmodell för Weichsel. Univ. of Lund, Dept. of Quat. Geol., Report 21, 120 p.
Lagerlund, E. 1987: An alternative Weichselian glaciation model, with special refrence to the glacial history of Skåne, South Sweden. Boreas, vol. 16, 433-459.
Larsen, G., Jørgensen, F.H. & Priisholm, S. 1977: The stratigraphy, structure and origin of glacial deposits in the Randers area, eastern Jutland. Danm. geol. Unders. II, nr. 111, 36 p.
Madsen, V. 1897: Beskrivelse til Geologisk Kort over Danmark, Kortbladet Samsø. Danm. geol. Unders. I, No. 5, 87 p.
Madsen, V. 1900: Beskrivelse til Geologisk Kort over Danmark. Kortbladet Bogense. Danm. geol. Unders. I, No. 7, 112 p.
Madsen, V. 1916: Ristinge Klint. Danm. geol. Unders. IV, vol. 1, No. 2, 32 p.
Madsen, V. 1928: Oversigt over Danmarks Geologi. Danm. geol. Unders. V, No. 4, 78-142.
Madsen, V. & Nordmann, V. 1940: Kvartæret i Røgle Klint ved Lillebælt. Danm. geol.Unders. II, No. 58, 142 p.
Mangerud, J. & Berglund, B. 1978: The subdivision of the Quaternary of Norden, a discussion. Boreas, vol. 7, 179-181.
Milthers, K. 1942: Ledeblokke og Landskabsformer i Danmark, Danm. Geol. Unders No. 69, 137 p.
Milthers, V. 1909: Scandinavian Indicator Boulders in Quaternary Deposits. Danm. geol. Unders. II, No. 23, 153 p.
Nielsen, J.B. 1987: Kvartærgeologiske observationer langs østsiden af Roskilde Fjord. Dansk Geol.Foren., Arsskrift for 1986, 41-47.
Nielsen, P.E. 1980: Kvartærgeologiske

undersøgelser i Korsør området. Dansk geol.Foren., Årsskrift for 1979, 55-62.
Overweel, C.J. 1977: Distribution and transport of Fennoscandian indicators. Scripta Geol. 43, 117 p.
Petersen, K.S. 1973: Tills in dislocated drift deposits on the Røsnæs peninsula, northwestern Sjælland, Denmark. Bull. Geol. Inst. Univ. Upps., vol. 5, 41-49.
Petersen, K.S. 1978: Applications of glaciotectonic analysis in geological mapping of Denmark. Danm. geol. Unders., Årbog 1977, 53-61.
Petersen, K.S. 1984: Stratigraphical position of Weichselian tills in Denmark. Striae, vol. 20, 75-78.
Rasmussen, L.Aa. 1973: The Quaternary Stratigraphy and dislocations of Ven. Bull. Geol. Inst. Univ. Upps., 5, 37-39.
Rasmussen, L.Aa. 1975: Kinetostratigraphic glacial drift units on Hindsholm, Denmark. Boreas, 4, 209-217.
Rasmussen, L.Aa. & Petersen, K.S. 1980: Resultater fra DGU's genoptagne kvartærgeologiske kortlægning. Dansk geol. Foren.,Årsskrift for 1979, 47-54.
Richter, K. 1929: Studien über fossile Gletscherstruktur. Zeitschr. für Gletscher kunde, Vol. XVII 1929.
Rosenkrantz, A. 1944: Nye Bidrag til Forstaaelsen af Ristinge Klints Opbygning. Meddr. dansk Geol. Foren., Vol. 10, 431-435.
Selsing, L. 1981: Stress analysis on conjungate normal faults in unconsolidated Weichselian glacial sedimants from Brorfelde, Denmark. Boreas, vol. 10, 275-279.
Sjørring, S. 1977: The glacial stratigraphy of the island of Als, southern Denmark. Z. Geomorph. N. F., Suppl. Bd. 27, 1-11
Sjørring, S. 1981: Pre-Weichselian Till stratigraphy in Western Jutland, Denmark. Meded Rijks Geol. Dienst., vol. 34, 62-68.
Slater, G. 1927: The Disturbed Glacial Deposits in the Neighbourhood of Lønstrup, near Hjørring, North Denmark. Transact. Roy. Soc. Edinb., Vol. LV, Part II (No. 13).
Thamdrup, K. 1970: Klinterne ved Mols Hoved, en kvartærgeologisk undersøgelse. Meddr. dansk geol. Foren., Årsskrift for 1969, 2-8.

Glaciotectonics: Forms and Processes, Croot (ed.), © 1988 Balkema, Rotterdam. ISBN 90 6191 848 0

Large-scale glaciotectonic deformation of soft sediments: A case study of a late Weichselian sequence in western Iceland

Olafur Ingólfsson
Lund University, Sweden

ABSTRACT: An investigation of the glacial geology of the Melabakkar-Asbakkar coastal cliffs in the lower Borgarfjördur region, western Iceland, revealed that the sequence has been subject to extensive glaciodynamic deformations. The stratigraphical and sedimentological evidence suggests that the sequence was deposited in an isostatically depressed fjord basin. The glacial stratigraphy of the sequence is briefly described and a model for the development of the depositional basin is outlined. A brief description of deformational structures is given, and the chronological control of glacial events is discussed. It is concluded that the glaciotectonic features developed in pro- and subglacial unfrozen soft sediments.

1 INTRODUCTION

As a part of a regional study of the glacial geology of the lower Borgarfjördur region, western Iceland (Ingólfsson 1984, 1985, 1987), the Melabakkar-Asbakkar coastal cliffs (Fig. 1) were studied to identify glacial episodes. Stratigraphical relations of the sediments and their sedimentological and structural properties were used to define a depositional model for the sequence, and a chronology for glacial events was established on the basis of radiocarbon dated shell samples (Ingólfsson 1988). A variety of "exodiamict glaciotectonic structures" (Banham 1975), ranging from small scale fracturing and shearing to large scale overfolding and thrusting, are superimposed on the sedimentary strata. Despite increased studies of glaciotectonic structures in recent years (Berthelsen 1978, Aber 1982, 1985, Hicock & Dreimanis 1985) and increasing evidence from both ancient sequences and recent glaciers, many fundamental problems regarding the development of glaciotectonic deformations remain to be satisfactorily explained. Controversial explanations have been suggested with regard to frontal/basal conditions for the development of large scale dislocations, i.e. whether they take place mainly in front of advancing glaciers (Berthelsen 1979) or below advancing glaciers via compressive flow in the basal zone (Moran 1971), and whether glaciotectonic deforma-

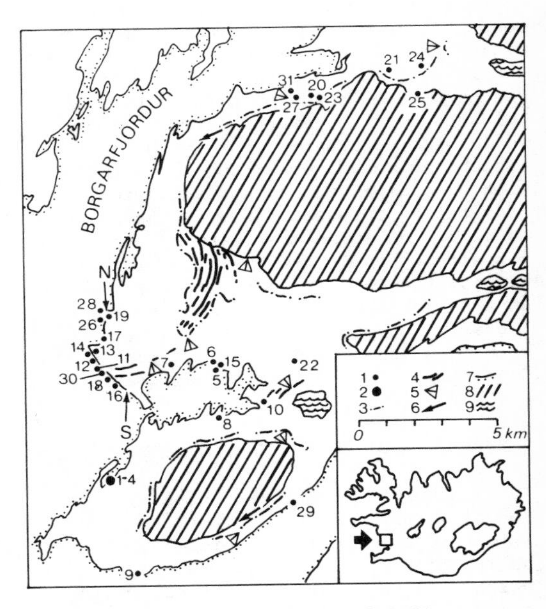

Fig. 1. Location map. The Melabakkar-Asbakkar section runs between points N and S. Legend: (1) location and no. of radiocarbon dated shells, (2) radiocarbon dated peat samples, (3) marine limit at 60-70 m a.s.l., (4) moraine ridges, (5) outwash deposits, (6) kame deposits, (7) present coastline, (8) area above 100 m a.s.l., (9) lakes.

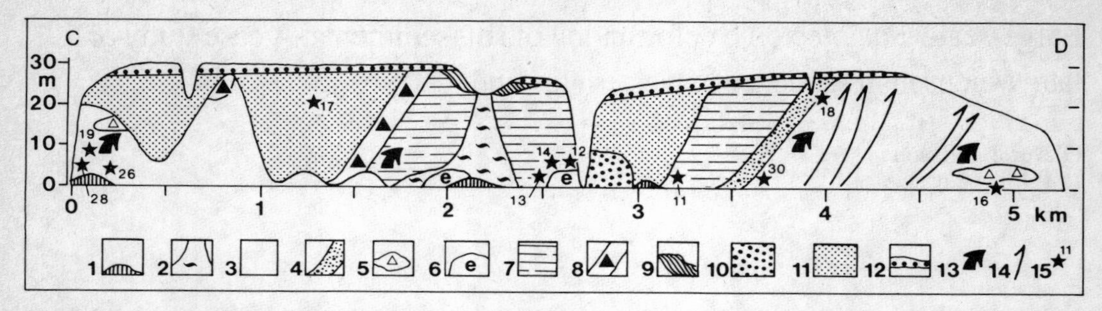

Fig. 2. Diagrammatic profile section of the Melabakkar-Asbakkar coastal cliffs. Legend: (1) bedrock, (2) section covered, (3) Asbakkar diamicton (glaciomarine sequence), (4) As beds (morainal bank sediments), (5) interbedded debris flow- and outwash sediments, (6) esker-fan sediments, (7) Látrar beds (glaciomarine sequence), (8) Melar diamicton (lodgement till), (9) Landhólmi sands (ice-marginal delta), (10) Asgil gravels (ice-marginal outwash), (11) Melabakkar silts and sands (glaciomarine and sub-littoral sequence), (12) Melagil gravels and sands (gravel lag and beach sediments), (13) glaciodynamic deformations, (14) large scale thrust faults, (15) location and no. of radiocarbon dated shells from the section.

tions form in frozen or unfrozen sediments (Banham 1975, Moran et al. 1980, Aber 1982, 1985, Thomas 1984, Sharp 1985). The role of basal meltwater, pore-water pressure and permeability of the substratum, differential loading etc. in the development of glaciotectonic deformation is not clear, as the information from recent glaciers on this subject is fragmentary. The purpose of this paper is to present evidence for the development of large scale glaciotectonic deformations in unfrozen sediments, occurring in connection with glacial advances into a glacio-isostatically depressed fjord/bay basin. As there are no structural methods available yet with which to get objective information on the basal conditions of a glacier at the time of glaciotectonic deformation, I base my conclusions on a good stratigraphical and chronological control of the development of the depositional basin.

2 STRATIGRAPHY AND BASIN DEVELOPMENT

The Melabakkar-Asbakkar coastal section is roughly 5 km long, and up to 30 m high (Fig. 2). The section is eroded by the sea during high tides and westerly storms, which results in almost vertical cliffs and good exposures along most stretches. The total thickness of the sedimentary strata exposed in the cliffs is 140-145 m, of which three glaciomarine sequences constitute about 85 m, ice-proximal and ice-contact outwash sediments, debris flows and tills 40-45 m, and littoral and beach sediments about 15 m.

The base of the Melabakkar-Asbakkar sequence is marked by a surface of glacially striated bedrock, indicating a glacial event when the glaciers reached beyond the present coast some time before 12.500-13.000 BP. The upper surface is a lag surface of wave erosion at 15 - 30 m a.s.l., marking the post-glacial isostatic rebound of the basin and subsequent marine regression. The sequence development may be divided into nine stages (Figs. 2 and 3): During the first stage (stage A), mollusc supporting glaciomarine sediments (the Asbakkar diamicton) accumulated from suspension, random underflows and ice-rafted debris. A glacial advance down the Borgarfjörður valley/fjord (stage B) caused the molluscs to disappear, and subaqueous ice-marginal/ice proximal stratified sediments (the As beds) were deposited from meltwater streams and glacigenic debris flows to form proglacial shoals or banks. Continued advance of the glacier (stage C) caused large-scale proglacial and subglacial deformations. Subsequently the glacier overran the whole sequence (stage D). Later the glacier retreated to a frontal position somewhere north of the section (stage E). In the course of retreat, esker fan sediments were deposited from meltwater streams (lower facies of the Látrar beds), and later, rhythmic glaciomarine sediments supporting a limited mollusc population (upper facies of the Látrar beds) accumulated from suspended fines, underflows and ice rafting.

During a glacial readvance out of the Borgarfjörður valley/fjord (stage F), lodgement till (the Melar diamicton) and ice-marginal delta sediments (the Landhólmi sands) were deposited in the sequence. The glacier advanced about halfway across the section, and a transverse ridge of stratified gravels and sands (the Asgil gravels) marks the position of the ice margin during the latest part of this advance (stage G). Glaciotectonic deformation of the strata also occurred in connection with this second glacial advance, but the processes were less intense than before. During stage H, some time before 10.000 BP, the glacier retreated from the coastal areas, and a third sequence of glaciomarine sediments and later sublittoral sediments (the Melabakkar silts and sands) were deposited. Finally (stage I) the postglacial marine regression caused the deposition of beach gravels and sands (the Melagil gravels and sands).

It is my conclusion that the Melabakkar-Asbakkar sequence contains a fairly continuous record of glacial episodes in the lower Borgarfjördur region for the later part of the Late Weichselian. The development of the depositional basin with regard to lithofacies distribution and stratigraphic associations, as outlined above and described in more details in Ingólfsson (1987, 1988), is fairly compatible with models for subarctic glaciomarine fjord glaciation as modelled by e.g. Domack (1983), Eyles et al. (1985), Mode et al. (1983), Molnia (1983) and Powell (1981, 1984).

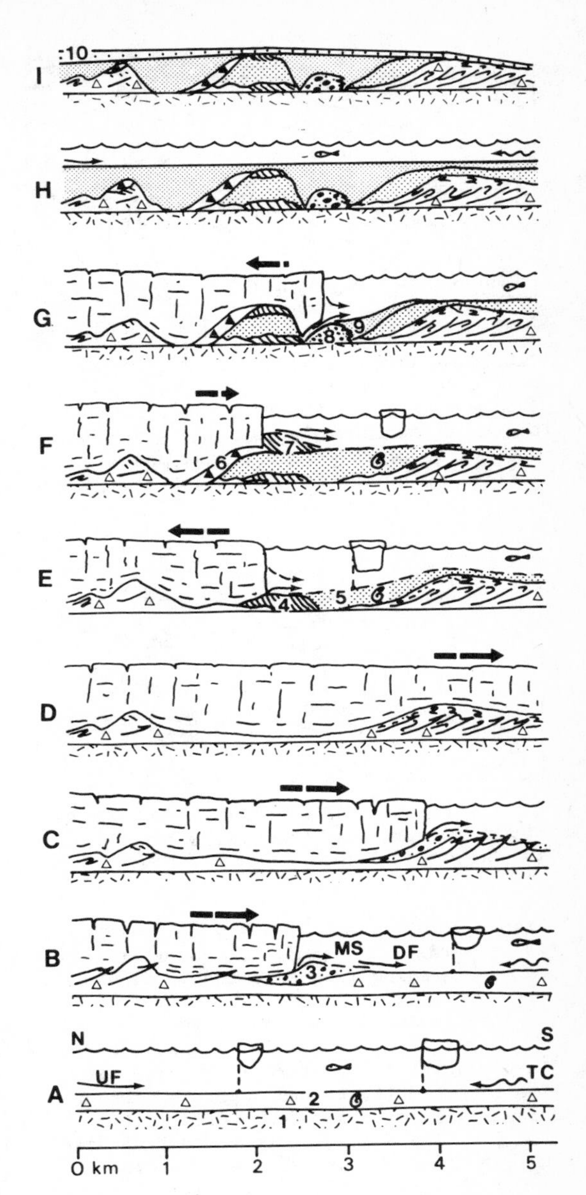

Fig. 3. Stages in the development of the Melabakkar-Asbakkar cliffs. Legend: (UF) underflows, (TC) tidal currents, (MS) meltwater streams, (DF) subaquatic glacigenic debris flows, (1) bedrock, (2) Asbakkar diamicton, (3) As beds, (4) Látrar esker-fan facies, (5) Látrar glaciomarine facies, (6) Melar diamicton, (7) Landhólmi sands, (8) Asgil gravels, (9) Melabakkar silts and sands, (10) Melagil gravels and sands. See text for discussion.

3 GLACIOTECTONIC STRUCTURES

Glaciodynamic deformations occur in two poorly defined zones in the stratigraphical sequence: a lower zone, mainly affecting the lowest glaciomarine sequence (the Asbakkar diamicton) and the morainal bank sediments (the As beds); and an upper zone of less intense deformations, mainly affecting the Látrar beds glaciomarine sequence and the Landhólmi sands ice-marginal delta deposits. The following is a brief description of some of the deformation features, which emphasises those structures which may be regarded as good structural indicators of glaciotectonism and the direction of movement.

3.1 Folds

Fig. 4 shows a large-scale isoclinal fold belonging to the lower deformation zone. The fold is a 8-10 m high anticline, truncated by a rising thrust plane that carries the bulk of the overlying sediments across the axis of the fold. Sediments on either side of the thrust plane are sheared and folded (Fig. 5). The fold axis

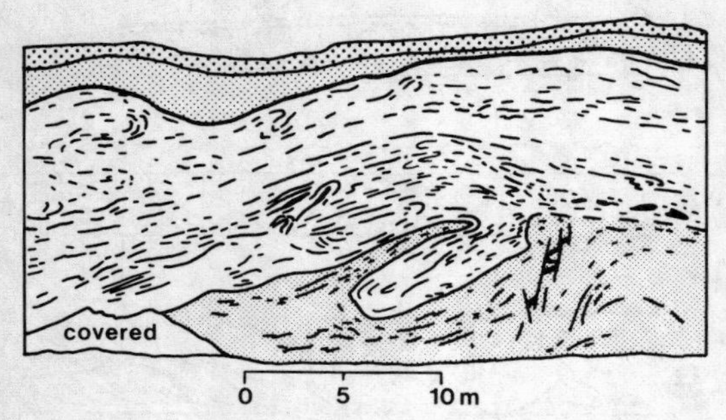

Fig. 4. A large isoclinal fold belonging to the lower deformation zone, at around 3700 m in Fig. 2. The fold affects the Asbakkar diamicton (glaciomarine sequence) and the As beds (morainal bank sediments). Glacial push from left to right

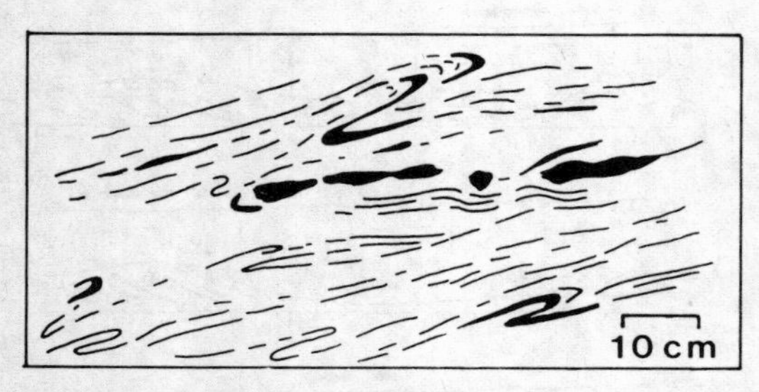

Fig. 5. Heavily sheared and folded structures in silty-sandy sediments belonging to the morainal bank sequence.

trends 75°-255° and the axial plane dips 28° towards 345°, indicating a deforming force operating from a northerly direction.

A large-scale recumbent structure with an almost horizontal axial plane, occurs in the lower deformation zone (Fig. 6). The inverted strata and a part of the anticlinal hinge zone are exposed in the section. The deformed strata is truncated by a horizon of glacial shear, approximately coinciding with the axial plane. The structure is overlain by a heavily sheared gravel-diamicton admixture ("deformation till"). The fold axis trends 100°-280°, indicating a southerly direction for the dislocating force.

3.2 Faults

A number of low angled reverse faults, commonly termed "thrust faults" (Dennis 1967) occur in the cliffs (Fig. 2). The fault planes are usually slightly concave upwards and measurements of 10 fault planes revealed apparent dips between 28° and 42° towards NW-N. The large-scale thrust faults appear to form a poorly developed imbricated dislocation pattern with overthrusts of blocks towards the south. On one occasion, fault drag below a major thrust fault had overturned a body of stratified sand, with a minimum displacement of about 5 m along the thrust plane. The displacement strike of the sand body measured 80°-160°, indicating thrust towards SSE-S.

In the substratum of the lodgement till recognized in the cliffs (Fig. 2) there is an example of a small-scale, asymmetrical, angular fold which has been transformed into a thrust fault when the sharp trough hinge fractured and the inverted limb was replaced by a slide plane (Fig. 7). The slide plane has an apparent dip of 26° towards NNW.

The thrust faults and the large-scale folded structures belong to the same dislocation pattern, indicating a direction of maximal stress during deformation from approximately north.

Other structures of glaciotectonic origin in the cliffs, not described further here, are boudins, joints, clastic dikes and wedges.

4 CHRONOLOGY AND CHANGES OF SEA LEVEL

The chronology for glacial events is based on 27 radiocarbon dated samples of subfossil molluscs from glaciomarine sediments throughout the lower Borgarfjördur region (Table I, Fig. 1). 11 samples were collected in the Melabakkar-Asbakkar section (Fig. 2). The radiocarbon dates fit the stratigraphic division of the Melabakkar-Asbakkar sediments reasonably well, given that the time intervals being dated are

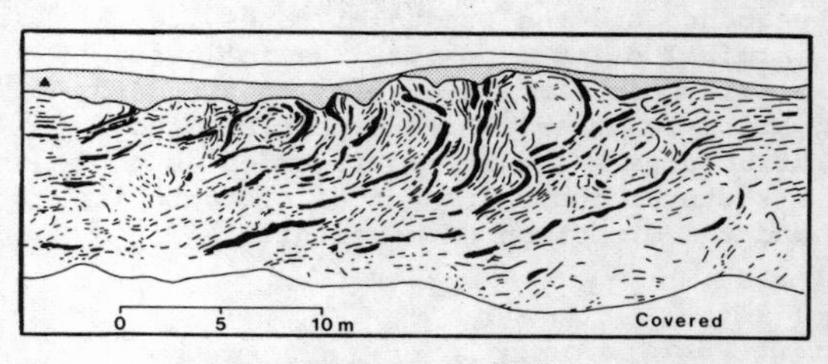

Fig. 6. Structural outline of a heavily sheared and folded strata at around 4700 m in Fig. 2. The deformed strata is truncated by a horizon of glacial shear, approximately coinciding with the axial plane of the folded structure, and overlain by a heavily sheared diamicton. Deforming shear from left to right. Height of section about 20 m.

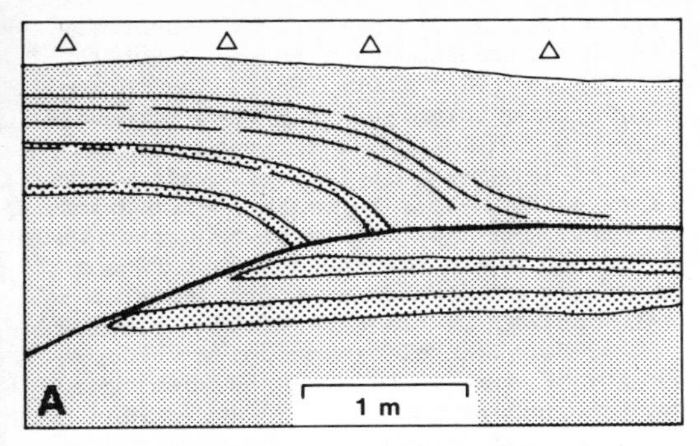

Fig. 7. A thrust fault developed from an angular fold in the Asbakkar diamicton, below the Melar diamicton lodgement till at around 1500 m in Fig. 2.

small and overlapping error limits can be expected. Three glaciomarine sequences are recognized and dated: the lowest and regionally most extensive sequence is dated by 16 radiocarbon dates to between 12.800±220 and 11.885±102 years BP. The stratigraphical control of this sequence is good: It is always the lowest sedimentary unit where it is found in a section, often found directly on striated bedrock and it is fairly homogeneous in both sedimentology and fossil fauna content. It has often been subject to intense glaciodynamic deformation or is found below a horizon of glacial shear. There is evidence that the relative sea level about 12.300 years BP was higher than 65-70 m, as mollusc supporting marine sediments were deposited in the present Skorradalur valley which opens at 62 m a.s.l. Sample 25 in Table I, was collected from glacial till at about 125 m a.s.l. in the mouth of the Skorradalur valley (see a discussion of this sample in Ingólfsson 1984).

The second glaciomarine sequence, corresponding to the rhythmic facies of the Látrar beds in the Melabakkar-Asbakkar section, is of more limited extent, not being found north of the Melabakkar-Asbakkar sequence. Stratigraphical control is not as good for this sequence as for the lowest glaciomarine sequence. In terms of lithology, stratification, glaciotectonic deformations, fossil content and extent/continuity it varies considerably, depending on proximity to ice-marginal features in the area. Additionally it may be bounded either conformably or unconformably by surrounding sediments. Nevertheless it is always stratified with dropstones, and contains high-arctic mollusc species such as Portlandia arctica (found in two localities). The sequence is dated

Table 1. Radiocarbon dates from the lower Borgarfjördur region.

Sample no.	^{14}C date, years BP	Lab no.	Sample
1	1.910±70	Lu-2394	Peat
2	6.060±80	Lu-2395AII	Peat
3	6.130±80	Lu-2395	Peat
4	6.300±120	Lu-2395AI	Peat
5	10.005±90	Lu-2197	Mollusc
6	10.155±150	Lu-2375	Mollusc
7	10.965±80	Lu-2056	Mollusc
8	10.985±100	Lu-2340	Mollusc
9	11.065±140	Lu-2338	Mollusc
10	11.155±100	Lu-2524	Mollusc
11	11.465±100	Lu-2376	Mollusc
12	11.545±140	Lu-2373	Mollusc
13	11.615±130	Lu-2196	Mollusc
14	11.715±120	Lu-2372	Mollusc
15	11.885±100	Lu-2374	Mollusc
16	11.945±110	Lu-2377	Mollusc
17	11.985±120	Lu-2375	Mollusc
18	12.015±110	Lu-2379	Mollusc
19	12.095±120	Lu-2192	Mollusc
20	12.100±150*	S-290	Mollusc
21	12.100±250*	S-289	Mollusc
22	12.470±110	Lu-2055	Mollusc
23	12.145±140	Lu-2371	Mollusc
24	12.240±200*	I-1825	Mollusc
25	12.270±150*	I-1824	Mollusc
26	11.925±160**	U-641	Mollusc
27	12.465±110	Lu-2194	Mollusc
28	12.465±110	Lu-2193	Mollusc
29	12.475±110	Lu-2339	Mollusc
30	12.505±112	Lu-2195	Mollusc
31	12.800±220*	S-291	Mollusc

Samples (*) from Ashwell (1975) and (**) from Olsson et al. (1969). All samples are referred to 0.95 NBS oxalic acid standard. The base year is 1950. Corrections for deviation from standard $^{13}C/^{12}C$ ratio and apparent age of living marine organisms (Håkansson 1983) applied, except for samples marked (*).

by 8 radiocarbon dates to 11.715±122 - 10.965±82 years BP. The sea level during its deposition was higher than 35 m above present sea level.

The third and youngest glaciomarine sequence, which corresponds to the Melabakkar silts and sands, is regionally extensive. It relates to the most recent marine limit in the region at 60-70 m a.s.l. (Fig. 1), and the postglacial marine regression is registered in the sequence as marine silts give way to sublittoral sands, later to be truncated by gravel lag horizon of wave erosion and overlying beach sediments. It is poor in subfossil molluscs, but 2 samples obtained date it to 10.155±151 years BP and 10.005±92 years BP.

5 DISCUSSION

My conclusions are that the glaciotectonic structures exposed in the Melabakkar-Asbakkar section developed in unfrozen pro- and subglacial sediments due to a combination of glacial- and hydrodynamic processes and mechanisms, in connection with glacial advances into a submerged fjord/bay basin.

The two glacial advances which caused glaciotectonic deformation of sediments in the Melabakkar-Asbakkar sequence, probably occurred during the time intervals 12.000 - 11.700 BP and 11.000 - 10.300 BP. The relative sea level before the former and after the later ice advance was above or at 60-70 m a.s.l. The sea level during the ice advances is not known. The sedimentological and stratigraphical evidence suggests that the glacier advanced into a submerged basin, and there are ample indications of meltwater input and calving activity. The conditions for pro/subglacial permafrost in connection with the ice advances, i.e. a marine regression, a very thick ice, or a combination of both, are not met for the Melabakkar-Asbakkar sequence. Also, the rate of isostatic rebound necessary to achieve marine regressions during the time spans available is in the order of 10 - 15 m x 100 y^{-1}. For comparison, an average postglacial isostatic rebound rate of 2 - 2.5 m x 100 y^{-1} can be inferred for the lower Borgarfjõrdur region on the basis of radiocarbon dating of submerged peat deposits (Fig. 1 and Table I).

Many researchers have regarded permafrost to be a pre-condition of glaciotectonism (e.g. Kupsch 1962, de Jong 1967, Moran 1971, Berthelsen 1979, Moran et al. 1980, Aber 1982), although current research is beginning to show that glaciotectonism can take place in a non-frozen sediment sequence (e.g. Funder & Petersen 1980, Aber 1985, Sharp 1985). The permafrost concept implies that high pore water pressure and low shear strength at the base of a permafrost layer provides a plane of décollement for the deformation, and that the sediments must be frozen when deformed. Their thickness thus reflects the depth of permafrost at the time of deformation. In the case of the Melabakkar-Asbakkar glaciotectonism the opposite appears to be true: the bedrock surface underlying the unfrozen strata has acted as a décollement plane. The comparatively coarse grained units of glaciofluvial outwash and till in the sequence have largely been competent and supported the tectonic forces, while the incompetent silty-clayey glaciomarine sediments have deformed and folded more easily, probably via fluidized flow. At this stage I cannot with certainty decide the sequence of deformation, but I tentatively suggest that the thrust faults developed at a late stage relative to the folding of the strata.

REFERENCES

Aber, J.S. 1982. Model for glaciotectonism. Bulletin of the Geological Society of Denmark 30:79-90.

Aber, J.S. 1985. The character of glaciotectectonism. Geologie en Mijnbouw 64: 389-395.

Ashwell, I.Y. 1975. Glacial and Late Glacial processes in western Iceland. Geografiska Annaler 57:225-245.

Banham, P.H. 1975. Glaciotectonic structures: a general discussion with particular reference to the contorted drift of Norfolk. In A.E.Wright & F.Moseley (eds.), Ice Ages: Ancient and Modern. Geological Journal, Special Issue No. 6, p.69-94. Liverpool, Steel House Press.

Berthelsen, A. 1978. The methodology of kinetostratigraphy as applied to glacial glacial geology. Bulletin of the Geological Society of Denmark 27:25-38.

Berthelsen, A. 1979. Recumbent folds and boudinage structures formed by subglacial shear: an example of gravity tectonics. Geologie En Mijnbouw 58:253-260.

Dennis, J.G. 1967. International tectonic dictionary. American Association of Petroleum Geologists Memoir 7:1-196.

Domack, E.W. 1983. Facies of late Pleistocene glacial-marine sediments on Whidbey Island, Washington: An isostatical glacial-marine sequence. In B.F.Molnia (ed.), Glacial-Marine Sedimentation, p. 535-570. New York, Plenum Press.

Eyles, C.H., N. Eyles & A.D. Miall 1985. Models of glaciomarine sediments and their application to the interpretation of ancient glacial sequences. Palaeogeography, Palaeoclimatology, Palaeoecology 51:15-84.

Funder, S. & K.S. Petersen 1980. Glacitectonic deformations in East Greenland. Bulletin of the Geological Society of Denmark 28:115-122.

Håkansson, S. 1983. A reservoir age for the coastal waters of Iceland. Geologiska Föreningens i Stockholm Förhandlingar 105:64-67.

Hicock, S.R. & A. Dreimanis 1985. Glaciotectonic structures as useful ice-movement indicators in glacial deposits: four Canadian case studies. Canadian Journal of Earth Sciences 22:339-346.

Ingólfsson, O. 1984. A review of Late Weichselian studies in the lower part of the Borgarfjördur region, western Iceland. Jökull 34:117-130.

Ingólfsson, O. 1985. Late Weichselian glacial geology of the lower Borgarfjördur region, western Iceland: a preliminary report. Arctic 38:210-213.

Ingólfsson, O. 1987. Investigation of the Late Weichselian glacial history of the lower Borgarfjördur region, western Iceland. Lundqua Thesis 19. Lund, Lund University.

Ingólfsson, O. 1988. The Late Weichselian glacial geology of the Melabakkar-Asbakkar coastal cliffs, Borgarfjördur, W-Iceland. Jökull 37:1-25.

de Jong, J.D. 1967. The Quaternary of the Netherlands. In K.Rankama (ed.), The Quaternary, p.301-426. New York, John Wiley & Sons.

Kupsch, W.O. 1962. Ice-thrust ridges in western Canada. Journal of Geology 70: 582-594.

Mode, W.N., R.W. Nelson & J.K. Brigham 1983. A facies model of Quaternary glacial-marine cyclic sedimentation along eastern Baffin Island, Canada. In B.F. Molnia (ed.), Glacial-Marine Sedimentation, p.459-533, New York, Plenum Press.

Molnia, B.F. 1983. Subarctic glacial-marine sedimentation: a model. In: B.F. Molnia (ed.), Glacial-Marine Sedimentation, p.95-144. New York, Plenum Press.

Moran, S.R. 1971. Glaciotectonic structures in drift. In: R.P. Goldwaith (ed.), Till/a Symposium, p.127-148. Ohio State University Press.

Moran, S.R., L. Clayton, R.LEB. Hooke, M. Fenton & L. Andriashek 1980. Glacier-bed landforms of the Praire region of N-America. Journal of Glaciology 25: 451-476.

Olsson, I.U, S. El-Gammal & Y. Göksu 1969. Uppsala Radiocarbon measurements 9. Radiocarbon 11 (2):515-544.

Powell, R.D. 1981. A model for sedimentation by tidewater glaciers. Annals of Glaciology 2:29-134.

Powell, R.D. 1984. Glacimarine processes and inductive lithofacies modelling of ice-shelf and tidewater glacier sediments based on Quaternary examples. Marine Geology 57:1-52.

Sharp, M. 1985. Sedimentation and stratigraphy at Eyjabakkajökull - An Icelandic surging glacier. Quaternary Research 24:268-284.

Thomas, G.S.P. 1984. The origin of the glacio-dynamic structure of the Bride Moraine, Isle of Man. Boreas 13:355-364.

Glaciotectonics: Forms and Processes, Croot (ed.), © 1988 Balkema, Rotterdam. ISBN 90 6191 848 0

Sedimentation and deformation of the North Sea Drift Formation in the Happisburgh area, North Norfolk

J.P.Lunkka
Subdepartment of Quaternary Research, University of Cambridge, UK

ABSTRACT: The North Sea Drift Formation is currently under investigation in the Happisburgh area of north east Norfolk.
The nature, geometry and interrelationships of the major sedimentary units exposed in the coastal cliffs have been determined by lithofacies mapping, textural and structural properties of the exposed sediments have also been investigated.
The sequences studied show laterally continuous diamicton facies interbedded with clays, silts and sands. Various deformation structures are found throughout the sequence.
This paper discusses the environment and processes of deposition of the diamictons, sands and fines together with the origin of the deformation structures.

1. INTRODUCTION

The glacial sediments of Norfolk have been lithologically divided into three major groups: 1. The North Sea Drift 2. The Marly Drift and 3. The Chalky Boulder Clay or Lowestoft Till (e.g. Boulton et al.,1984).
On the basis that the North Sea Drift overlies the Cromer Forest Bed Formation (e.g. West, 1980) and that Hoxnian temperate Stage deposits overlie the Lowestoft Till and Marly Drift (Ehlers et al.,1986) these glacial sediments are regarded as having been deposited during the Anglian Stage (Bowen et al., 1986 and references therein).
A complex stratigraphy has been developed for the Anglian glacial sediments in north east Norfolk from a series of studies during the last hundred years (e.g. Reid, 1882; Solomon, 1932; Baden-Powell, 1948; West and Donner, 1956; Dhonau & Dhonau, 1963; Kazi & Knill, 1969; Banham, 1966, 1968, 1970, 1975, 1977a,b; Perrin et al., 1979). However, the interrelationships of the different diamicton units and related sorted sediments between different areas are still not fully understood.
Nowadays it is widely accepted by most authors that during the Anglian Stage there were at least two major ice lobes advancing into north east Norfolk. At the Anglian stratotype at Corton, Suffolk (Mitchell et al.,1973) the Cromer Till (Banham, 1968) i.e. the diamicton of the North Sea Drift Formation (Perrin et al., 1979) is overlain by Corton Sands and Lowestoft Till. This implies that the Lowestoft Till Formation (Perrin et al., 1979) postdates that of the North Sea Drift Formation Ice (Perrin et al., 1979) at Corton. The stratigraphical position of the Marly Drift has recently been questioned and has been shown to be a chalky variety of the Lowestoft Till (Ehlers et al., in press).
The present investigation is restricted to the North Sea Drift Formation (Gibbard & Zalasiewicz, 1988) in the Happisburgh area (fig.1). There this Formation comprises three different diamicton units which are interbedded with sorted sediments. The coastal sequence in the area was previously studied by Reid (1882), Solomon (1932) and more recently in a series of papers by Banham (1966, 1968, 1970, 1977a,b), Kazi & Knill (1969) and Hart & Boulton (in press).
Banham (1968, 1970, 1977b) proposed a lithostratigraphical division for the area (Table 1); his stratigraphy being virtually the same as that outlined by Reid (1882). Banham concluded that the so-called 'Cromer ice' (responsible for the deposition of the Cromer Tills) advanced into the area on three distinct occasions. The Intermediate Beds and the Mundesley Sands were deposited in a shallow water environment respectively after retreat of the first and second ice lobes (Banham, 1977b). He further suggested that these sorted sediments were supplied with material derived from land towards the south and west (Banham, 1977b).
The lower part of the sedimentary sequence in coastal cliffs between Happisburgh and Cromer were also studied by Kazi & Knill (1969). They concluded that the Intermediate Beds at Happisburgh could be divided into two subunits, which include various deformation structures. Further they suggest that many strata in the Intermediate Beds are the result of rapid deposition from turbidity currents.
More recently Hart (1987) and Hart and Boulton (in press) have presented a rather complex picture of glaciation in the Anglian Stage. They proposed that

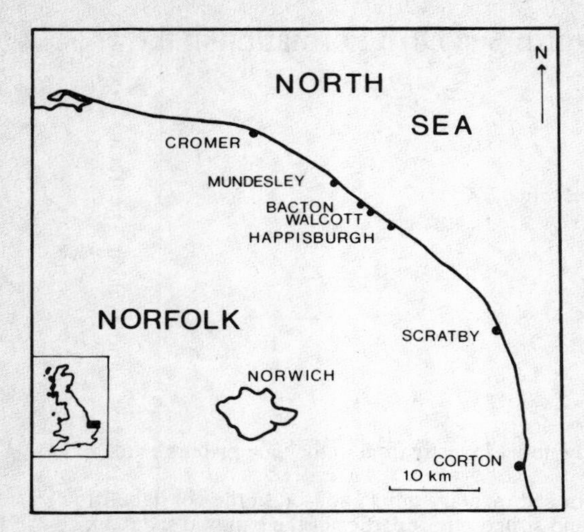

Fig. 1. Location map showing the position of the cliffs studied, between Happisburgh and Walcott.

there were at least four ice advances into North Norfolk; their First and Second North Ice Advances reached the Happisburgh area. They suggest that the diamictons were produced by subglacial deformation. (Hart & Boulton, in press).

The coastal cliffs at Happisburgh have been selected for the present study since they comprise exposures in the North Sea Drift Formation. The sedimentary sequence at these localities are laterally continuous and less strongly deformed than further NW on the North Norfolk coast (e.g. Boulton et al., 1984). These exposures provide an important opportunity for palaeoenvironmental reconstruction by sedimentological interpretation of the sediments. Arising from the contraversy between the results of previous studies the main purpose of this paper is to concentrate on the evolution of the sedimentary environments during deposition, and on the process of sedimentation of both the diamictons and sorted sediments.

2. METHODS

For the purpose of this study the cliff exposures have been recorded in detail using lithofacies logging techniques modified for the present work following Eyles et al. (1983). Lithofacies logging was used to define the lateral extent and vertical variation of the units within each section to give information on the nature of sedimentation and through this the contemporary sedimentary environments.

Clast fabric measurements have been made on suitable sediments to provide information on the genesis of diamictons (Dreimanis, 1969) and also to define the general ice movement direction (e.g. Holmes, 1941). The a-axis orientation and dip of at least 50 pebbles with a-axes of 1-10 cm long and, a:b ratio of 3:2 were measured. Samples were drawn from a 20x20 cm area. The data were recorded on a polar equidistant plots, and resultant vectors were calculated (Andrews, 1971) and their significance defined (Curray, 1956).

Structural measurements (strike and dip) from fault and fold planes, and hinge lines of folds were measured in order to define the general orientation of shear stresses in the sediments investigated. The orientation of current- induced structures were measured to define general flow directions during deposition. All these were measured using a palaeocurrent disc, compass and clinometer in three dimensional sections. The measurements obtained from faults and folds are presented as poles to planes or poles to lines on a Schmidt net, and palaeocurrent directions are plotted on rose diagrams.

3. LOCALITY AND GENERAL LITHO-STRATIGRAPHY

The area investigated consists of a 3 km long cliff section that extends from Happisburgh Lighthouse to the sea wall at Walcott (TG 388307-TG 364326) (Fig. 2.).

The coastal cliffs in this area reach 20 m in height, but are usually less than 15 m. The highest, well exposed sections are in the central part of the area, and in the south-eastern and north-western parts the sections are only about 5 m high.

A formal terminology for these sediments was proposed by Banham (1970) and this has been used to equate them with other exposures in the region. However, in order to avoid possible confusion in this paper, the author will use a descriptive terminology (Table 1).

4. DESCRIPTION OF THE SEDIMENTARY UNITS

4.1. Pre-Anglian deposits

In the vicinity of section 4 (fig. 2) 1 m of Pastonian estuarine sediments (West, 1980) are exposed for 20 m immediately beneath the first diamicton. These estuarine sediments consist of interbedded silts and fine sands and occasionally contain quantities of organic, mainly lignite fragments. The nature of the primary sedimentary structures in these sediments resemble wavy and lenticular bedding, although they are slightly displaced by minor extensional faults. Orientations of the fault planes were measured throughout the lateral extent of the exposure and they show a consistent strikes of 270° and a dips of 25-80° degrees towards the south (fig. 3). The throw of the faults is between 0.5-10.0 cm.

4.2. The first diamicton

The first diamicton is well exposed throughout the area investigated. It has previously been described as being a brown or grey, sandy or silty diamicton with rare clasts. It is mainly massive but in places is highly laminated (West & Banham,1970; Boulton et

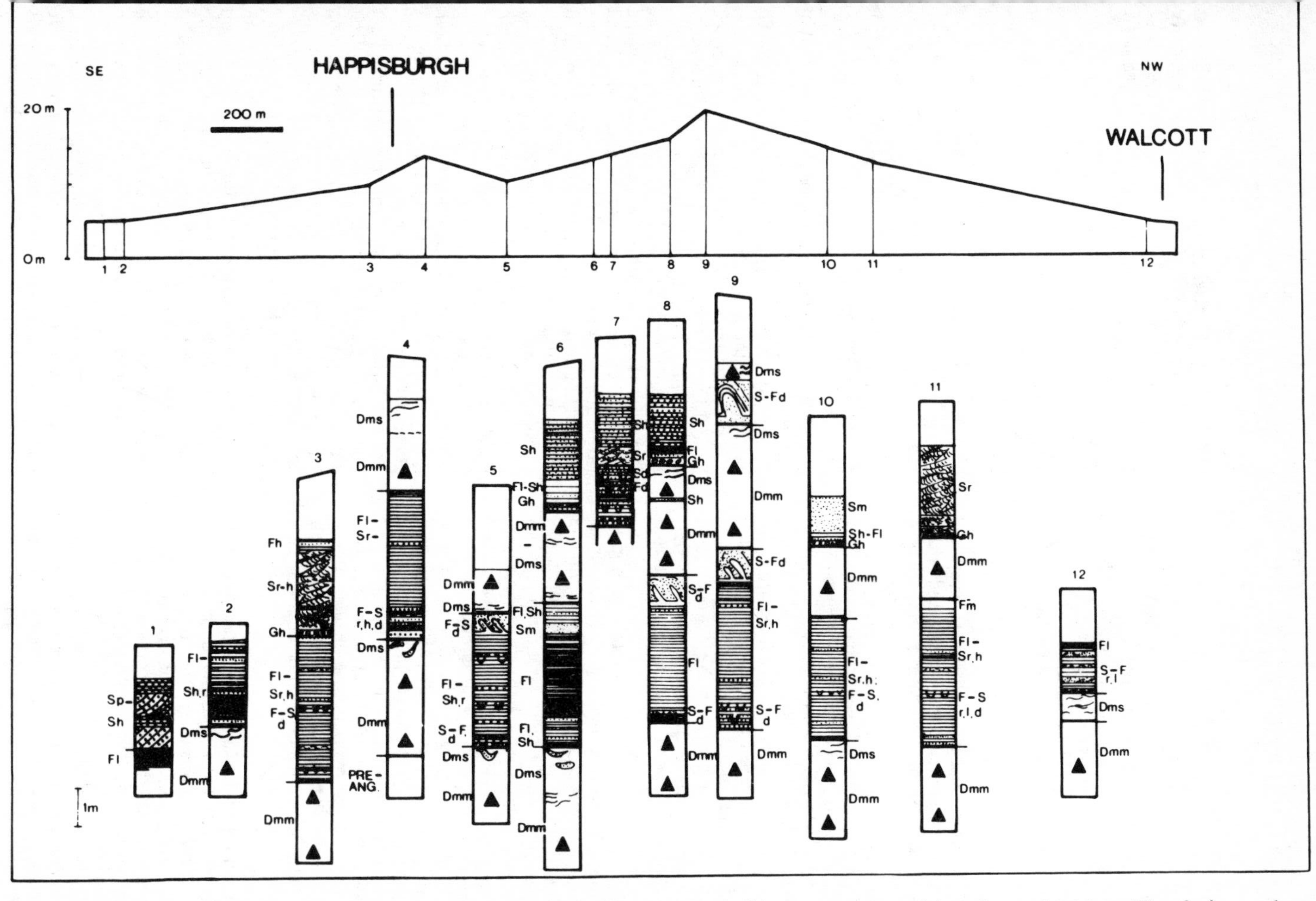

Fig. 2. Sedimentary sequence exposed in the coastal cliffs and the facies log of sites between Happisburgh and Walcott. The facies codes are modified for present work after Eyles et al. (1983). Key: D= diamicton, G= gravel, S= sand, F= fines (silt, clay), Dmm= diamicton matrix supported massive, DMS= diamicton matrix supported stratified, PRE-ANG.= pre-Anglian deposits, h= horizontal bedding, p= planar cross-bedding, r= ripple structure, m = massive, l= laminated, d= deformed.

Table 1. Comparison of the terminology used in this paper with that of Banham (1970).

Banham (1970)	This paper
6. Third Cromer Till	third diamicton
5. Mundesley sands	rippled, cross-bedded and horizontally bedded sands
4. Second Cromer Till	second diamicton
3. Intermediate Beds	massive, laminated and rippled fines and sands
2. First Cromer Till	first diamicton
1. Cromer Forest Bed Formation	pre-Anglian deposits

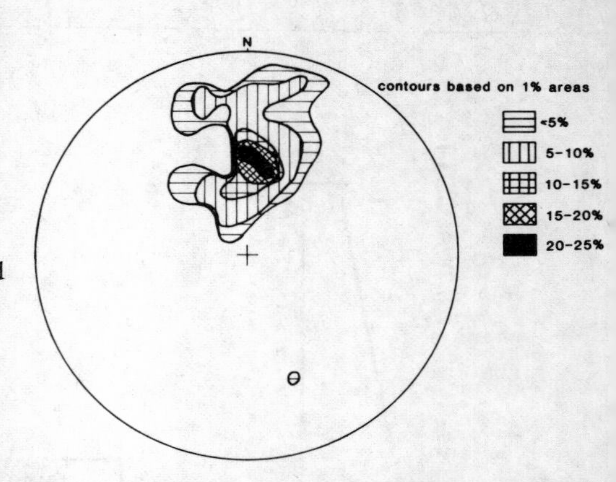

Fig. 3. Stereographic polar plot (poles to planes) of the small extensional fault planes in pre-Anglian sediments exposed around section 4 (39 measurements).

al., 1984). In this area it is silty, dark grey and massive with occasional sand lenses and chalk bands, particularly in the middle and upper part of the unit.

The contact between the first diamicton and the underlying pre-Anglian sediments near section 4 is sharp, but shows no clear evidence of erosion. The lower 30 cm of the diamicton is brown in colour, sandier than the overlying sediment, and includes scattered sand and lignite clasts. The latter are a few millimetres thick and a few centimetres long and are derived from the underlying deposits. Clast macrofabric data obtained from the extreme basal part and from the lower middle part of the first diamicton do not indicate any significant preferred orientation (fig. 4.).

The upper surface of the diamicton is highly undulating, comprising upstanding ridge-like areas separated by shallow depressions. The sediments within the 1.0-2.5 m high ridge-like areas charasteristically exhibit clear stratification, with sand lenses and chalk bands. Fabric measurements made in these upper parts show a significant clast orientation, in general NE-SW at three sites and a less significant peak from NNE-SSW at another (fig. 4.). The ridge-axes at 20 sites along the coast show generally an orientation parallel to that obtained from fabric data (fig. 5.).

In the north western part of the area, the massive dark grey diamicton is overlain by a distinct, highly stratified and much more sandy diamicton which shows no significant clast orientation (fig. 6.). This unit occurs about 100 m south of the sea wall at Walcott.

The genesis of the diamictons in the North Sea Drift Formation has not been considered in detail in previous studies, moreover the depositional environment of these diamictons has not been considered until recently.Hart & Boulton (in press) have proposed that the first diamicton, that forms part of their Lower Diamicton Member, consists of subglacially-deformed sediment. However, Gibbard (1988) and Gibbard & Zalasiewicz (1988) have proposed that at least part of the diamictons in the North Sea Drift Formation have been deposited in water. A waterlain origin has also suggested for the genesis of the lowermost till at Corton by Hopson & Bridge, (1987). This till has been correlated with the first diamicton at Happisburgh by Banham, (1970) and Hopson & Bridge, (1987).

According to the results presented here, it seems evident that the deposition of the first diamicton at Happisburgh results from operation of a variety of depositional processes in different sedimentary environments. Initially the deposition took place in a glaciolacustrine or glaciomarine environment. The lack of strong preferred clast orientation, which could be expected immediately above the basal interface (Banham, 1977a), as well as palaeogeographical setting of north east Norfolk at that time (Gibbard, 1988) suggest that the first diamicton was deposited subaquatically. The processes such as 1. rain-out of fines in suspension with an ice-rafted component 2. down-slope resedimentation of debris, and 3. current reworking (Eyles & Eyles, 1983) could have operated in both the proximal and distal glaciolacustrine or glaciomarine subenvironments.

The consistent orientation of minor extensional fault planes and flutings on the surface of the first diamicton, with a parallel preferred clast orientation (cf. Shaw, 1977) suggest that in the later part of this advance ice then became in contact with its bed and overrode the underlying waterlain diamicton. The depositional processes at this stage was both subglacial deformation (cf. Boulton, 1987) and lodgement. Subglacial deformation gave rise to the extensional faults in the pre-Anglian sediments. But the lower part of the diamicton must have escaped internal deformation since it has no significant fabric and was apparently moved en-bloc. However, there appears to have been an upward transition into a rapidly deforming zone (Boulton, 1987) where clasts became orientated. Although flutings are formed by subglacial deformation the debris transported by the ice could have contributed to the deposition by lodgement. It has been suggested that the ice movement at this stage was from NW to SE (Hart & Boulton, in press). However, the consistent

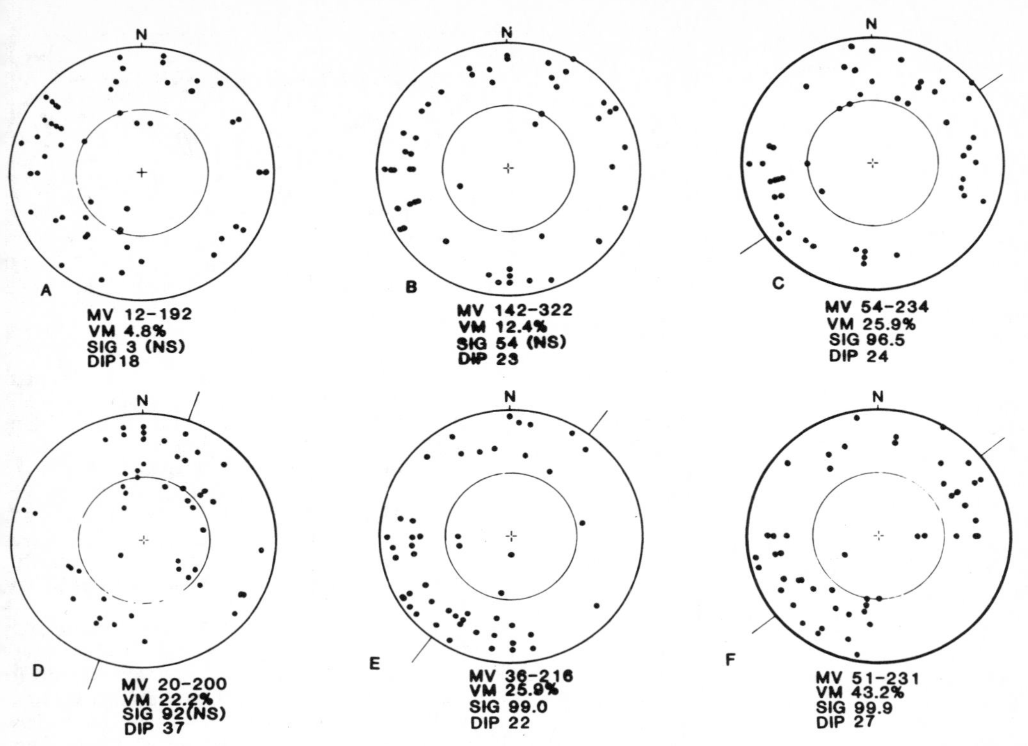

Fig. 4. Polar equidistant plots of the three dimensional fabric data from the first diamicton. A-B= the basal part around section4. C= the upper middle part of section 12. D= the top of the first diamicton in section 2.E= the upper part ofsection 3. F= the upper part of section 11.
Concentric circles represent dips of 0 degree (outer) and 45 degrees (inner). RV= resultant vector, VM= vector magnitude, SIG= Rayleigh test of significance, DIP= mean dip.

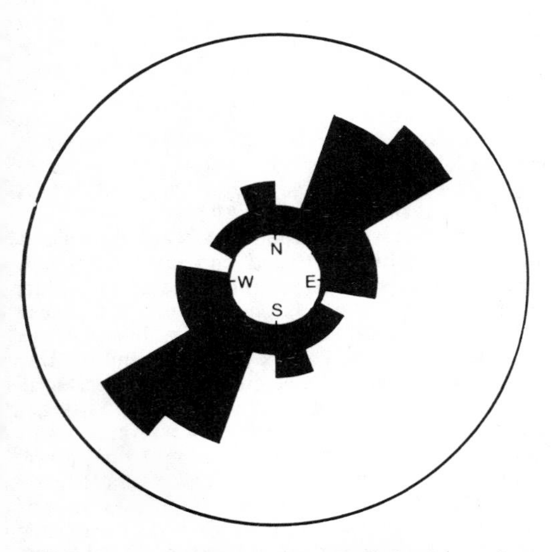

Fig. 5. The orientation of the 20 main ridges from the upper surface of the first diamicton in the study area.

orientation of the ridges on the upper surface of the diamicton, together with the fabric data from its upper parts, as well as occurrence of exotic erractics (such as larvikite: Banham, 1970; Gibbard & Zalasiewicz, 1988) supports Banham's (1970) conclusion that the ice movement at this stage was from NE to SW.

4.3. Massive, laminated and rippled fines and sands

These sediments, which occur between the first and second diamicton, have been termed the Intermediate Beds by Reid (1882) and Banham (1968).They are well exposed throughout the area investigated and are several metres in thickness. Banham (1977b) noticed that they both fine and thin towards the north.
The unit often starts with 0.5-2 m of deformed clay, silt and sand which infills the small basins in the upper surface of the first diamicton.This unit is deformed by listric faults and disharmonic fold structures. Diamicton also is commonly involved with these structures. In the sites investigated the deformation structures show no consistent orientation, but the fault structures are usually related

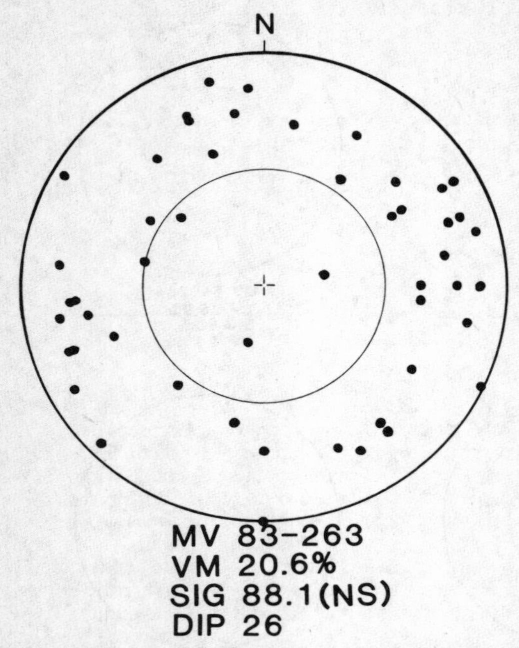

Fig. 6. Polar equidistal plot of the three dimensional fabric data from the stratified diamicton that overlies the massive diamicton near south eastern end of the Walcott sea wall.

to sediment sliding and slumping towards the centre of the small basins (fig.7).

The surface layers of the first diamicton are usually also deformed and mixed with the clay, silt and sand. At least part of the stratified structures, especially the sand lenses in the upper part of the first diamicton, continue into the overlying unit. The fabric measurement in section 2 (fig. 4) was made in the uppermost zone of the first diamicton where deformation structures occurred. These structures may explain the less significant orientation of the clasts recorded at this site.

In some sections particularly in the south-eastern part of the area, massive, light grey clays occupy the basins and rest on an eroded surface of the pre-existing laminated, and rippled fines and sands, and diamicton (fig. 8).

All the evidence assembled indicates that most of these deformation structures are related to mass movements of laminated fines and sands down the flanks of the ridges on the upper surface of the first diamicton. As these sediments moved, the unconsolidated and water-saturated diamicton may have been injected into them in places and become involved in the deformation.

Kazi and Knill (1969) concluded that the orientation of small scale ripple laminae in the lower subunit of the Intermediate Beds at Happisburgh indicate two distinct directions corresponding to NW and SE

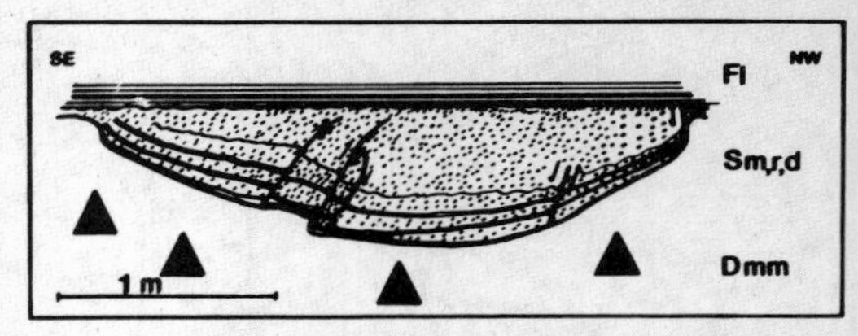

Fig. 7. Listric fault planes in the lower unit of the Intermediate Beds. Fault strikes are from 150-23(degrees with dips 45-60 SE. (For full key see caption to fig. 2)

sediment sources. However, in section 9 the original sedimentary structures of the lower unit are well preserved and here the fine sands include both asymmetric and symmetric ripples with foreset laminae charasteristic of wave ripples (Boersma, 1970; Reineck & Singh, 1980). These have a ripple index of 3.2-8.0. In other sites the sand or fine sand ripples were asymmetrical and represent either wave or small current ripples formed by a generally constant current and/or wave direction from SE to NW.

Above the basal subunit of the deformed sandy and silty facies, there are mainly laminated and massive clays and silts with rippled and horizontally-bedded sands as minor constituents. On a large scale this unit has distinct beds of clay and silt, but when studied in detail, shows regular sedimentation patterns (fig. 9) with penecontemporaneous deformation structures (Reineck & Singh, 1980).

Dropstones were found in the lower part of this subunit in the north western part of the area near Walcott, where the stratified sandy diamicton also occurs above the dark grey massive diamicton.

The fine sand ripples studied were wave or small current ripples with a wavelength of 10-14 cm and a height of 1.9-2.5 cm and gave a ripple index 3.6-11.2. They indicate a general flow direction from SSE to NNW.

The characteristics of this upper facies indicate that these sediments are turbidites, deposited in a glaciolacustrine or possibly glaciomarine environment, as proposed by Kazi and Knill (1969) and Gibbard and Zalasiewicz (1988).

The sedimentary succession (fig. 9) reveals cycles of coarse sediments (sand and silt) which are overlain by massive clays. The sequence represents density underflow conditions. In some cases these massive clays are overlain by laminated clays probably representing quiet periods when overflow or interflow conditions persisted between turbidite underflow events (Smith & Ashley, 1975). What triggered the turbidite underflows is still under investigation. Whether they arise from quasi-continuous currents from a river or rivers flowing into the lake, or whether they indicate slump-generated surge currents caused by mass movements in an unstable glacial delta or lake margin sediments (Smith & Ashley, 1975) cannot be determined at present. However, the general current flow direction obtained from the results indicates perhaps a single

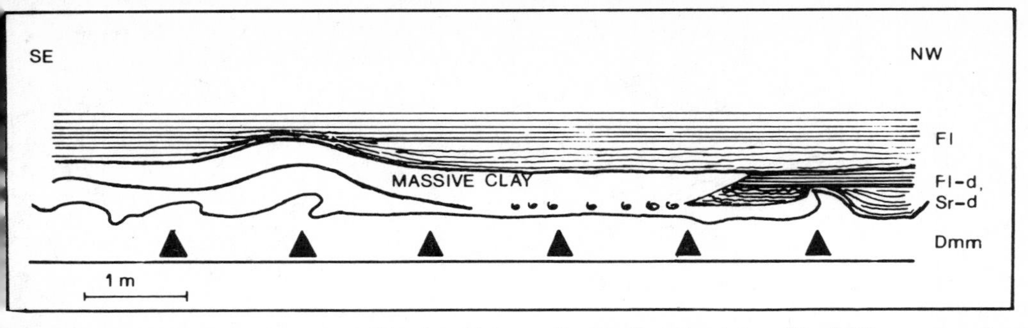

Fig. 8. Massive clay occupying a small basin which overlies the first diamicton. (For full key see caption to fig. 2)

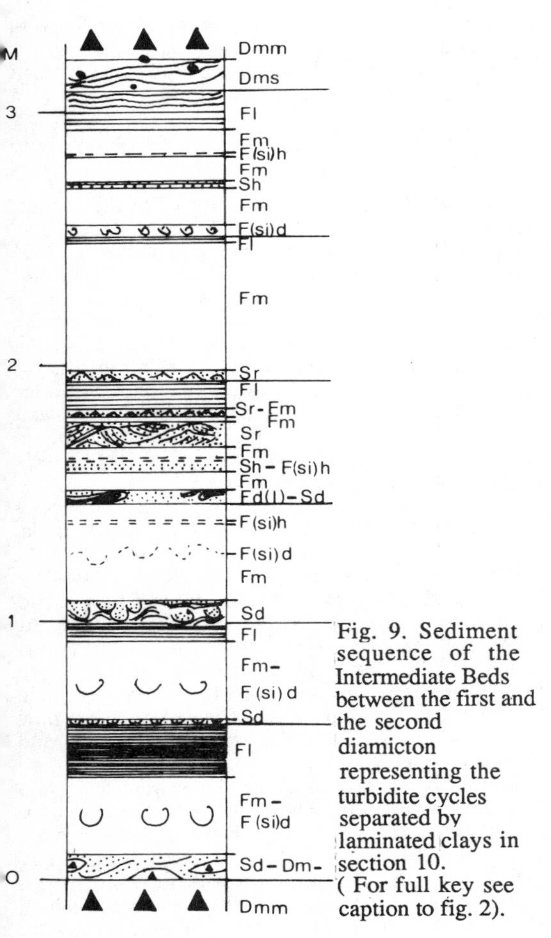

Fig. 9. Sediment sequence of the Intermediate Beds between the first and the second diamicton representing the turbidite cycles separated by laminated clays in section 10. (For full key see caption to fig. 2).

persistent source from the S or SE in the Happisburgh area. Penecontemporaneous deformation structures in the upper facies are mainly 'ball-and-pillow' structures of various types (cf. Kuenen, 1965), where the sand and silt is broken into more or less ellipsoidal masses within massive clay. These masses are occasionally present in a clay matrix and their size range from 1-10 cm. 'Ball-and-pillow' structures are normally associated with rapid sedimentation and have been reported from shallow water environments, as well as from beneath deeper water turbidites (Reineck & Singh, 1980).

This upper subunit grades into a more sandy facies which has been deformed by overriding ice associated with deposition of the second diamicton.

The whole unit, i.e. the Intermediate Beds, represents a rising water level of the basin ultimately infilling with sediments. Initial deposition in shallow water is indicated by the wave ripple structures, together with the occurrence of the lower subunit in small basins, bounded by the fluted surface of the first diamicton. Deformation structures arising from unequal loading indicate rapid sedimentation, evidently contemporary with the ice retreat from the area.The presence of the diamicton in deformation structures, and the occurrence of stratified diamicton and dropstones beneath and within the Intermediate Beds around section 12 indicate the proximity of the ice. The upper part of the unit was deposited into deeper water and represents rapid deposition from turbidite currents, as described by Kazi & Knill (1969), with consistent flow from SSE. The Intermediate Beds grade upwards into coarser sediments, but were eroded in many sections eroded during the second ice advance.

4.4. The second diamicton

The second diamicton is usually dark grey in colour, although it often grades into a brown colour upwards. The latter is attributed to weathering above the water table (Banham & West, 1968). The second diamicton contains silt and distinctly more chalk than the first diamicton (Banham, 1968; Perrin et al. 1979).

In the area investigated the second diamicton is mainly massive, but is occasionally distinctly stratified. This structure arises from chalk bands and laminae especially in the upper part of the unit. It is normally rather thin but in some places reaches over 3 m in thickness. This unit is well represented from section 4 to section 12, but is not preserved south east of section 3. It has also been completely

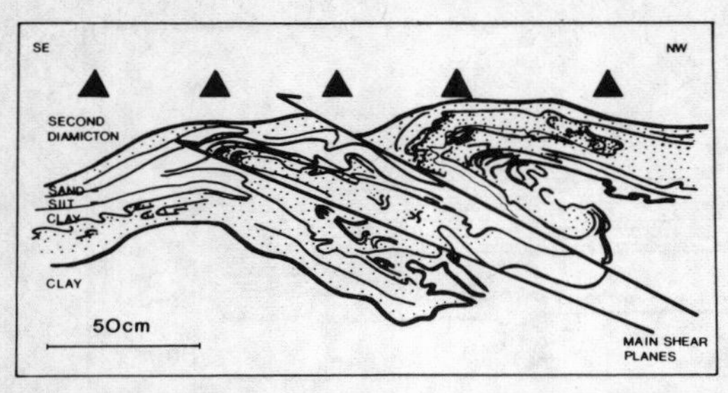

Fig. 10. Glaciotectonic structures below the second diamicton in section 7.

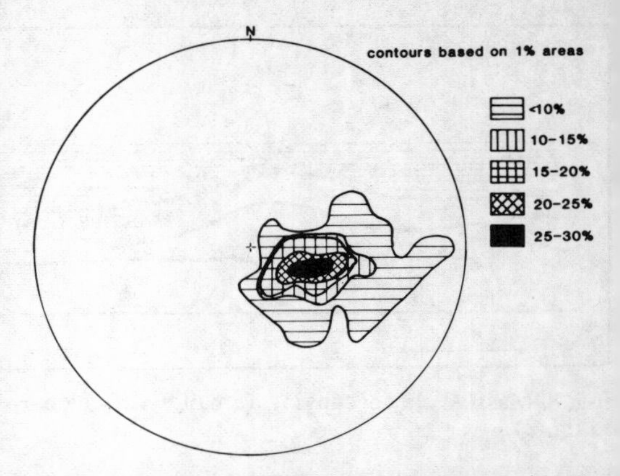

Fig. 11. Stereographic polar plot (poles to planes) of the fold and shear planes under the second diamicton in the area investigated (44 measurements)

removed by subsequent erosion at one site between sections 8 and 9.

At many localities along the coast the beds beneath the second diamicton are deformed, as reported by Banham (1976). Deformation normally occurs in the underlying beds where sand overlies silt or clay, but in those sites where the second diamicton rests on the massive or laminated clay, there is little or no sign of deformation.

Deformation structures are predominantly disharmonic recumbent folds which are often separated from each other by thrust faults (fig. 10) which cut through the overturned limb of the fold.

The general trend of the fold axial planes and thrust fault planes measured beneath the second diamicton show a remarkably consistent orientation from NE to SW (fig. 11). This indicates that the shear stress was applied from NW to SE. The clast fabric measurements from the basal, middle and upper part of the second diamicton also show very strong pebble a-axis orientations from NW to SE (fig. 12).

In the SE part of the area (near section 3) the second diamicton terminates in a fold system which is a tight, disharmonic and anticlinal and appears to have formed when the ice bulldozed the sands at the margin and also deformed the underlying bed from NW (fig. 13).

Banham (1977b) has presented the results of measurements on the glaciotectonic structures mainly from the area north of that investigated here and presented a rather complex picture with the pebble long axis direction parallel to the fold axis present below the second diamicton. He regarded the existing fold axis orientation from beneath the diamicton to be the best evidence of the direction of the last ice movement. In the Happisburgh area however, the orientation of the clasts and fold axis are transverse to that described by Banham (1977b).

The consistent clast fabric found in the massive and stratified parts of the second diamicton, as well as the orientation of the glaciotectonic structures, indicate that the ice was in contact with its bed and the ice movement during this phase was from NW to SE as also noted by Banham (1970b).

The sharp erosional contact between the diamicton and the underlying deformed beds, together with the strong preferred clast orientation throughout the diamicton, indicates that the till was probably deposited by lodgement.

4.5. Rippled, cross-bedded and horizontally bedded sands

These sands occur immediately SE of the second diamicton and also overlie it. Banham (1970) has equated the sands above the second diamicton, NW of section 3, with the Mundesley Sands. He further suggests (Banham, 1977a), that the sands SE of section 3 are so-called 'Valley Gravels and Sands', unrelated to the glacial sequence. This conclusion is mainly based upon the evidence of Phillips (1976) who argues that the so-called 'Mundesley River Bed' contains vertebrate- and plant -bearing silts and peat of Ipswichian age (Phillips, 1976). Banham (1977a) correlates this bed with his 'Valley Gravels and Sands' at Happisburgh.

SE of section 3 the sands overlie massive and laminated fines and sands of the Intermediate Beds. As described above (fig. 13), the sands adjacent to the second diamicton have been glaciotectonically deformed and must therefore predate the second diamicton, at least between sections 3 and 2. At section 3, the unit comprises a thin (10 cm) gravel bed at the base, which passes upwards into deformed sands that previously had a climbing ripple structure;.only the lee sides of these ripples are now preserved (climbing ripple type 'A') (Jopling and Walker, 1968).

Further SE of section 2 the sands were deposited into a large basin. These sands in the basin are composed of 0.3-0.5 m thick planar cross-bedded sets, interbedded with 0.2-0.3 m thick horizontally bedded sands. The planar cross-bedded sets have a highly tangential (concave) foreset contact, which alternates with angular contacts. However, it has yet to be determined whether this apparently logical sedimentary pattern results from the influence of depth ratio, sediment type and/or velocity which are the main variables controlling the shape and the slope of the foreset laminae (Jopling, 1963; 1965).

Palaeocurrent indicators observed adjacent to section 1 show a highly variable current direction in

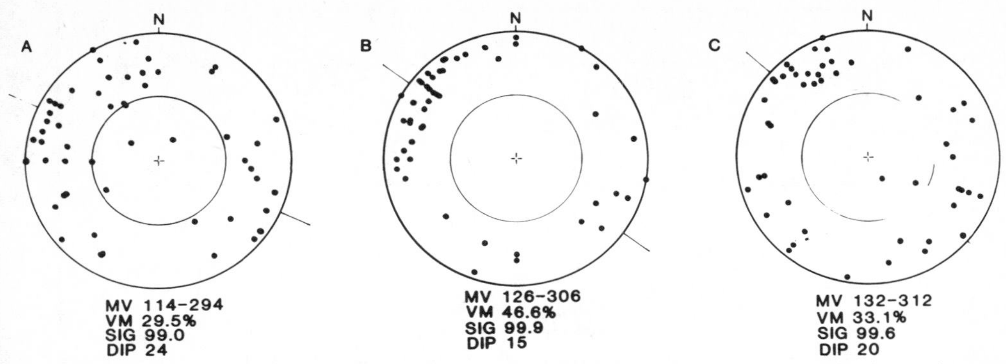

Fig. 12. Polar equidistant plots of the three dimensional fabric data from the second diamicton. A= basal part in section 11. B= the middle part and C= the top part of the second diamicton in section 10. (For full key see caption to fig. 4)

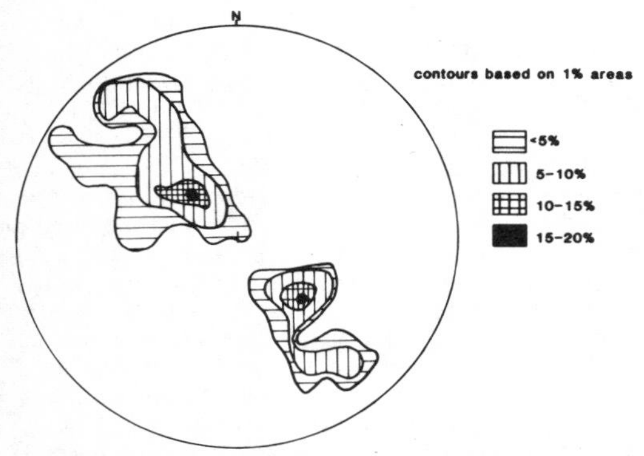

Fig. 13. Stereographic polar plot (poles to planes) of the fold planes of the tight anticlinical fold near section 3 (36 measurements) in adjacent to the second diamicton.

different sets. This ranges from N to SSE, although the most frequent direction is from the SE.
The age of these planar cross-bedded sands SE in section 2 is still uncertain. The palaeocurrent variations between the various cross-bedded sets argues against unidirectional flow from present land into a basin, although statistically the number of measurements are so few that a firm conclusion regarding the flow direction cannot be achieved at present for the whole unit.
NW of section 5 the sands are well exposed above the second diamicton. The sands show 'Type A' climbing ripples and in places they become horizontally bedded. Massive sands were also found, but they are normally at the top of the section where weathering processes have disturbed the primary structures. No large planar or trough cross-bedded sets were found in the sites investigated.
Palaeocurrent directions measured from the sands indicate a rather constant flow direction. The climbing ripple sets show directions mainly from the S or SE, but in section 7 the opposite direction was also measured in one set (fig 14).

All these sands, including the interbedded silts above the second diamicton, are horizontally bedded, although some sets show climbing ripple structures. Moreover, there is no evidence of large scale cross-bedded sands nor channel fillings. This indicates that these sands could probably represent delta bottomsets. The strong, preferred palaeocurrent direction from SSE or SE suggests a single persistent sedimentary input such as an extra-marginal river.
The sands between sections 8 and 9 contain different structures from those found elsewhere in the sites studied (fig 15). The upper and middle part of these sands contain deformation structures, although the basal part shows undisturbed primary structures, where the diamicton has been removed before emplacement of the gravels and sands.

Fig.14. 30 observations of palaeocurrent directions in different sets of the climbing ripple sands that overlie the second diamicton.
Flow direction towards NNW.

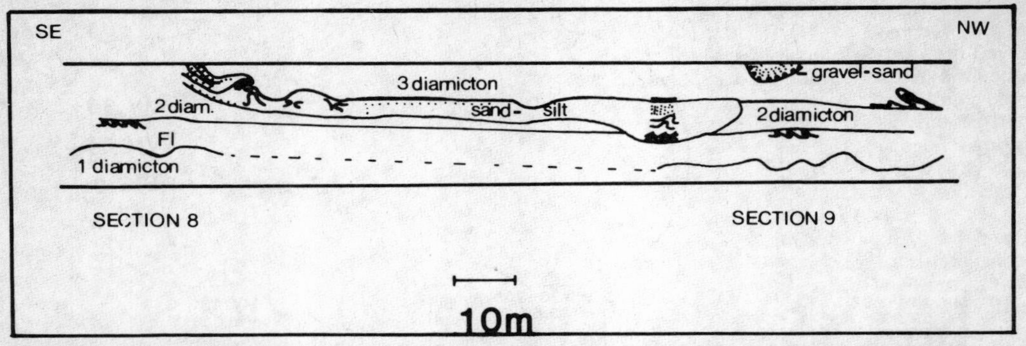

Fig. 15. The coastal sequence between sections 8 and 9 showing three different diamictons, the deformed upper part of the Intermediate Beds and deformed sand and silt unit below the third diamicton.

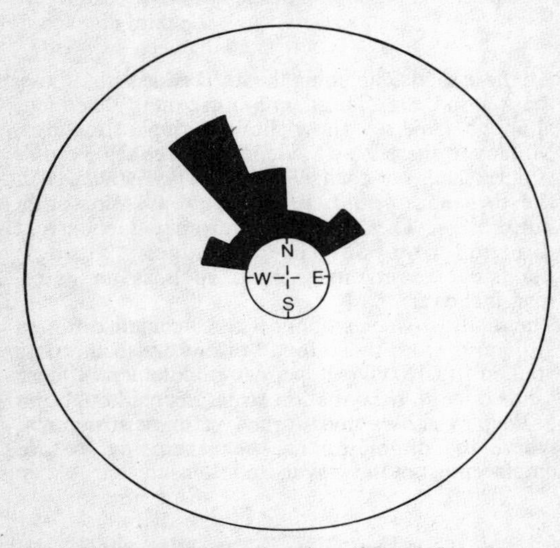

Fig. 16. Palaeocurrent directions from 1.5 m thick lower rippled sand SE of section 9 indicating a flow direction towards NW (15 observations).

This unit reaches its maximum thickness of 7m just SE of section 9. There a gravel bed 0.7 m thick was deposited at the base and cut out the second diamicton. Above this gravel bed, 1.5 m ripple cross-laminated sands occur. Palaeocurrent directions measured in the 15 sets indicate flow from the SE (fig. 16). Above these undisturbed subunits there is a sand and silt bed in which various deformation structures occur. These structures comprise folds which can best be seen in the deformed silt beds. Although the folds in this subunit often plunge in different directions, structural measurements reveal a general NW-SE trend of stress (fig. 17). The sands and silts are overlain by a thin silt band which is deformed in section 9 beneath the third diamicton.

The palaeocurrent directions from the SE in the lowermost sands and the deformed silt layers including variable fold-axis orientations (see fig. 17) indicate that this sand unit was deposited as a mass flow. This is further supported by deformed ripple structures in the sands adjacent to the flow. The flow also cut through the second diamicton. However, part of the internal folding in the sands took place after the mass flow when there is evidence of a third ice movement broadly from SW to NE.

According to this investigation the rippled, cross-bedded and horizontally bedded sands were deposited during at least two different episodes. The sands between sections 2 and 3 deposited above the Intermediate Beds could represent a upwards gradational change from deep water facies (turbidites) to deltaic bottomsets, as suggested by Hart & Boulton (in press). Planar cross-bedded sands above the Intermediate Beds SE of section 2 could be related to this deltaic succession, although they may not necessarily be of Anglian age, if Banham's (1977b) conclusion (see above) is correct. These sands have been pushed near section 3 by the second ice which entered the area from the NW. After or at the same time as the second ice was retreating, the sands resting on the second diamicton were deposited. They represent the bottomsets of a

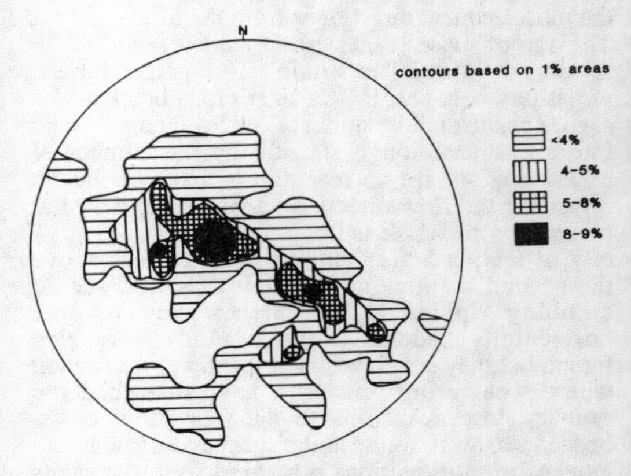

Fig. 17. Stereographic polar plot (poles to planes) of the fault planes in the sand-silt unit beneath the third diamicton between sections 8 and 9 showing general NE-SW trend (78 measurements).

deltaic environment where some mass flows could have also occurred on the floor of the basin, as observed between sections 8 and 9. The consistent palaeocurrent direction from the SSE or SE indicates that the sands above the second diamicton are not related to outwash sediments derived from the retreating second ice.

4.6. The third diamicton and sand and gravel basins

Banham (1970) suggested that in the upper parts of the coastal cliffs in the Happisburgh area there are remnants of the Third Cromer Till. In the area investigated these remnants are situated between sections 8 and 9, but they are too few to provide sufficient evidence to determine the origin of this unit.

The third diamicton is light brown, stratified and sandy with relatively few clasts and many shell fragments. In this area it appears quite dissimilar from the first diamicton, contrary to the views expressed by Perrin et al. (1979).

The third diamicton overlies thin and deformed laminated silt and sand in section 9. At one site,10 m NW of section 9, the sands and silts underlying this third diamicton have been deformed and the orientation of deformation structures indicates shear stress from WNW to ESE (fig. 18). Fabric measurements from basal part of this diamicton above a thick sand unit do not, however, show any significant orientation (fig. 19). The lack of preferred clast orientation could result from emplacement of the gravel and sand in a basin above (fig. 15). Modification of the fabric would have occurred as the sands and gravels sunk into the underlying unconsolidated diamicton during deposition.

In the sections investigated the third diamicton has local textural and structural properties which reflect derivation from the underlying sands. During this third advance the ice overrode these underlying sediments, and it was in contact with its bed. This caused some deformation of the sand and silt unit below which was expressed by the NW-SE trend of stress (fig. 17 and 18). The deposition of gravels and sands in a basin in the third diamicton may have caused the reorientation of the clasts in the diamicton.

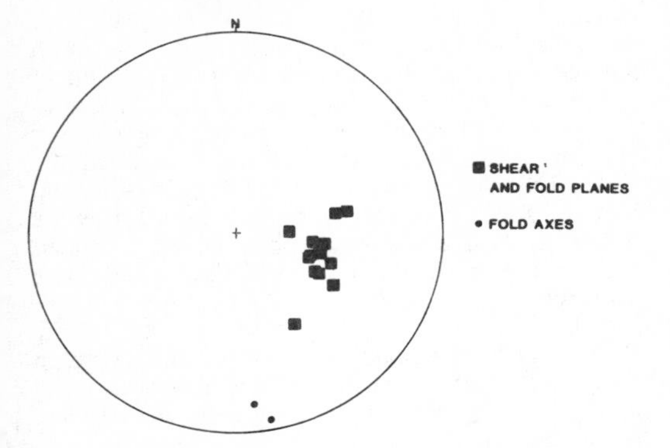

Fig. 18. Stereographic polar plot (poles to planes) of fold axes and fold planes of the fold formed by the third ice advance from WNW.

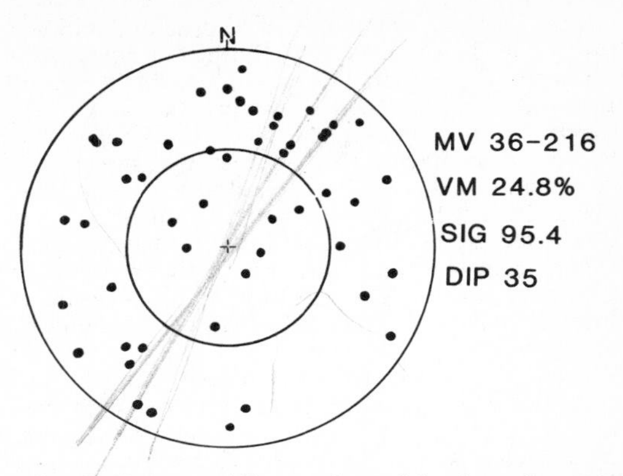

Fig. 19. Polar equidistant plots of the three dimensional fabric data from the basal part of the third diamicton below the gravel and sand basin between sections 8 and 9. (For full key see caption to fig. 4.)

5. CONCLUSION AND DISCUSSION

The general lithostratigraphic units recognised in the North Sea Drift Formation in the Happisburgh area are basically similar to those proposed by Banham (1968,1970), although the results presented here provide new evidence of the sedimentary environments and processes operating during deposition of the Formation.There are still many unresolved questions concerning the actual processes of the deposition of the diamictons, as well as precisely how these units should be correlated with those in other parts of North Norfolk. These points are still under investigation by the author. Nevertheless, from the results presented in this paper, the following conclusions can be drawn for the Happisburgh area.

Clast fabric measurements within the fluted surface of the upper part of the first diamicton and the occurrence of the exotic clasts in it suggest that the ice advanced into the area broadly from the NE. This conclusion has generally been accepted by most authors in the past but was recently doubted by Hart (1987) and Hart and Boulton (in press).

The lack of preferred clast orientation in the lower part of the first diamicton suggests that the first ice i.e. the Scandinavian Ice Sheet (e.g. Bowen et al., 1986) advanced from the NE into a large water body in the southern North Sea basin (Gibbard, 1988). Initial deposition of the diamicton resulted from different processes such as, rain-out of fines in suspension with ice-rafted component, downslope resedimentation of debris and current reworking (cf. Eyles & Eyles, 1983).

The minor deformation structures in the underlying pre-Anglian sediments, the upward gradational change from random to preferred clast orientation, parallel to the orientation of flutings on the upper surface of the first diamicton, indicate that this unit was initially deposited in water, and was later overriden by grounding ice. During this phase the ice was in contact with its bed so that subglacial deformation and probably also lodgement could take place. To judge from the parallel alignment of the ridges and the fabric orientations (cf. Shaw, 1977), it is evident that the surface of the first diamicton was most probably generated by fluting rather than from ice pushing as proposed by Hart & Boulton (in press).

Wave-ripple structures in the lower subunit of the succeeding Intermediate Beds, together with the location of this subunit in the small inter-ridge basins on the fluted surface of the first diamicton suggest that deposition initially occurred in shallow water, while the first ice was retreating, as also noticed by Hart & Boulton (in press). Deformation, mostly slump structures involving the first diamicton indicate rapid sedimentation when the underlying diamicton was water saturated. Contemporary with the retreat of the first ice, water level in the basin rose sharply and the turbidite units of the Intermediate Beds were deposited. These turbidite currents were broadly derived from the SSE, perhaps indicating a single persistent source, rather than meltwater and debris input from the retreating first ice. If the latter were the case, one would expect more variable palaeocurrent directions comparable to those encountered in the lower subunit (Kazi & Knill, 1969).

The ice responsible for the second diamicton deformed and eroded the upper part of the Intermediate Beds NW of the area investigated. But SE of section 3 the sands lying above the Intermediate Beds may represent delta bottomsets. Further SE, around section 2, the origin of cross- and horizontally bedded sands above the Intermediate Beds is still under investigation. They could have been deposited during the Ipswichian Stage as proposed by Banham (1977b), but could also represent Anglian deltaic sands.

The second diamicton overlying the Intermediate Beds, NW of the area (NW of section 3) has a sharp contact with the underlying beds; these beds are often deformed. The orientation of the fault and fold planes in the upper part of the Intermediate Beds, together with the transverse and strong preferred clast orientation in this second diamicton suggest that the deformation structures represent subglacial deformation. This occurred when the second ice overrode the underlying sediments from the NW. The sharp and erosional contact of the second diamicton with the underlying deformed beds, accompanied by strong preferred clast orientation suggest that the diamicton was deposited by lodgement.

The sands occasionally interbedded with silt above the second diamicton were deposited as delta bottomsets by water currents from the SE or SSE. The consistent palaeocurrent direction in these sands indicates a persistent sediment source, such as a river flowing into a glaciolacustrine or glaciomarine environment. Subaquatic sediment mass flows on the basin floor occurred in places.

The third diamicton is only found in the higher parts of the area investigated. Moreover remnants of this diamicton are too few to provide sufficient evidence regarding its origin. However, the underlying silt and sand have deformed beneath the third diamicton in places. The orientation of some of these structures indicates that the third ice advance came from the WNW. Emplacement of the gravel and sand basin above this diamicton may have resulted in disorientation of clast fabric since the former may have sunk into the underlying unconsolidated diamicton.

On the basis of the results presented here some general conclusions can be drawn about the depositional sedimentary environments in the Happisburgh area, and their relation to neighbouring areas.

During the Anglian (e.g. Bowen, 1986) the Scandinavian Ice Sheet entered this area of the southern North Sea Basin from the NE. Initially, the first diamicton was deposited in a standing water body, but as the ice advanced and became grounded, subglacial deformation and lodgement also occurred. In the NW of the area the massive diamicton formed by this process is overlain by stratified diamicton including with dropstones. This probably indicates that the waterlain sedimentation returned during retreat of this first ice possibly as a consequence of ice thinning.

The first diamicton at Happisburgh has been correlated with the Norwich Brickearth and with the First Cromer Till of the Corton and Scratby areas (Banham,1970; Bridge & Hopson, 1987). However, this first diamicton is not preserved at Mundesley (Banham, 1970). This implies that the first ice advance reached the area around Corton. It did not however reach Mundesley, unless the unit was removed by erosion prior to or during the second ice advance.

During retreat of the first ice, rapid sedimentation initially into shallow and later into deeper water occurred. The lower shallow water subunit could have resulted from meltwater and sediment input derived from the retreating ice. However, the upper turbidite subunit of the Intermediate Beds indicates deposition from the SE or SSE probably originating from an extra-marginal river which entered the basin. Thus the rise of water level in the basin could have resulted from the river being dammed by the retreating first ice. Later when this ice retreated further NE, regression of the lake could have occurred.

The coarse sediments in the upper part of the Intermediate Beds could then represent the migration of the delta bottomsets towards the NW or infilling of the basin as Hart and Boulton (in press) have suggested.

The Intermediate Beds have not previously been recorded in the Scratby area although sands above the first diamicton there (Hopson & Bridge, 1987) could conceivably represent the lateral equivalent of this Member.

The second ice advance from the NW caused subglacial deformation and deposited the second diamicton by lodgement, at the same time the ice eroded and pushed the underlying sediments. This second diamicton can be correlated with the Second Cromer Till at Mundesley (Banham, 1970) and also in the Scratby area (Hopson & Bridge, 1987), but it has not been reported from around Corton (e.g. Banham, 1971; Pointon, 1978). The lowest sedimentary unit in the Corton Sands, which overlie the Norwich Brickearth i.e. the First Cromer Till has palaeocurrents from the N and NW and contains chalky material (Hopson & Bridge, 1987). This 'lower cycle' of the Corton Sands could have been derived from melting of the ice responsible for deposition of the second diamicton at Happisburgh. This diamicton contains distinctly more chalk than the other diamictons in the North Sea Drift Formation (Perrin et al., 1979).

While the second ice was retreating from the Happisburgh area, the sands above the second diamicton were deposited from SE or SSE. It seems obvious since this ice retreated broadly towards the north, that the overlying sands cannot represent outwash derived from this ice. The consistent orientation of the palaeocurrent directions, together with the bedding structures of these sands, indicate that they represent delta bottomsets deposited into a large water body probably by extra-marginal river.

These sands have been correlated with so-called the Mundesley Sands (Banham, 1970) at Mundesley itself and also with the sands lying above the Second Cromer Till at Scratby (Hopson & Bridge, 1987). However, palaeocurrent directions from the Mundesley Sands at the type locality above the Second Cromer Till show that the sands were deposited from the SW, W or NW as outwash from the second ice (Banham, 1970). This implies that the sands above the second diamicton at Happisburgh are not equivalent to those at Mundesley.

The ice responsible for the deposition of the Third Cromer Till at Mundesley entered the area from the west (Banham, 1977b). In the Happisburgh area the authors results indicate broadly the same ice movement direction from the WNW. The genesis of this diamicton remains unclear. However, subglacial deformation occurred as this third ice overrode the underlying sediments.

Overall, the complex suite of diamictons and interbedded sorted sediments indicates that there were three ice advances into the Happisburgh area in the Anglian Stage (cf. Banham,1968). During that time the Intermediate Beds, equivalent sands and the sands above the second diamicton were deposited from SE or SSE. This suggests that during that time there was probably extramarginal drainage flowing into basin, such as the proto-Waveney (Hopson & Bridge, 1987), its tributaries or possibly the Thames (Gibbard, 1988). The waterlevel in this basin was undoubtedly influenced by the North Sea Drift Ice (Perrin et al., 1979) which throughout the period represented interacted with glaciolacustrine and/or glaciomarine environments in the southern North Sea Basin.

Future work will be directed towards an extension of these investigations both to the north and south of the area presented here.

ACKNOWLEDGEMENTS: I thank my supervisors Dr. P.L. Gibbard and Dr. M.J. Sharp for invaluable direction and discussion during this project. I wish to acknowledge the support of the Academy of Finland and also Jenny ja Antti Wihurin rahasto, Emil Aaltosen Säätiö and Hämäläisten Ylioppilas-Säätiö for funding this research.

REFERENCES

Andrews, J.T. 1971. Techniques of till fabric analysis. British Geomorphological Research Group Technical Bulletin No. 6: Geo Abstracts: Headley Bros. Ltd, Ashford, Kent.

Baden-Powell, D.F.W. 1948. The Chalky Boulder Clays in Norfolk and Suffolk. Geological Magazine 85, 279-296.

Banham, P.H. 1966. The Significance of Till Pebble lineations and relation to Folds in two Pleistocene Tills at Mundesley, Norfolk. Proceedings of the Geologists' Association 77, 469-474.

Banham, P.H. 1968. A preliminary note on the Pleistocene stratigraphy of north east Norfolk. Proceedings of the Geologists' Association 79, P.507-512.

Banham, H. 1970. North Norfolk. In G.S. Boulton (ed.): Quaternary Research Association. Field Guide.

Banham, P.H. 1971. The Pleistocene Beds at Corton,Suffolk. Geological Magazine108, 281-285.

Banham, P.H. 1976. Glaciotectonic structures: a general discussion with particular reference to the Contorted Drift of Norfolk. In Wright, A. E. & Moseley, F. (eds.): Ice Ages: Ancient and Modern. Geological Journal, Special Issue 6, 69-94. Seel House Press, Liverpool.

Banham, P.H. 1977a. Glaciotectonites in till stratigraphy. Boreas, 6, 101-105.

Banham, P.H. 1977b. North Norfolk. In R.G. West (ed.): X INQUA Congress: East Anglia, Excursion Quide. Geoabstracts; Norwich.

Banham, P.H., Davies, H. & Perrin, R.M.S. 1975. Short field meeting in North Norfolk. Proceedings of the Geologists' Association 86, 251-258.

Boersma, J.R. 1970. Distinguishing features of wave-ripple cross-stratification and morphology. stratification and morphology. Doctoral Thesis, University of Utrecht. 65 pp.

Boulton, G.S. 1987. A theory of drumlin formation by subglacial sediment deformation. In Menzies, J. & Rose, J. (eds.) 1987: Drumlin Symposium. Balkema; Rotterdam.

Boulton, G.S., Cox, F., Hart, J., & Thornton, M. 1984.The glacial geology of Norfolk. Bulletin of the Geological Society of Norfolk 34, 103-122.
Bowen, D.Q., Rose, J., McCabe, A.M. & Sutherland, D.G. 1986. Correlation of the Quaternary glaciations in England, Ireland, Scotland and Wales. Quaternary Science Reviews 5, 299-340.
Curray, J.R. 1956. The analysis of two dimensional orientation data. Journal of Geology 64, 117-131.
Dhonau, T.J. & Dhonau, N.B. 1963. Glacial structures on the north Norfolk coast. Proceedings of the Geologists' Association 75, 433-439.
Dreimanis, A. 1969. Selection of genetically significant parameters for investigation of tills. Zesk. Nauk. UAM Geografia 8, 15-29.
Ehlers, J., Gibbard, P.L. & Whiteman, C.A. 1987. Recent investigations of the Marly Drift of northwest Norfolk, England. In Van der Meer, J.J.M. (Ed). Tills and Glaciotectonics. Balkema; Rotterdam-Boston,
Ehlers, J., Gibbard, P.L. & Whiteman, C.A. In press. North Norfolk. In Ehlers, J., Gibbard, P.L. & Rose, J. (Eds). Glacial deposits in Great Britain and Ireland. Balkema; Rotterdam.
Eyles, C.H. & Eyles, N. 1983. Sedimentation in a large lake: A reinterpretation of the late Pleistocene stratigraphy of Scarborough Bluffs, Ontario, Canada. Geology 11, 146-152.
Eyles, N., Eyles, C.H. & Miall, A.D. 1983. Lithofacies types and vertical profile models; an alternative approach to the description and environmental interpretation of glacial diamict and diamictite sequences. Sedimentology 30, 395-410.
Gibbard, P.L. 1988. The history of the great northwest European rivers during the last three million years. Philosophical Transactions of the Royal Society of London B 318, 559-602.
Gibbard, P.L. & Zalasiewicz, J.A. 1988. Anglian Stage. In Gibbard, P.L. & Zalasiewicz, J.A. (eds.) 1988. Pliocene-Middle Pleistocene of East Anglia. Field Guide 27-31, Quaternary Research Association: Cambridge. 195 pp.
Hart, J.K. 1987. The genesis of the north east Norfolk Drift. Ph.D. Thesis, University of East Anglia.
Hart, J.K.,& Boulton, G.S. In press. The glacial drifts of Norfolk. In Ehlers, J., Gibbard, P.L. & Rose, J. (eds). Glacial deposits in Great Britain and Ireland. Balkema; Rotterdam.
Holmes, C.D. 1941. Till fabric. Bulletin of the Geological Society of America 52, No 9, 1299-1354.
Hopson, P.M. & Bridge, D.McC. 1987. Middle Pleistocene stratigraphy in the lower Waveney Valley, East Anglia. Proceedings of the Geologists' Association 98(2), 171-185.
Jopling, A.W. 1963. Hydraulic studies on the origin of bedding. Sedimentology 2, 115-121.
Jopling, A.W. 1965. Hydraulic factors and shape of laminae. Journal of Sedimentary Petrology 35, 777-791
Jopling, A.W. & Walker, R.G. 1968. Morphology and origin of ripple-drift cross lamination, with examples from the Pleistocene of Massachusetts. Journal of Sedimentary Petrology 38, 971-984.
Kazi, A. & Knill, J. 1969. The sedimentation and geotecnical properties of the Cromer Till between Happisburgh and Cromer, Norfolk. Quarterly Journal of Engineering Geology 2, 63-86.
Kuenen, Ph. H. 1965. Value of experiments in geology. Geologie en Mijnbouw 44, 22-36.
Mitchell, G.F., Penny, L.F., Shotton, F.W. & West, R.G. 1973. A correlation of Quaternary deposits in the British Isles. Special report of the Geological Society of London No 4.
Perrin, R.M.S., Rose, J. & Davies, H. 1979. The distribution, variation and origins of the pre-Devensian tills in eastern England. Philosophical Transactions of the Royal Society of London B. 287, 535-570.
Phillips, L. 1976. Pleistocene vegetational history and geology in Norfolk. Philosophical Transactions of the Royal Society of London B. 275, 215-286.
Pointon,W.K. 1978. The Pleistocene Succession at Corton, Suffolk. Bulletin of the Geological Society of Norfolk 30, 55-76.
Reid, C. 1882. The geology of the country around Cromer. Memoir of the Geological Survey of England and Wales.
Reineck, H.-E. & Singh, I.B. 1980. Depositional Sedimentary Environments. Springer-Verlag, Berlin-Heidelberg- New York. 549 pp.
Shaw, J. 1977. Till body morphology and structure related to glacier flow. Boreas 6, 189-201.
Smith, N.D. & Ashley G. 1975. Proglacial lacustrine environment. In Church, M.A., Gilbert, R. (Eds.) 1975. Glaciofluvial and glaciolacustrine sedimentation. Society of Economic Palaeontologists and mineralogists. Special Publication 23, 135-215.
Solomon, J.D. 1932. The glacial succession of the North Norfolk coast. Proceedings of the Geologists' Association 32, 241-271.
West, R.G. 1980. The pre-glacial Pleistocene of the Norfolk and Suffolk coast. Cambridge University Press, 203 pp.
West, R. G. and Donner, J.J. 1956. The glaciations of East Anglia and the East Midlands: a differentiation based on stone orientation measurements of the tills. Quarterly Journal of the Geological Society of London112, 69-91.
West, R.G. & Banham, P.H. (eds.) 1968. Short Field Meeting on the North Norfolk Coast. Proceedings of the Geological Association 79, 493-512.

Glaciotectonics: Forms and Processes, Croot (ed.), © 1988 Balkema, Rotterdam. ISBN 90 6191 848 0

Wet-sediment deformation of Quaternary and recent sediments in the Skardu Basin, Karakoram Mountains, Pakistan

Lewis A.Owen
Departments of Geology and Geography, University of Leicester, Leicester, UK

ABSTRACT: The terraces in the Skardu Basin, Karakoram Mountains comprise floodplain, fluvial, glaciofluvial and till sediments, as well as palaeosols. These sediments have been frost-heaved, thrust and folded. At some locations, up to five duplexes of stacked sediments with sheath-fold geometries occur. These structures result from glaciotectonic deformation by glacier ice from tributary valley glaciers entering the Skardu Basin and overriding the terraces. A sedimentological, palaeogeographical, and glaciotectonic model is developed for the formation of these deformed terraces.

1 INTRODUCTION

The Karakoram Mountains are one of the most extensively glaciated areas outside the polar regions. During the Pleistocene, the area underwent three extensive valley glaciations and during the Holocene, as many as five glacial stages can be recognised (Derbyshire et al, 1984). Quaternary and recent terraces and valley fills reach several hundred metres above river-level. These comprise mudflows, tills, debris-flows, alluvial fans, fluvial and lacustrine sediments. Many of the sediments have been tilted, faulted and folded. This results from neotectonics, landslide and glaciotectonic processes (Owen, in press). The distinction between these types of deformation is particularly important in elucidating the role of glacial processes and reconstructing glacial histories of high, dynamically-active mountain regions.

Deformed terraces in the Skardu Basin are used to provide examples of the type of deformation produced by glacial processes.

2 SKARDU BASIN, PAKISTAN

The Skardu Basin is some 30km long and 15km wide and lies at the confluence of the Indus and Shigar Rivers (Fig. 1). The basin trends SE-NW, with the Marshakala Range of the High Karakoram to the north and mountains of the Deosai Plateau to the south. Dainelli (1922) mapped the superficial deposits within the Skardu Basin and suggested that it had been extensively glaciated.

Cronin and Johnson (1988) studied the palaeomagnetism of the Bunthang Sequence, a 1300m thick sedimentary sequence, situated in the NW corner of the Skardu Basin. This comprises a basal diamicton overlain by lacustrine and glaciofluvial sediments. They showed the sequence to be magnetically reversed which suggests that the base of the sequence began to accumulate towards the end of the Matuyama chron (c. 0.72 Ma.). The diamicton at the base of the section was interpreted as a till and is correlated with glacial till exposed on a butte, Karpochi Rock, near the centre of the Skardu basin (Drew, 1873; Lydekker, 1883; Conway, 1894; Oestreich, 1906; Filippi, 1912; Dainelli, 1922). Cronin (1982) considered these to be the oldest sediments exposed in the basin, and Johnson (1986) suggested that they are related to the first glaciation in the northern and southern hemispheres which according to Berggren and Couvering (1974), occurred 3 Ma ago. The recent palaeomagnetic studies of Heller and Liu Tungsheng (1984) show that the initiation of loess accumulation on the Loess Plateau of Central China started about 2.4 Ma years ago. Given the apparent coincidence between the onset of hemispheric glaciation and the beginning of loess deposition, a maximum age of about 2.4 Ma for the first glaciation in High Asia is suggested.

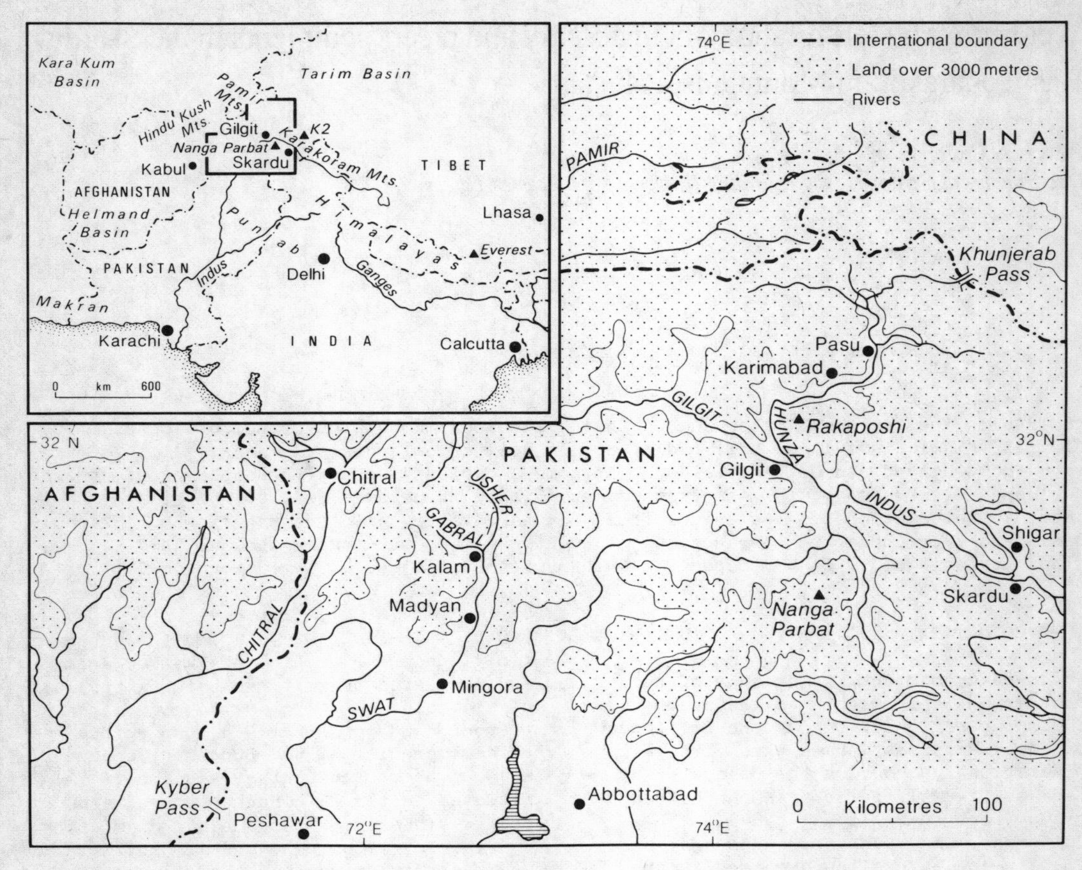

Figure 1. Map of the Western Himalayas showing location of the Skardu Basin.

This restricts the deposits in the Skardu Basin to less than 2.4 Ma years old. In the Kashmir basin Agrawal (1984) argued for a more modest date (0.7 Ma) for the first glaciation of the Himalaya. Others (Kusumger, 1980; Burbank, 1982) argue for much earlier glaciations but such suggestions are based on less reliable evidence.

A number of planation and glacial surfaces can be recognised throughout the Skardu Basin and Shigar valley. The highest of these is analogous to the Gipfelflur in the Alps and is at a height of between 5000-54000m (Fig. 2 & 3A). It is probably pre-Pleistocene in age. This can be correlated with a high surface in the Hunza valley which was recognised by Derbyshire et al (1984): this surface occurs at a height of 5200m and may correlate the Eocene surface at 5000m in Tibet noted by Li Jijun et al (1979).

Along the Shigar valley at a height of between 3200 and 3500m, glacially eroded benches with tills deposited on them can be recognised e.g. east of Shigar at about 3500m (Fig. 3B). These glaciated surfaces probably correlate with the surfaces at a similar height on the southern side of the Skardu basin. Together these may represent an early extensive glaciation of the Shigar valley and Skardu basin. A later glaciation eroded this glaciated surface, deepening the Shigar valley and producing an extensive surface (Fig. 3C). Later ice advance produced large mesa consisting of what is now Karpochi Rock, Blukro Rock, and Stronodoka Ridge (Fig. 3D). Ice over-topped this mesa depositing tills against its cliffs and on its surfaces. The deepened valleys were then infilled with sediment. Cronin and Johnson's (1988) basal till in the

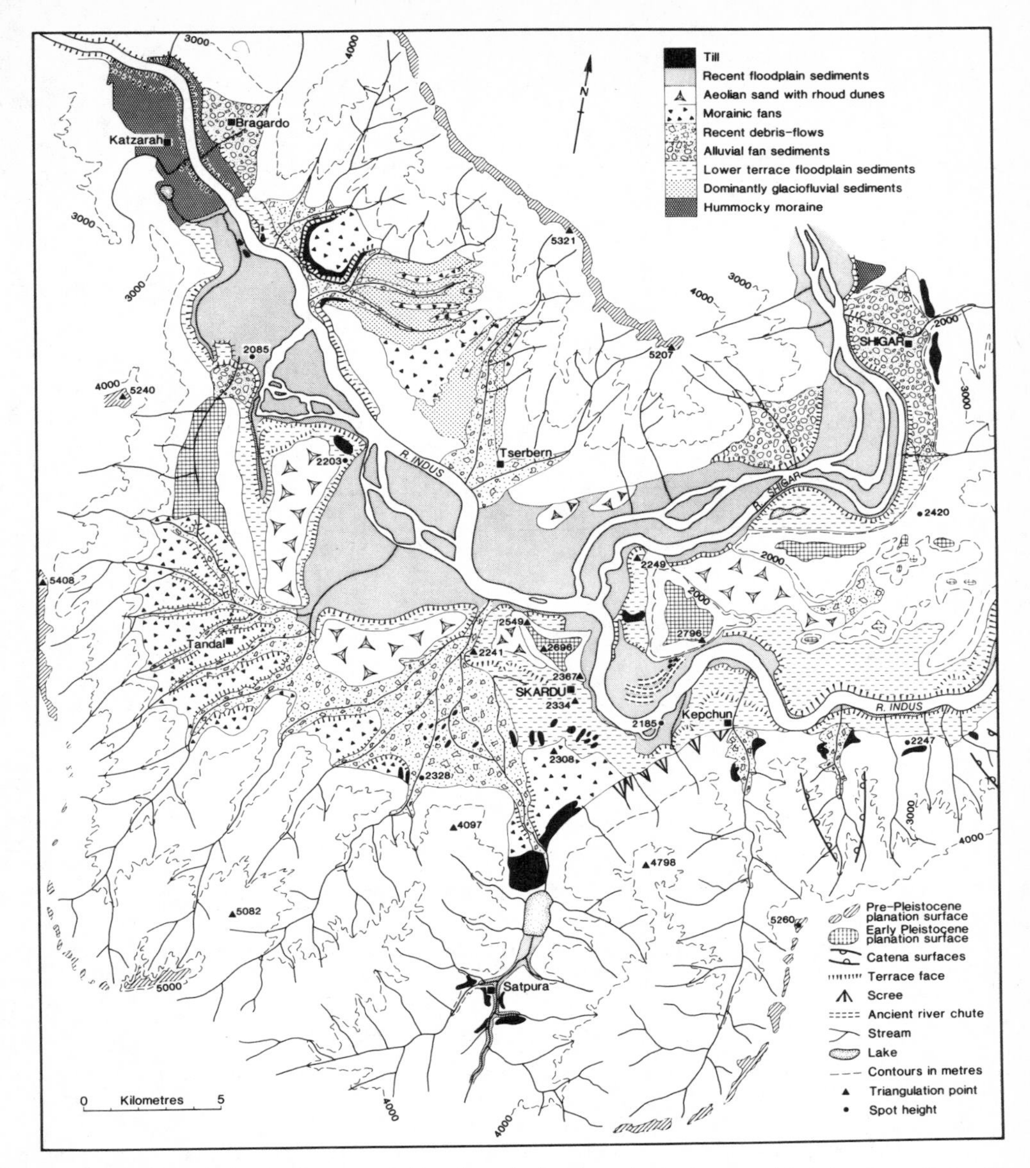

Figure 2. Sedimentological map of the Skardu basin adapted and added to from Dainelli (1922) and Cronin (1982).

Bunthang Sequence may represent till deposited in such a deep palaeovalley during this glaciation.

Valley glaciers from tributary valleys advanced into the Skardu basin during Holocene times. These blocked the Indus when they advanced into the Skardu basin diverting the rivers course over the Karpochi-Blukro-Stronodoka mesa, producing deep rock-cut channels, so dissecting the mesa into three blocks - Karpochi Rock, Blukro Rock and Stronodoka Ridge (Fig. 3E and Plate 1). Tills were deposited and now form inliers in valley fill sediments. To date thus far, no absolute age determinations have been made on these sediments but very dark desert varnishes on surface boulders of the tills suggest that they are early Holocene or possibly late Pleistocene in age.

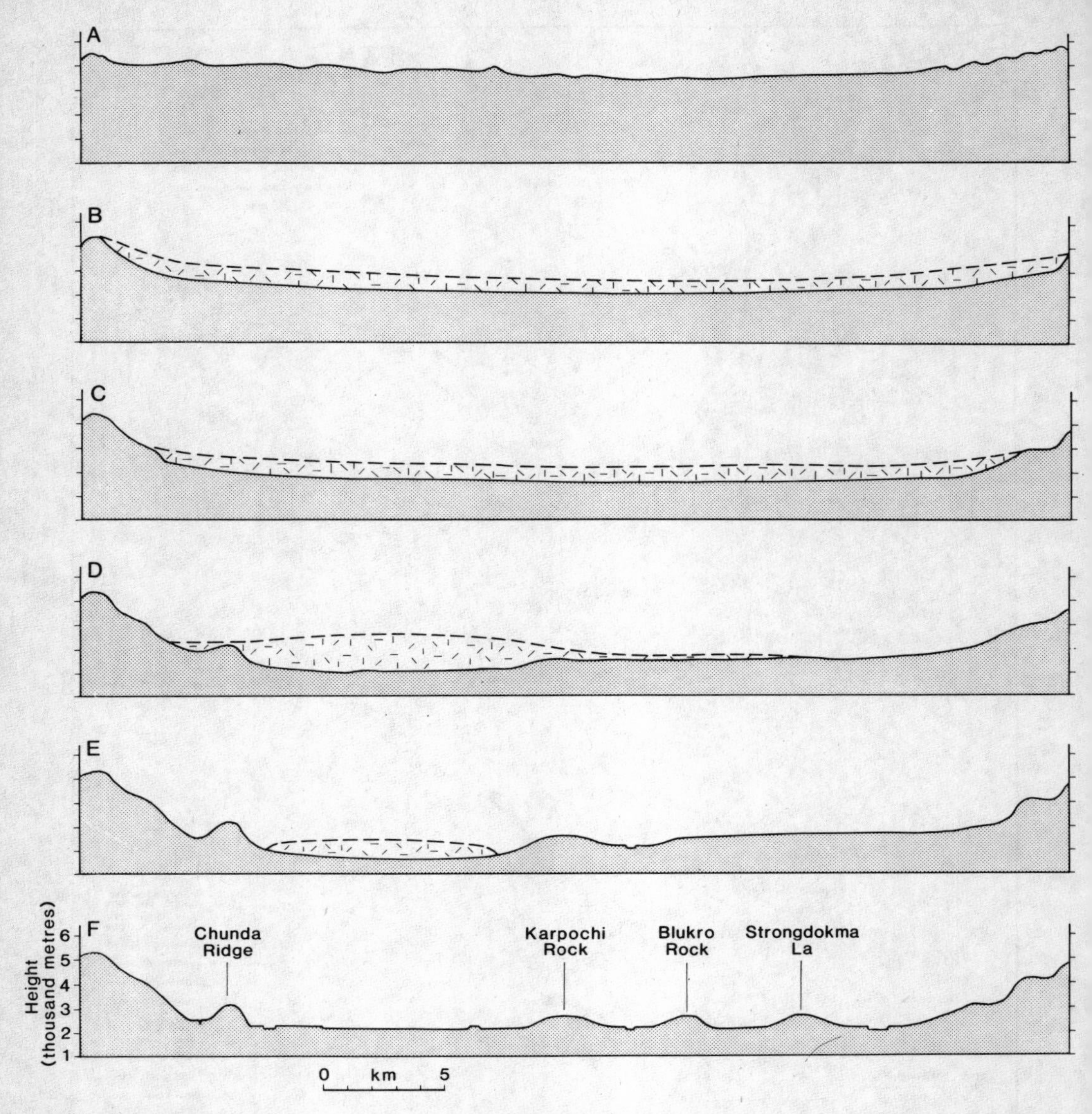

Figure 3. Cross-section across the Skardu basin indicating the planation and glaciated surfaces. See text for explanation.

The first part of this study is concerned with the sediments, which are present in the lower terraces around Skardu town. These comprise flooplain silts, fluvial and glaciofluvial sands and gravels, and tills all of which have been thrust and folded into complex geometries. Burgisser et al (1982) and Cronin (1982 & 1988) considered the deformation to be the result of glaciotectonism but, in fact, gave them minor consideration. The presence of lateral moraines in the glacially moulded Sapara Valley, immediately south of Skardu, strongly suggests a valley glacier extended down that valley into the Skardu Basin, depositing tills and deforming the main terraces. These deformed terraces provide good detailed examples of the nature and mode of formation of the glaciotectonic structures.

Plate 1. View eastwards across the Skardu basin with an early Pleistocene glaciated surface in the foreground and on the top of the buttes in the distance.

3 THE LOW SKARDU TERRACES

3.1 Sedimentology of the sediments

Three to four kilometres north of Skardu, the lowest terraces are undeformed. At their base they comprise planar bedded clayey silts and medium silts which have been channelled. The channels are filled with sandy silts and fine to medium sands (Fig. 4). These planar beds are overlain by silts and sands which pass up into poorly zoned clayey silts with cm to dm bedding and fossil rootlets. Interbedded with these are silts and sandy silts. The whole package of sediments is unlithified and generally very friable, though some are lightly consolidated.

Figures 5 and 6 show grain size plots for the clayey silt beds and the interbedded silts and scour fill sediments (Fig. 7). Table 1 shows the Folk and Ward (1957) mean, sorting, skewness and kurtosis statistics for these sediments. Both groups are poorly sorted and skewed towards the fines, but statistically they form two distinct groups. The interbedded silts have a coarser mean grain size (c. 12 µm cf. c. 5 µm), are slightly better sorted and are not so markedly skewed towards the fines. The mineralogy of the silt and clay fractions was constant throughout the section comprising, chlorite (Fe rich), illite, kaolinite, plagioclase and orthoclase feldspars, and quartz (Figure 8 : Brindley and Brown, 1980).

The clay-rich layers were very homogeneous, Laminations could not be discerned at SEM magnifications, but horizontal alignment of the platey clay and silt size grains was evident (Plate 2A). These horizontally-aligned clays are disrupted in some samples by a sub-vertical alignment of clays lying sub-parallel to vertical fissures with widths of the order of 10 µm (Plate 2B). The grains make up a relatively open framework and some of the clays form aggregates which are similar to clay peds observed in soils. Some samples contain sub-horizontal lenticular fissures several tens of µm long. In the more clay-rich samples, isolated sand grains are often conspicuous. The clays immediately above these sand grains are frequently aligned with their long axes vertical to sub-vertical, and, less frequently the matrix below them is compressed. Organic material is abundant in the form of rootlets with a well structured cellular form,

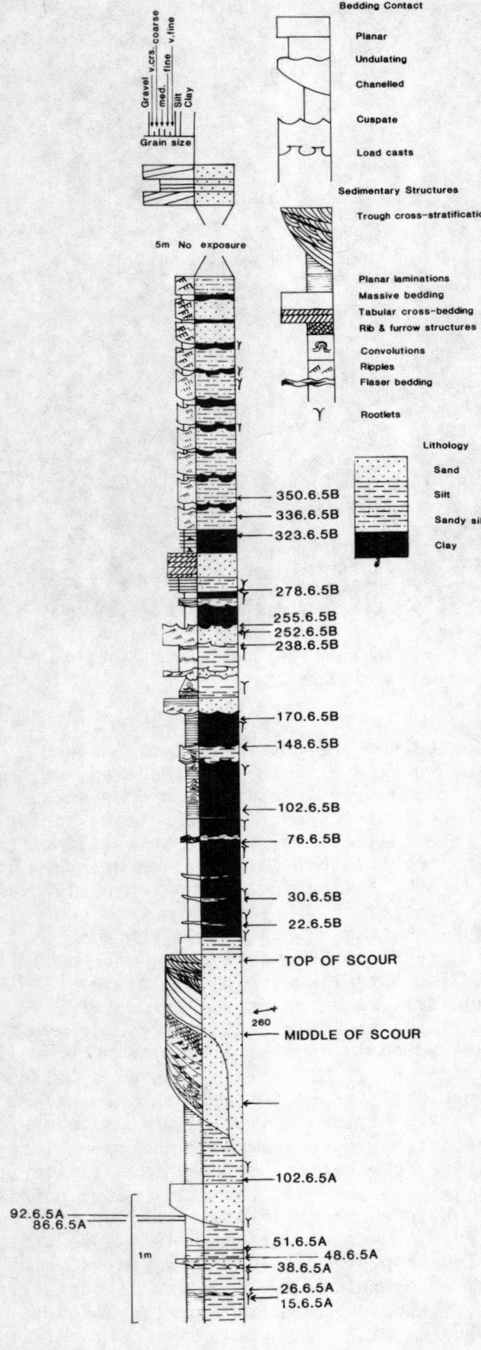

Figure 4. Typical sedimentary log of the sediments in the lower Skardu terraces.

spores or pollen grains, and very fine rootlet hairs (1 µm wide) often coated with calcium carbonate. Millimetre to cm thick layers of organics (dominantly woody stems and rootlets) are present along some horizons and may be traced laterally for several decimetres, often lying in depressions between hummocks.

Towards the top of the terraces, well sorted sands and gravels fill channels cut into them. Figure 9 shows the grain size distribution curves for some of these gravels: they compare quite closely with the glaciofluvial sand envelopes from the Upper Hunza Valley (Li Jijun et al, 1984) although they are slightly better sorted and finer-skewed. The clast lithologies vary from dominantly granite to mixtures of granites, metasediments, and greenstones. All clasts are sub-angular to rounded.

In the terraces SE of Kepchun village, a channel-fill at least 10m wide and 4m deep can be seen. It comprises metre-size rafts of silt (similar to the silt making up the terrace into which the channel has been eroded), boulders and pebbles of granite, greenstone and metavolcanics/ sediments. The matrix comprises pebbles and coarse sands of these lithologies. There is no evidence of any stratification and a number of large elongate boulders are disposed vertically (Plate 3 & 4).

The terraces are capped by alluvial fan gravels with a poor stratification parallel to the present (active) fan surface dipping at less than 5°. Thicknesses are usually of the order of 2 to 3m, although in some cases 6m was recorded. At several locations the gravels fill vertical-sided channels in the underlying silts and gravels.

Inliers of till project through the surface of the recent Skardu alluvial fan their frequency increasing towards the Sapara valley mouth (Plate 5).

3.2 Deformation structures

Near Skardu and particularly around the mouth of the Satpara valley, the terrace sediments are deformed.

Hummocky structures (Frost heave structures)

The clayey silt beds frequently have hummocky structures with cm to dm wavelengths and amplitudes (Figs. 10, 11 & 12). The crests of the hummocks are vertically stacked and equally spaced laterally (Figs 11 & 12). Commonly, the hummocks are

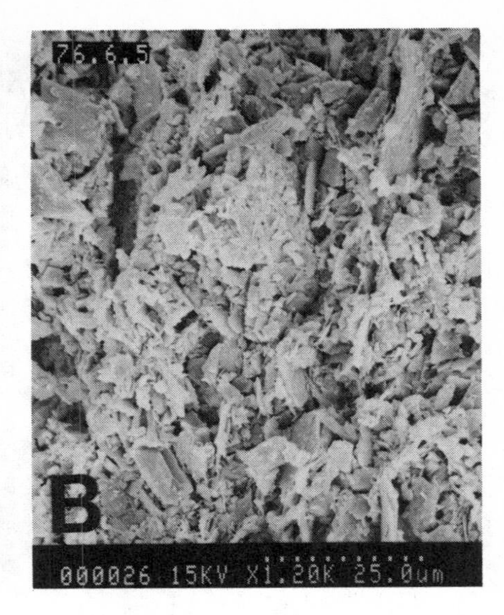

Plate 2, A) Scanning electron photomicrograph of a typical palaeosol with an horizontal alignment of clays, B) Scanning electron photomicrograph of a frost-heaved palaeosol, note the sub-vertical re-orientation of angular silts and clays.

Plate 3. Catastrophic flood channel fill east of Kepchun.

Plate 4. View of the catastrophic flood sediments.

Plate 5. Inliers of till on the lower terraces at Skardu.

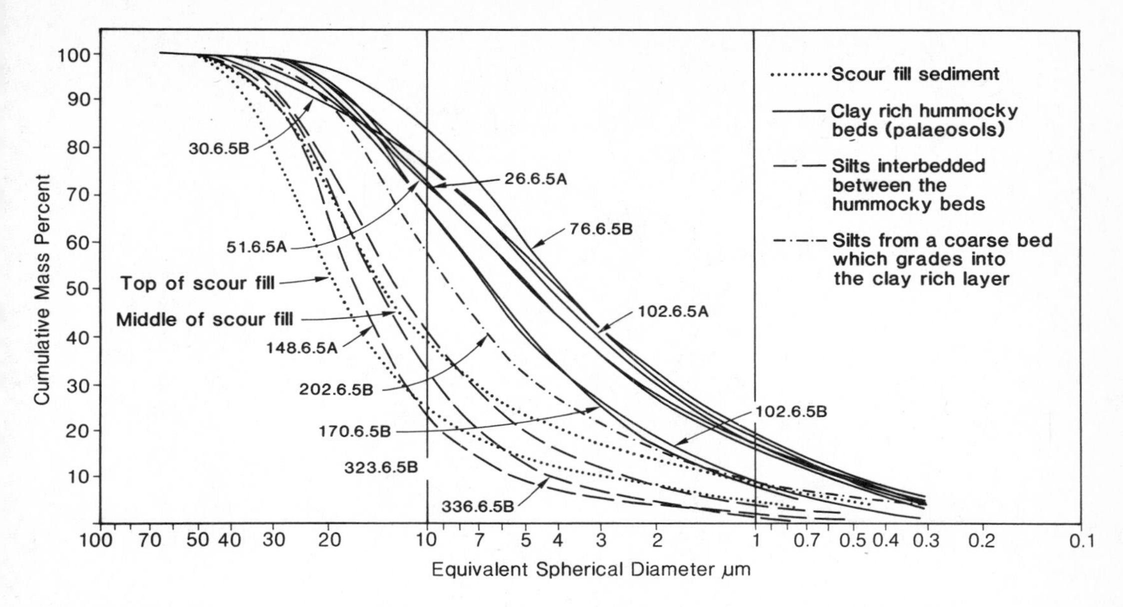

Figure 5. Grain size distribution curves for the silts in Fig. 4 analysed using a SediGraph.

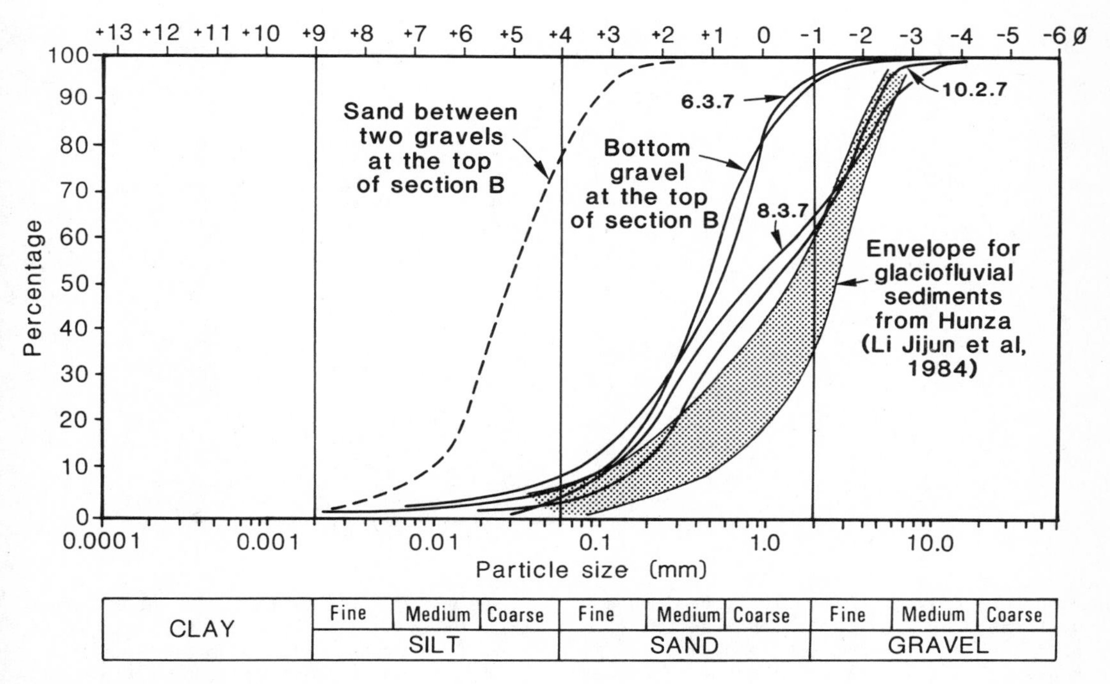

Figure 6. Grain size distribution curves for the silts in Fig. 4, analysed by wet sieving and SediGraph.

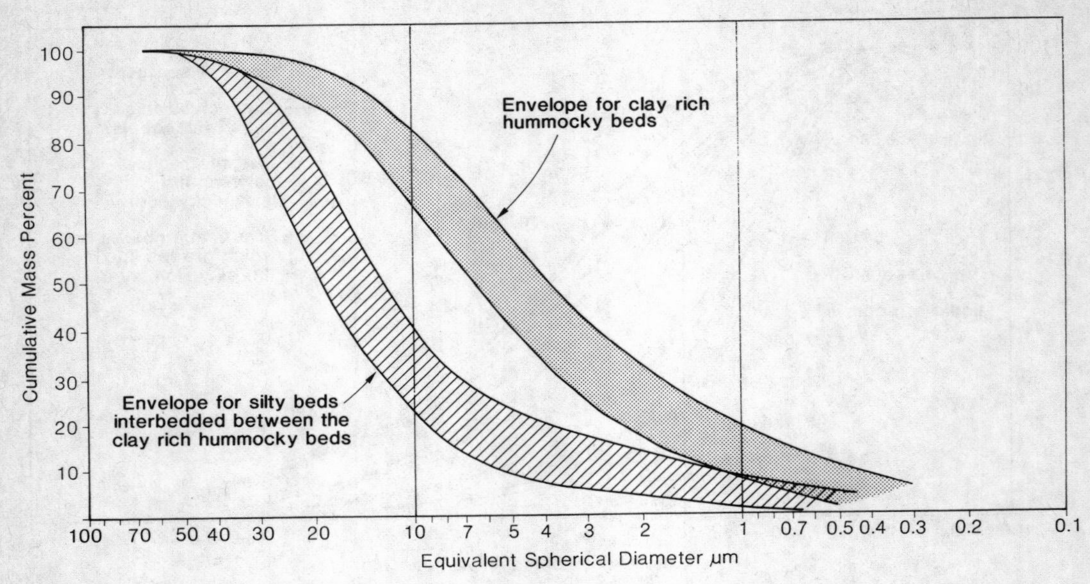

Figure 7. Grain size distribution envelopes for the silts in Fig.4.

Table 1.

Sample code	Mean Grain size um	Sorting,	Skewness	Kurtosis
Palaeosol				
26.6.5A	4.10	1.95	0.34	0.56
51.6.5A	4.34	1.92	0.30	0.55
102.6.5A	3.91	2.01	0.27	0.51
30.6.5B	3.56	2.11	0.15	0.58
76.6.5B	3.25	1.79	0.30	0.60
102.6.5B	5.70	1.60	0.36	0.62
170.6.5B	5.80	1.56	0.30	0.65
Interbedded silts				
148.6.5A	15.50	1.13	0.47	1.10
323.6.5B	10.71	1.37	0.42	0.83
336.6.5B	13.82	1.12	0.33	1.05
Middle of scour fill	9.80	1.87	0.58	0.79
Top of scour fill	15.71	1.47	0.53	1.05

cuspate in section (Fig. 11). The more competent clay and clayey silt beds have suffered brittle deformation fracturing normal to the surface of the hummock. The coarse silt and sand beds are not fractured but are frequently injected into the clay beds. Many of the silt injections exploit the jointing in the overlaying clays and infill the joints. Small scour and fill features occur less than 150cm in width and 30-50cm deep, fine to medium sands making up most fills. Silts onlap the hummocks along some horizons and, in some cases, hummock surfaces have clearly been truncated before the deposition of the next silt or sand bed.

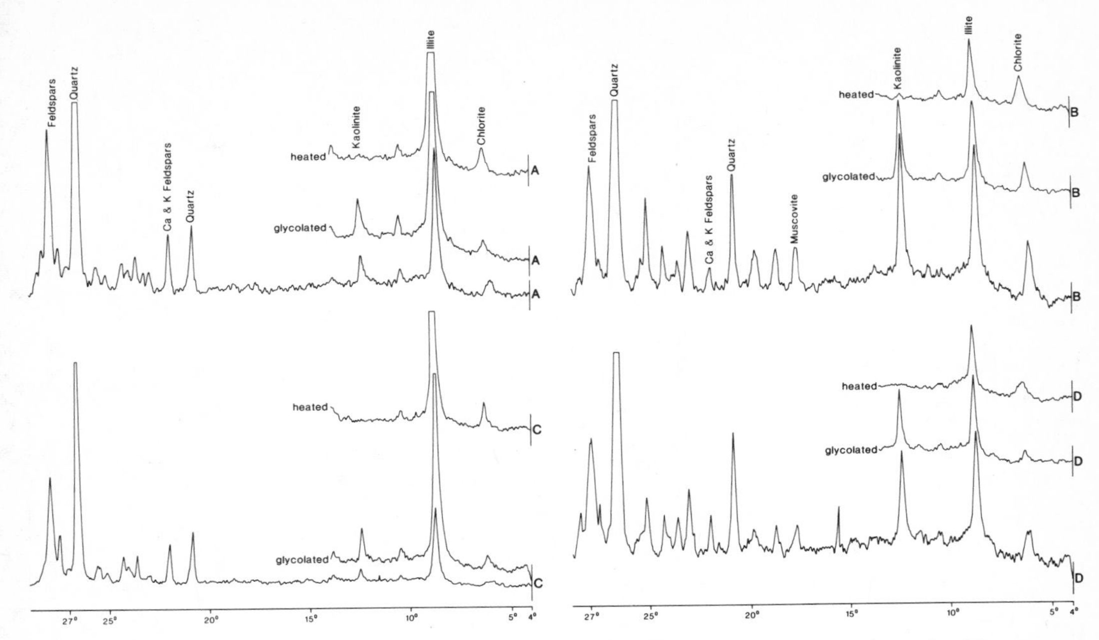

Figure 8. X-ray diffraction traces for the silts in samples in Figure 4, note these samples have been heated and glycolated.

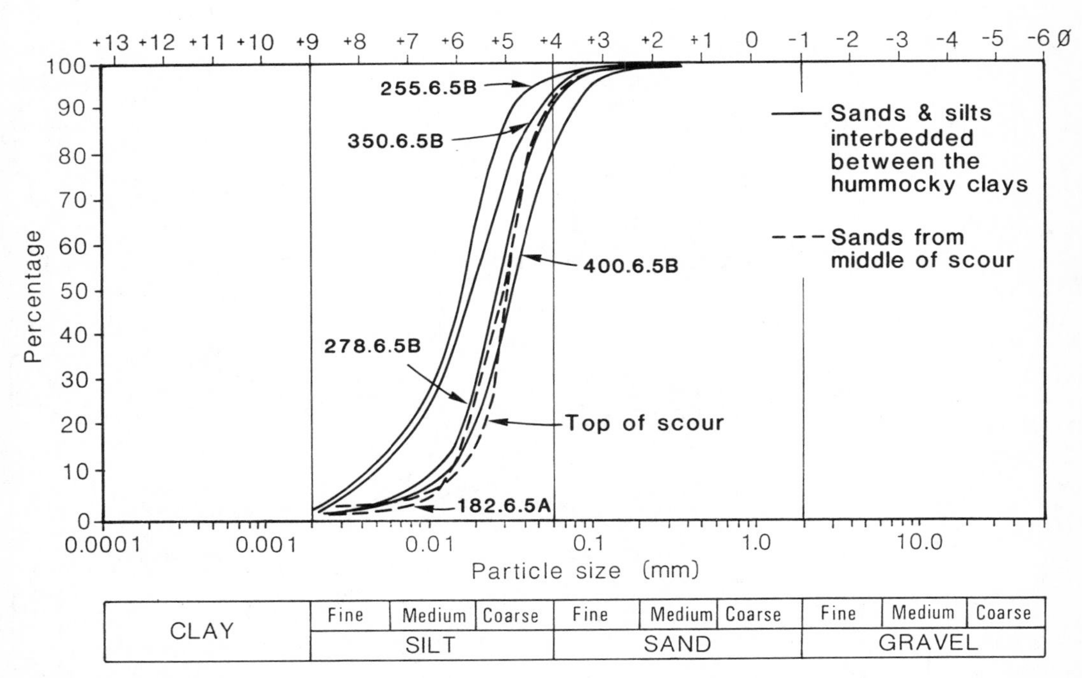

Figure 9. Grain size distribution curves for the gravels towards the top of the terrace (see Fig. 4 for sample numbers).

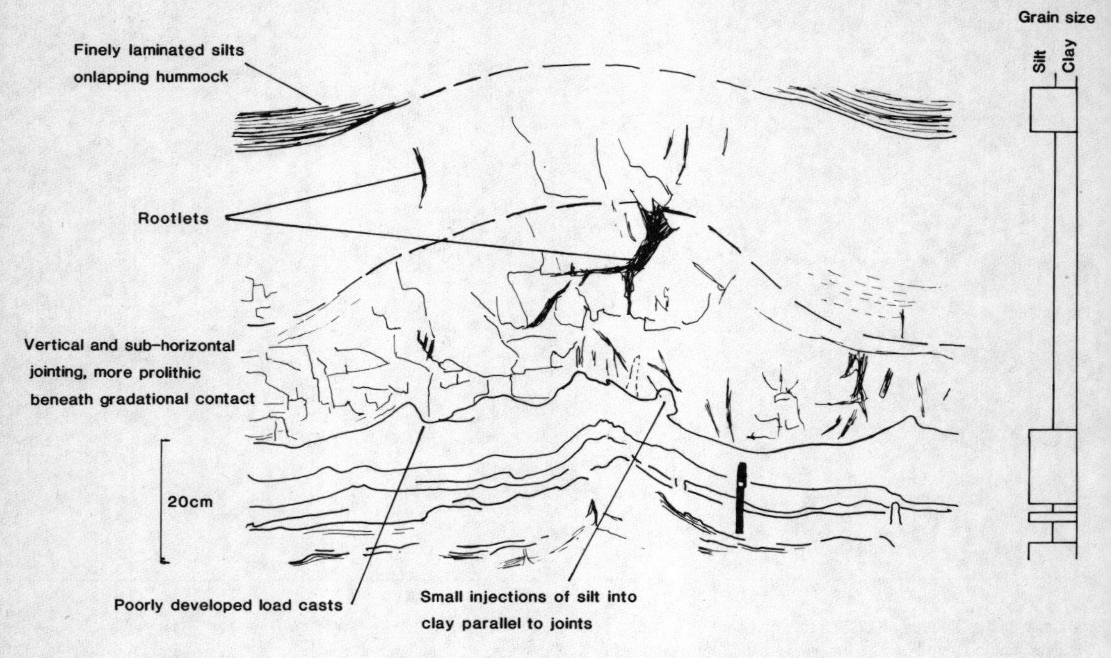

Figure 10. Enlarged view of a hummock in the palaeosols.

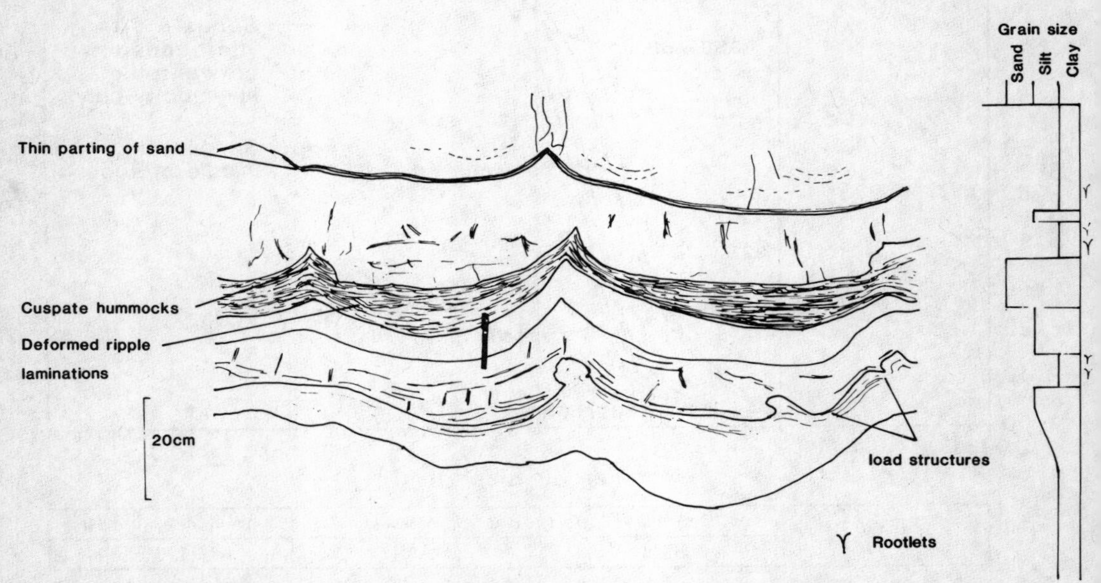

Figure 11. View of parallel cuspate structures developed in the palaeosols.

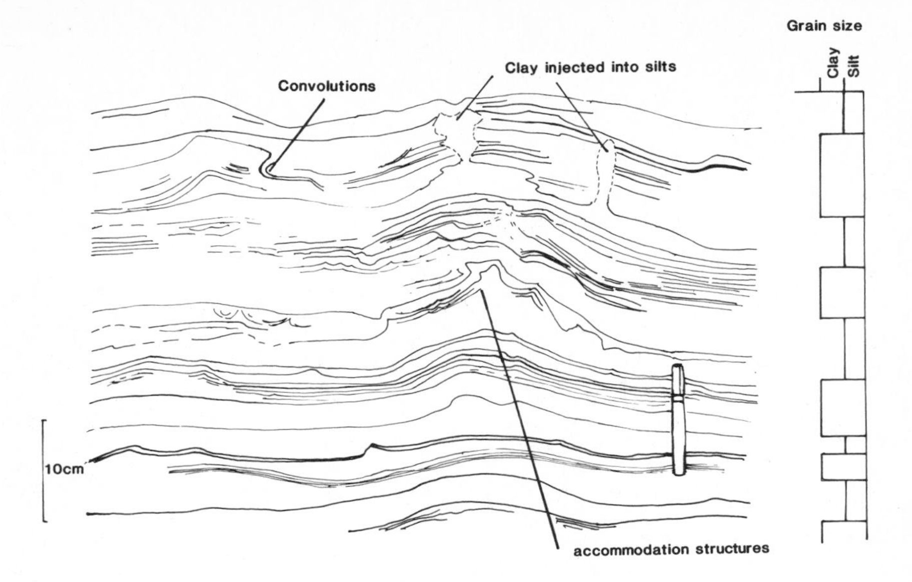

Figure 12. Parallel hummocky structures in the palaeosols.

Glaciotectonic deformation

Folding

Gentle folds with wavelengths of the order of a few tens of metres are present. The style of minor folding varies from simple symmetrical open folds in the more homogeneous beds to box folds in the well laminated beds of alternating sand and silt, and from tight isoclinal and asymmetrical folds in the noses of first order folds and open gentle folds on their limbs.

Near the top of the terraces sand and imbricated pebbly gravel channel-fills of metre depth and widths from 1-10m dominate the fill and are commonly folded (Fig. 13). The channel walls are frequently vertical when incised into lacustrine silts. In most sections, folding within the underlying silts and clays is discordant with that of the overlaying channel-fill sediments. Gravels and sand channel fills are also thrusted and folded into crude sheat-fold geometries.

Normal faulting

Folds are cut by conjugate sets of small normal faults with displacements dying out within a few cm to dm on their limbs (Fig. 14). Often, the fault planes are lines with silts, while in other places there is a segregation of silt or clay within the coarser sediments and parallel to the normal faults. Listric faults are present within the more deformed terraces and are sometimes associated with small thrusts.

Thrusts

Thrusting is common and varies from cm to m scale. Tills are incorporated within some of the terraces although they generally constitute only a minor componant (Fig. 15). Irregular folding of the till and its mixing with silts and sands, associated with the thrusting, makes identification difficult. Figure 16 shows two typical grain size distribution curves for the till. Large rafts of silts also occur within these thrusted admixtured (Fig.15).

Where thrusting was more intense at least five duplexes can be found stacked vertically above each other (Fig. 17). Thrusting frequently exploits bedding usually along the upper bedding plane of a clay bed. Individual thrust planes may be traced for several tens of metres. Scanning electron microscopy of microthrusts indicates that deformation was either confined to discrete shears, shear, zones of anastomosing shears often creating micro-duplexes or shear bands with a rotation of clay and silt size grains normal to the maximum principal

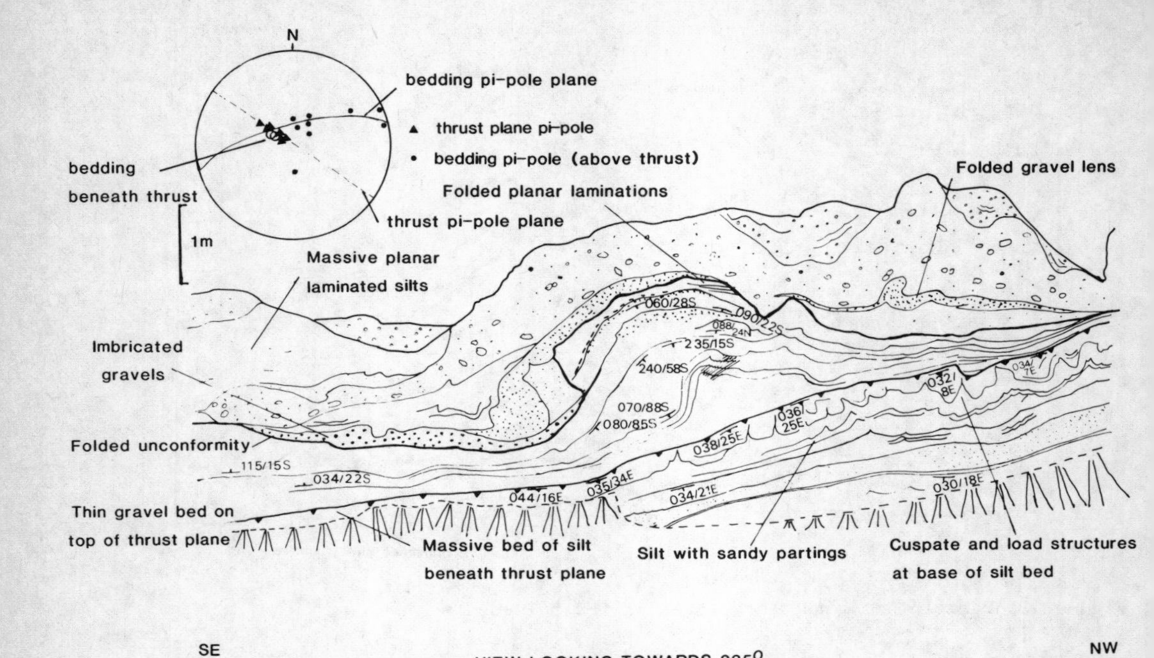

Figure 13. Field sketch of a deformed terrace section showing folded channel sands and gravels.

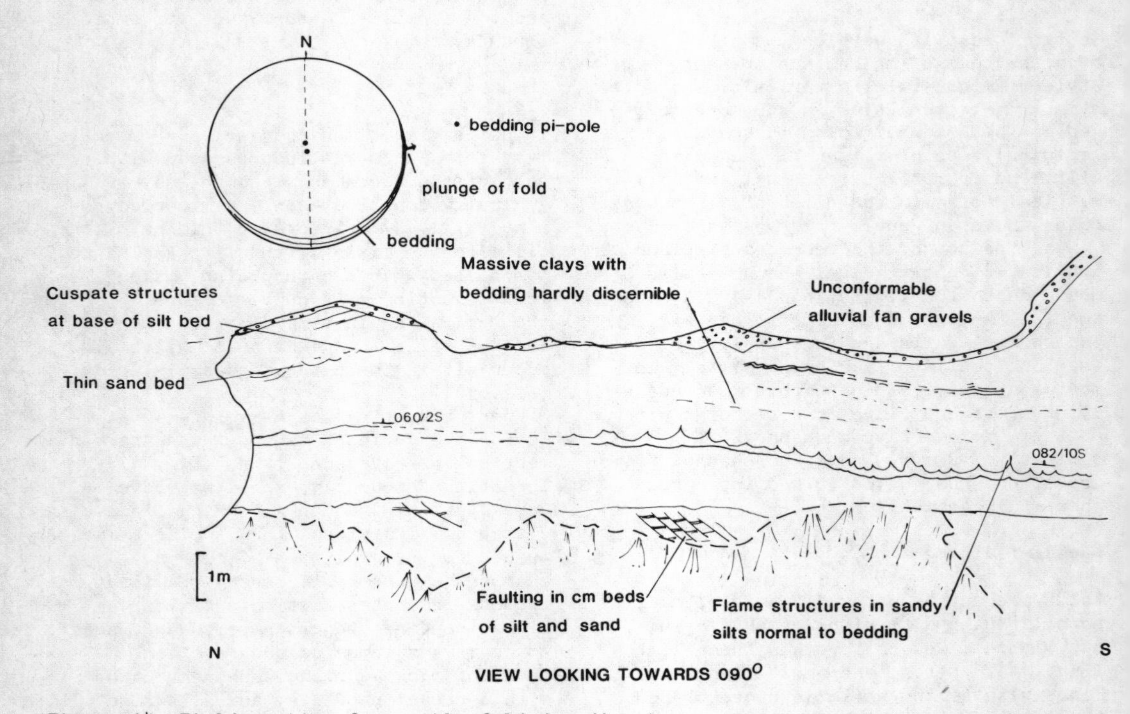

Figure 14. Field section for gently folded sediments.

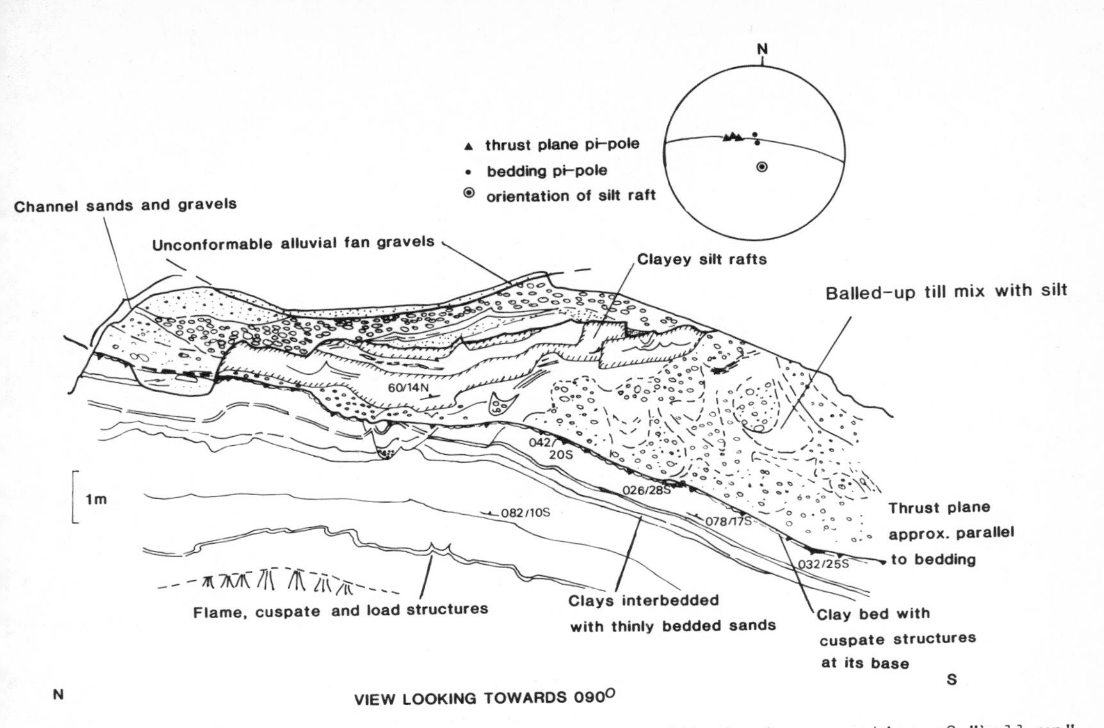

Figure 15. Field section showing a major thrust zone with the incorporation of "ball-up" tills in the section.

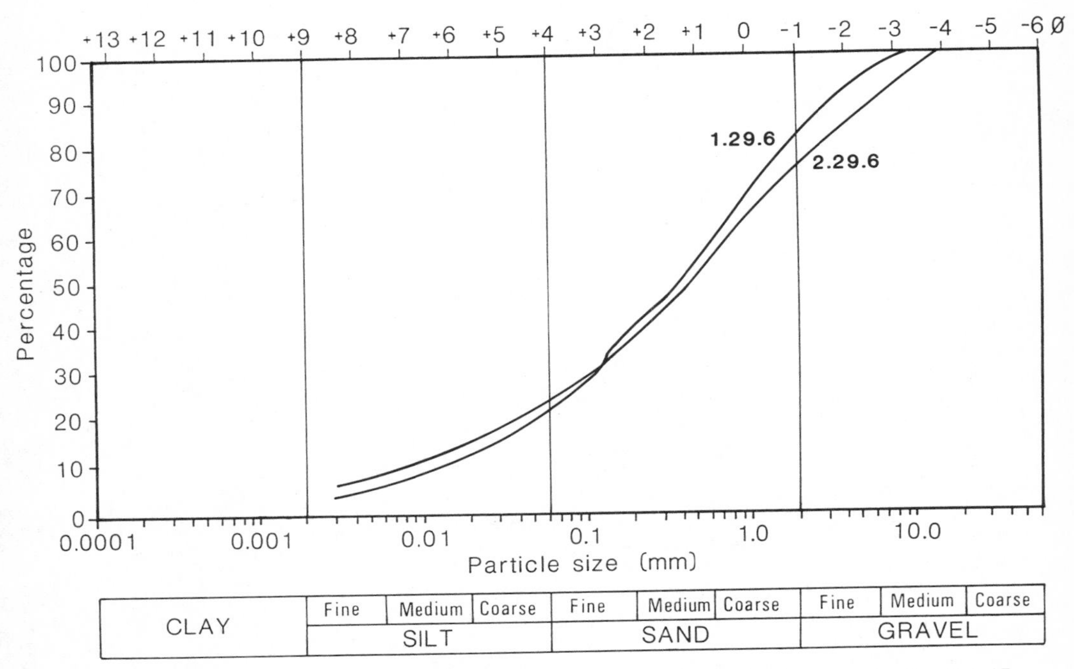

Figure 16. Grain size distribution curves for the "balled-up" till seen in Figure 15.

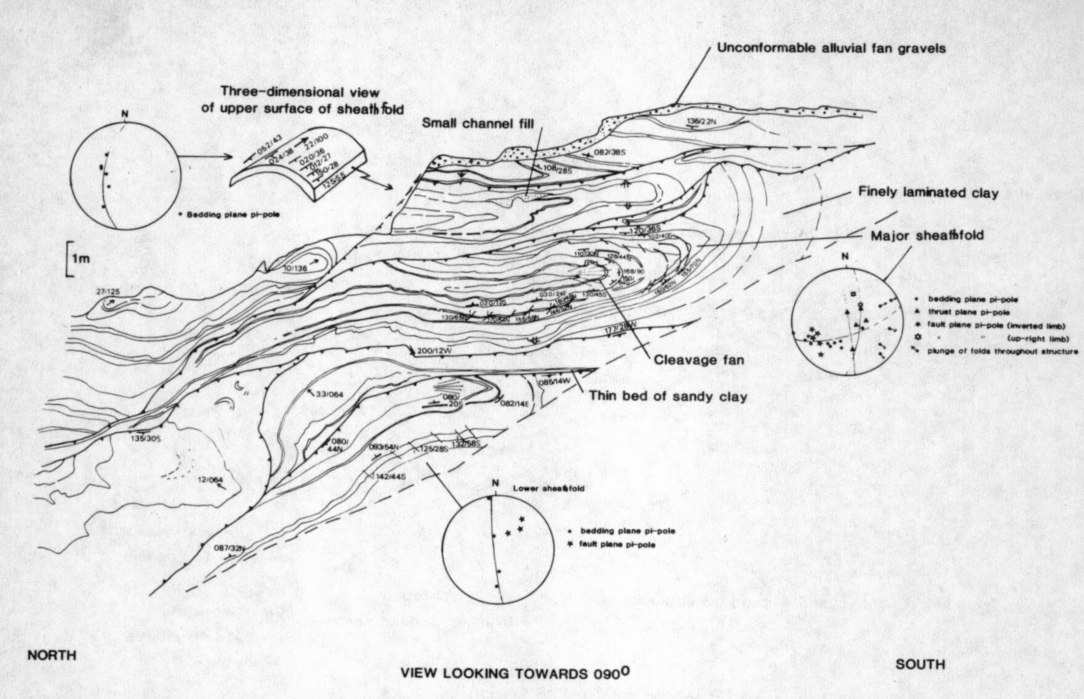

Figure 17. Field sketch of five duplexes stacked on top of one another enclosing large sheath-folds.

stress direction (Fig. 18). The thrust planes within these micro-thrusts are very narrow (of the order of a grain or two thick) with a re-orientation of grains along the plane parallel to its length. Thrust planes within the first order thrusts are also narrow, usually less than 2cm.

Sheath folds

Duplexes commonly comprise sheath folds (Fig. 17, 19 & 20) within which lower order structures can be recognised; small-scale conjugate faulting is present on the limbs of the sheaths striking sub-parallel to the plunge of the sheath; small thrusts within the duplex are less common, other faults are concentric and mimic the sheath-fold's geometry, some of these faults extending into overlaying nappes. It is not known how pervasive these concentric faults are. Cleavages are present in some of the fold noses but their development is restricted to the clays and clayey silt beds, although some have exploited the flame structures in the coarser beds (Fig. 17 & 19; Plate 6). Hummocky and dewatering structures have been folded with the beds although some dewatering structures face upwards in inverted beds (Fig. 19). Primary sedimentary structures are remarkably well preserved, retaining ripple and planar laminations after folding.

3.3 Model of development

A sedimentological and glaciotectonic model to account for the structures and sediments found in the Skardu terraces is presented in Figure 21.

Sedimentology

The unfolded sediments northeast of Skardu represent fluvial and floodplain sediments, deposited adjacent to the palaeo-Indus River. The clay-rich silt layers are interpreted as palaeosols deposited adjacent to the palaeo-Indus and frequently flooded or partially flooded by a few cm of water. Although plants were apparently

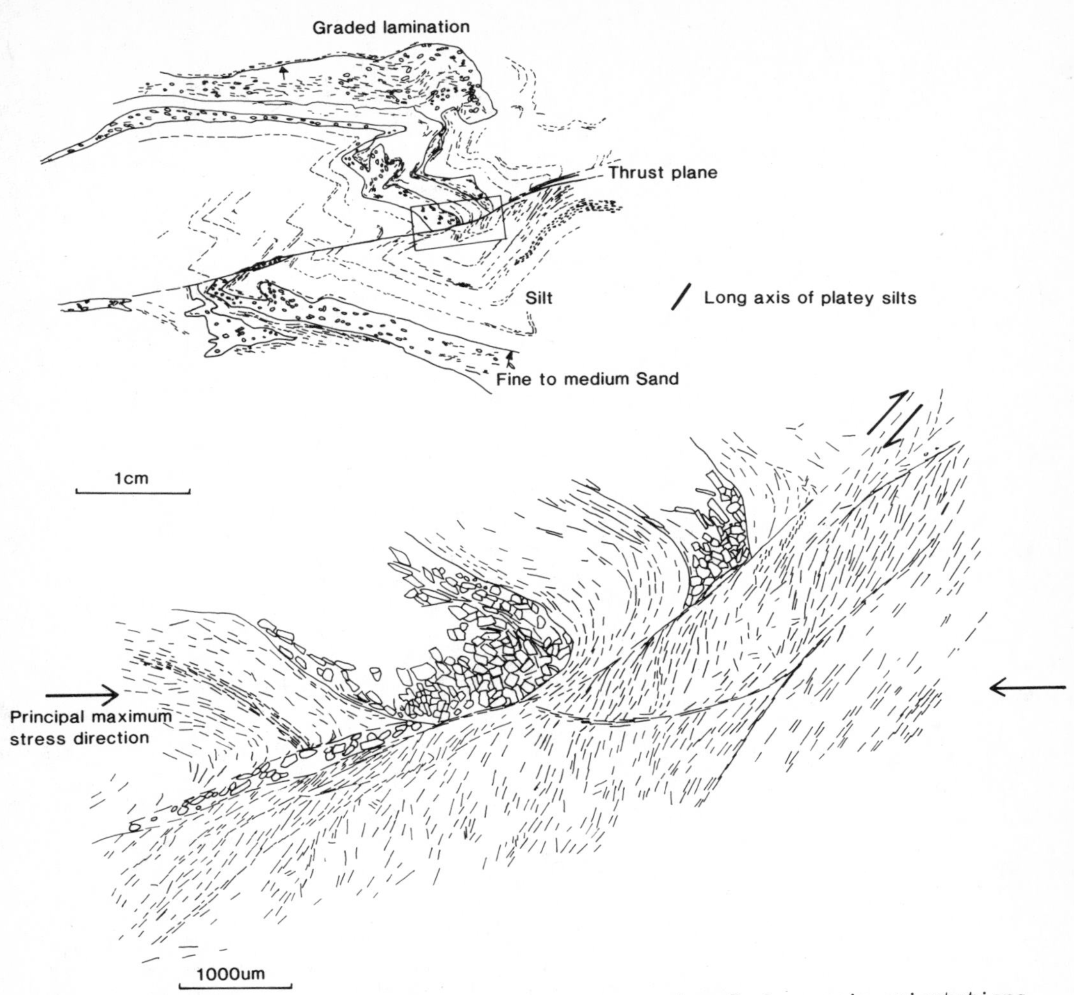

Figure 18. Sketch drawn from scanning electron micrographs of clay grain orientations within a shear zone.

able to grow, and are represented by a few organically-rich mm thick horizons which were present at some locations, a well developed soil failed to develop. The alternation of incipient palaeosols with silts and sands may have been allocyclic or seasonally influenced. The similarity of mineralogies of all the sediments examined suggests a constant source, with kaolinite, quartz and feldspars being derived from the weathered granites, and the illite and chlorites from the metasediments and metavolcanic rocks. The broadly similar grain size distribution of the palaeosols and the interbedded silts suggests that this alternation of lithologies was a result of changes in flow regime. Both the silts and sands appear to have been deposited in a large floodplain-type environment under higher flow regimes during flooding by shallow sheet water or during changes in the position of the palaeo-Indus. Similar depositional conditions can be observed in the present Indus River, the regime of which is seasonally controlled. Sediment output and river discharge may have increased 500 to 1000 and 20 to 50 times respectively during the day in the summer because of the ablation of glacial ice which supplies the waters for the Indus catchment (Ferguson, 1984 and Ferguson et al, 1984). Comparison of Landsat images taken in different seasons in 1979 shows the extent

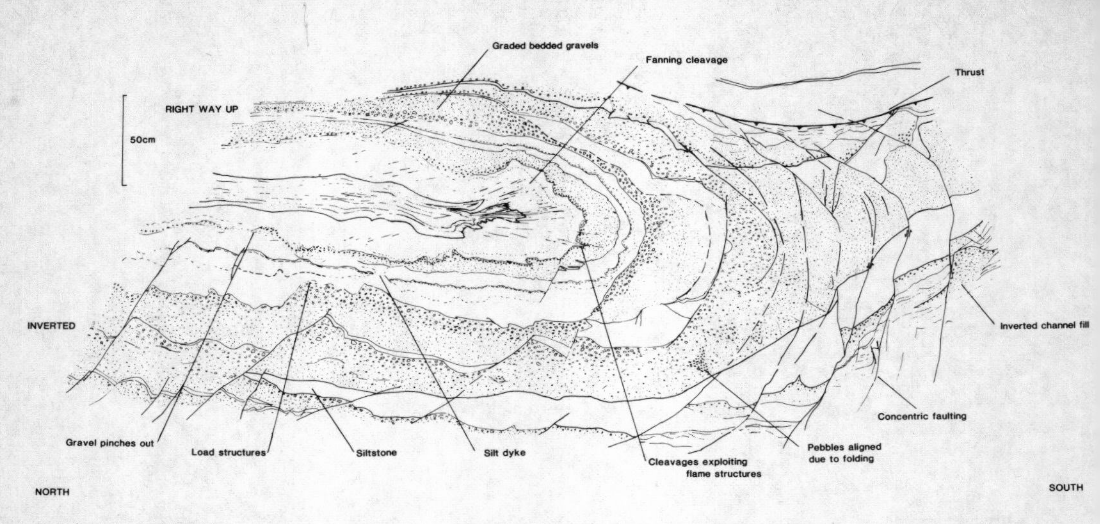

Figure 19. Enlarged view of the fold nose within a sheath-fold, seen in Figure 17, showing cleavages and concentric faulting.

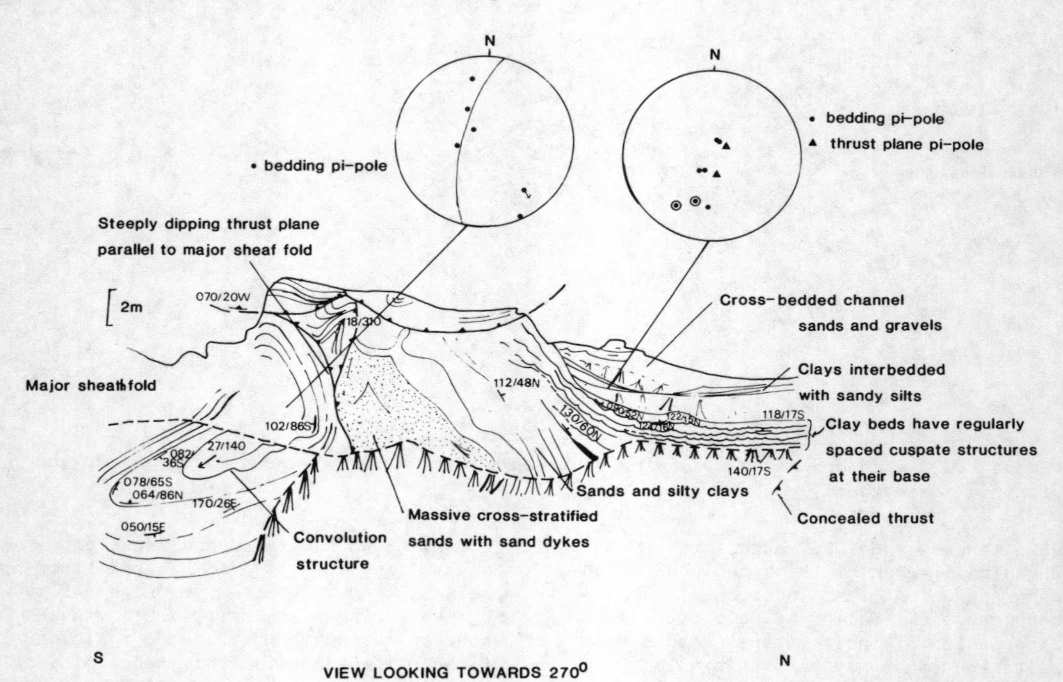

Figure 20. Field sketch of a deformed terrace section showing the geometry of a sheath-fold.

of changes in the flow regimes due to seasonally-controlled melting of glacial ice today (Plate 7). The grain size distribution of the silts is poorly sorted and skewed to the fines which is also characteristic of floodplain environments (Brown, 1984). The palaeosols are poorly zoned probably because the formative period was so brief prior to burial by coarser sediment. The isolated sand grains in some of the clay-rich layers are interpreted as small drop-stones which

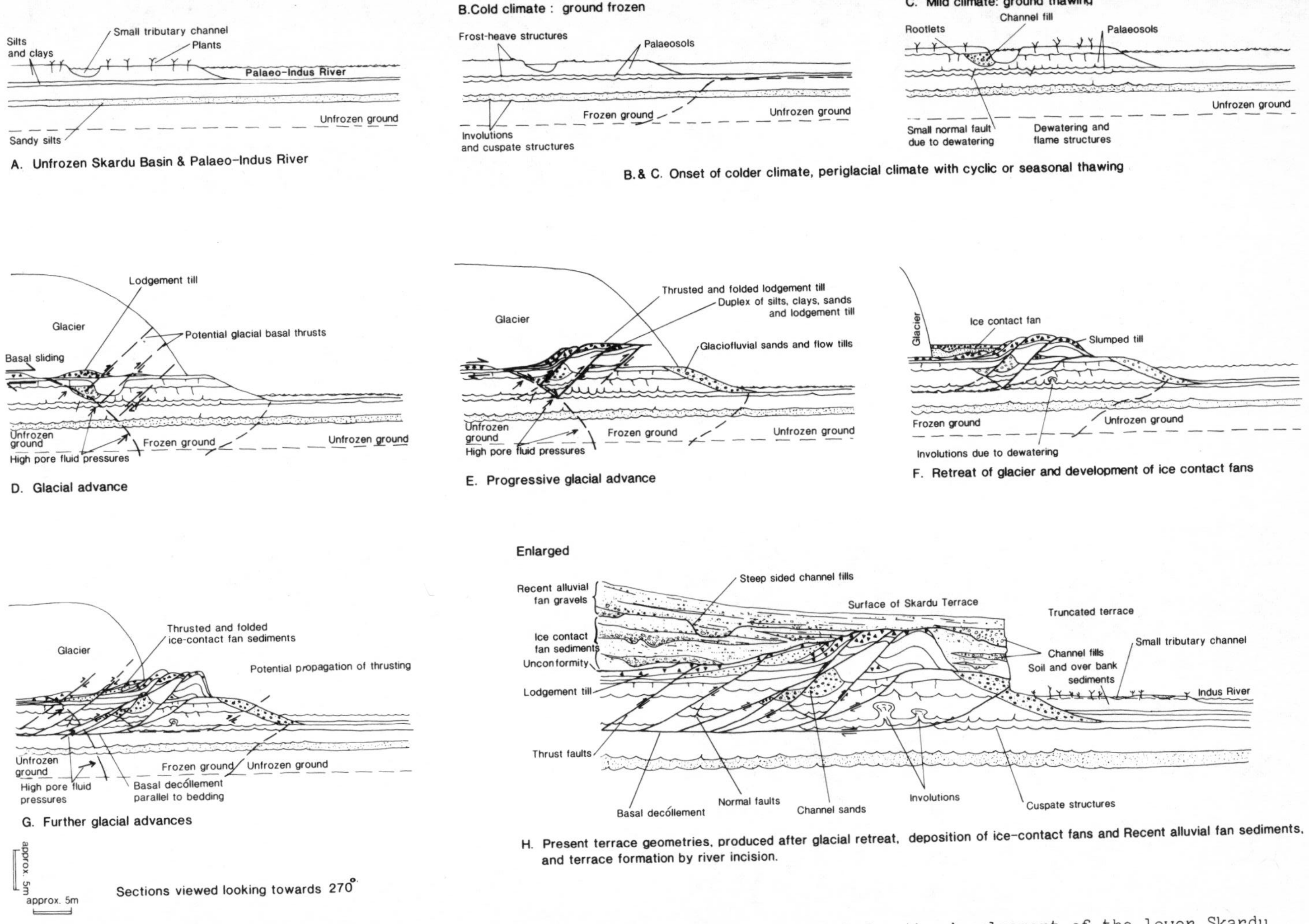

Figure 21. Summary diagram showing the sedimentological and glaciotectonic model for the development of the lower Skardu terraces.

Plate 6. Fanning cleavage in the nose of a sheat-fold at Skardu.

either dropped from pieces of floating ice or fell through the clay when sands were deposited above them.

Frost heaving

The hummocky structures are very characteristic of the palaeosols in this area and are interpreted to be the result of frost-heave. Their vertical stacking is typical of frost effects in soils and sediments with the segregation or concentration of ice at one level producing a lowering of the temperature gradient over a vertical zone throughout the soil/ sediment system (Vliet-Lanoe 1985). The vertical alignment of clays seen under the scanning electron microscope is consistent with this explanation. The presence of sub-horizontal lenticular voids probably represents the growth of minute ice lenses within the sediment.

The regular arrangement of the hummocks suggests a regular and constant freezing of the ground, Muller (1947) showed that, with freezing and increasing moisture content, clayey sediments initially increase in strength but soon weaken again as water content increases and pores become progressively infilled with ice. As a consequence of such freezing, the strength of the clays was increased and probably caused them to behave in a more brittle fashion at lower stresses and, being more competent than the sands, they developed fractures. In contrast, the strength of a sand body increases to a maximum with progressive freezing. This suggests that the moisutre content of the sand may have been less than that of the clays which did not allow the clay to exceed or reach the same strength at the sand. Alternatively both may have been relatively depleted in moisture, but this seems unlikely in a floodplain environment. The injection of silt into the clays may similarly be explained by differential melting of the sediment, the clay retaining its strength while ice melted first in the sand beds and was aided by dewatering allowing it to be injected into the strengthened clay often into the already developed fractures. The cuspate form of many of these hummocks may be explained by the great anisotropy of the sediments resulting from the rapid alternation of lithologies. The truncation of hummocks at some horizons was probably due to fluvial erosion. This indicates that they may have developed subaerially adjacent to small streams or rivelets or in shallow water with a high flow regime. Upon melting the frozen sediment dewatered producing "flame-like" structures, large involutions and associated normal faults. In summary the field data is best explained in terms of a periglacial climate with cyclic or seasonal thawing and freezing of the ground ice in a floodplain environment.

Glacial deformation

With the onset of more severe climate, glaciers advanced from the Deosai Ice Field and the Marshakala Range down side valleys and down the Shigar Valley into the Skardu Basin eroding the valley and producing lateral moraines (Fig. 22). Cronin (1982) suggests that the glaciers that occupied the Satpara Valley originated in cirques on Oltingblok and moved down into and eroded the lower reaches of the Satpara valley depositing tills in the lower valley only. The same glacier

A

B

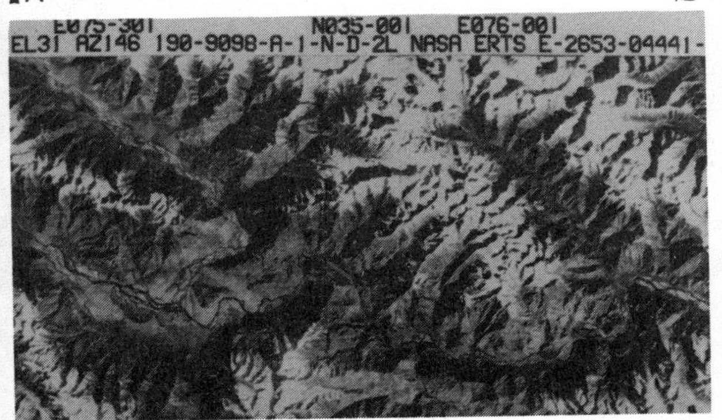

Plate 7. Landsat image of the Skardu basin, A. during the Summer; B. during the Winter.

was responsible for the deposition of till, but only in the lower valley. The ice advanced at least 5km into the Skardu basin where the younger tills are now exposed as inliers. As glaciers advanced into the Skardu Basin they were no longer confined by valley walls and may have converged to form piedmont glaciers. Dainelli's (1922) map suggests that glaciers from the side valleys advanced into the basin throughout the area.

The movement of a cold-based glacier over wet and probably frozen floodplain sediments would have allowed the ice to adhere to them by freezing to the wet substrate (Fig. 21D & E; Banham, 1975; Boulton and Jones, 1979; Moran et al. 1980; van der Wateren, 1985; Boulton, 1986). Temperate glaciers and isothermal ice would tend to slide over the sediment though they may exert high frictional drag along the ice/sediment contact and may initiate shortening of the substrate by horizontal compression or may transmit ice movement into the substrate (Dredge and Grant, 1987). With the initial movement of the glacier over the sediments, the underlying silts were folded into large open or gentle folds possibly being pushed in front of the glacier snout. These folds appear to have been related to the concurrent development of ramping thrusts (Morley, 1986). Tension in the limbs of the folds was accommodated by sets of conjugate normal faults, now lined with clays. Dewatering probably accompanied the folding and faulting, and with further movement of ice over the substrate, thrusting was produced beneath or adjacent to the glacier. This thrusting was probably controlled primarily by differences in pore fluid pressures between frozen and unfrozen ground (Boulton, 1970 & 1971; Harrison, 1957; Moran, 1971; and Virkkala, 1952). Bedding plane thrusts are particularly common suggesting that lithology may have been an important control on thrusting. The thrust décollement surface ccommonly lies along the upper bedding plane of clay-rich beds and, at some locations, the crests of hummocks have even been truncated by thrusting (Figure 13). This suggests that the clays and sands may have been frozen, since clay is weaker than sand in the frozen state (Muller, 1947; Banham, 1975) while in the unfrozen state sands would be the weaker of the two, having an essentially zero plastic limit; thus they would have acted as the décollement surface. The preservation of many fine laminations within the sandy silts after they had been folded may also indicate that the ground may have been frozen to a depth of at least 3m when deformation occurred.

Tills were incorporated with the sediments during thrusting. Where tills are seen in the sections they are balled-up and mixed with the silts. The grain size distribution curves in Figure 16 demonstrates the polymodal grain size distribution resulting from this mixing. Thrusting incorporated large rafts of internally undeformed silt, along with an admixture of coarser sediments. Displacements on minor thrusts were often small enough to be accommodated by compression of the overlying beds.

The common duplexing and sheath-fold geometries may be explained with reference to the movement of modern glaciers emerging from confined valleys. Much of their movement is controlled by longitudinal crevasses, which may have a spacing from a

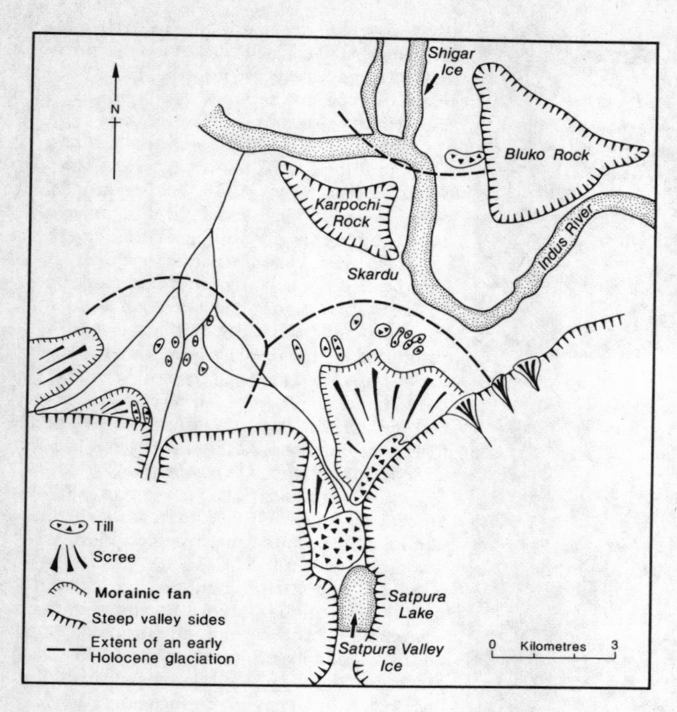

Figure 22. Map showing the extent of glacial ice in the Skardu area probably during an early Holocene advance.

few metres to several hundred metres, e.g. Blaisen Glacier in Norway (Derbyshire pers. comm.). A contemporary example of this was described by Hewitt (1967) near the snouth of the Baifo Glacier due to glacial push resulting from seasonal fluctuations in the position of the snout. This results in differential movement and thus provides a mechanism where zones of zero longitudinal stress occur adjacent to zones of intense stress. A thrust produced beneath the ice may thus be "pinned" laterally with maximum displacements in the centre of the thrust sheet, this would allow the development of a sheath-fold. Croot (1987) describes groups of push moraine ridges in lobate shaped complexes several hundreds of metres across marginal to the snout of the Eyjabakkajokull glacier in Iceland. These represent discrete structural units resulting from differential movement of the ice. These may be analogous to the deformation within the Skardu sediments.

Small-scale conjugate normal faults are found in areas of both extension and compression, suggesting that some of the faulting occurred before the sediment was folded into a sheath. Some of these small normal faults are associated with small thrusts which formed within the larger duplex. Hummocky and associated dewatering structures which bear no relationship to the sheath's geometry are also present, indicating that they developed prior to folding. During folding, areas of higher stress developed within the tightened nose of the fold, leading to an alignment of pebbles in the gravel units and a fanning cleavage in the finer sediments. Scanning electron microscopy has shown that the cleavages resulted purely from a rotation of platey minerals; pervasive cleavage is rare and is confined to laminations dominated by platy minerals. The style of minor folding varies and includes box folds, tight almost isoclinal folds, to asymmetrical open folds. These are interpreted to be a result of sediment anisotropy and their position within the larger fold. A more open framework at the base of some of the laminations, seen under the scanning electron microscopy, suggests that small-scale bedding plane flexural slip took place during the folding possibly as a result of a concentration of pore fluid pressures at the base of these laminations. However, most of the deformation within the folds was taken up as tangential longitudinal slip. Concentric faulting mimicking the sheath-fold geometry produced a "cone-in-cone" structure which developed late in the process with some faults extending into a later overlying nappe; these are clearly, not folded normal faults. A shift from ductile to brittle deformation of the sheath-fold is indicated, possibly as a result of increased strength of the fold with time.

The preservation of many of the primary sedimentary structures within the sheath folds initially suggests that the ground was frozen. However it may be argued that porosity and rate of propagation of pore fluids through the sediment is the most important control on the preservation and deformation of these sedimentary structures. Thus it may not be necessary for the sediment to have been frozen to preserve these structures. Thus it may not be necessary for the sediment to have been frozen to preserve these structures. The trend from dutile to brittle deformation within the structural units may indicate that the freezing of these sediments occurred late in their deformation helping to produce brittle structures as sediment was thrust from depth into colder zones nearer the surface. The process must have been relatively slow, however, in order for it to cool before being finally emplaced.

Dewatering structures in inverted limbs are occasionally directed upwards indicating that dewatering occurred after inversion; in such cases, the dewatering structures must have occurred later than the concentric faulting suggesting that the sediment may have been frozen and started to thaw after the sediments had been brittly deformed. Whether thrusting took place beneath the ice or adjacent to it has not been fully evaluated. However, the absence of till in the majority of terraces north of the outcrops mapped by Dainelli (1922) suggests that the ice did not advance beyond these sites. Additionally, had the ice extended beyond Dainelli's limit any ice supported structures would have collapsed with glacial retreat. There is little evidence of this however and the most logical hypothesis is that all the structures described developed in front of the glacier.

Climatic amelioration caused glacial retreat and associated thawing of the proglacial structures developed. These sediments were then eroded by glacial meltwater streams depositing new bodies of sediment in channels the banks of which were frequently very steep probably because the sediments were frozen in the surface layers (Fig. 21F). This pulse of meltwater sediment was, in turn, channelled and filled by sands and gravels their geometries were typical of small braided channels.

A subsequent glacial advance caused a recurrence of the whole process of folding, and deposited more glaciofluvial sediments which have also been deformed (Fig. 21G). The number of advances of the Satpara glacier is not known, but at least two can be recognised on the basis of the two sets of folded glaciofluvial sands and gravels. Whether these two advances represent major glaciations or minor fluctuations has yet to be resolved.

With the final retreat of the Satpara Valley glacier, sands and gravels were deposited, infilling steep-sided gullies (Fig. 21H) and forming the surface of the present alluvial fan. A recessional moraine now blocks the Satpara Valley and impounds the Satpara Lake. The morainic barrier was probably breached in the past, the consequent catastrophic flooding channelling these Skardu terraces and depositing an admixture of sediments within the channels as seen in the terraces and depositing an admixture of sediments within the channels as seen in the terraces SE of Kepchun. Similar events have been described from the Khumbu Himalaya (Vuichard and Zimmermann, 1987).

4 CONCLUSIONS

The terraces around the town of Skardu provide good examples of the mode of deformation and geometry produced by glaciotectonic disturbance of sedimentary sequences in an intermontane basin. Frost-heave structures are also present in these deformed sediments in the region around Skardu town; these formed in frozen ground before the advance of glacial ice from the Satpara Valley. The glaciotectonic structures are dominated by thrust-related geometries and sheath folds are one of the most common structures found. As till is not present in most of the sections along the Indus river, and ice-contact structures have not been recognised, an ice-front origin is favoured. At least two phases of ice advance have been recognised on the basis of two groups of glaciofluvial sediments and a structural discontinuity between the structures in each group in the Skardu region.

This model provides a comparative framework for the study of deformed sediments in other high, active mountain belts with particular relevance to the trans-Himalayan regions. It might usefully be tested in the central Andes, the Rockies and the Southern Alps of New Zealand, with special referecne to the distinction between glaciotectonic and neotectonic processes.

ACKNOWLEDGEMENTS

E. Derbyshire and D. Croot for useful and constructive comments on the manuscript and S. Haywood for typing the document. This research was funded by the Violet Cressey Marcks Fisher Travelling Scholarship, awardeded by the Royal Geographical Society.

REFERENCES

Agrawal, D.P. 1984. Palaeoclimatic studies in Kashmir: a summary. East Asian Tert. Quat. Neslett., 1, 17-21.

Banham, P.H. 1975. Glaciotectonic structures - a general discussion with particular reference to the contorted drift of Norfolk. In: Wright, A.E. & Moseley, F. (ed.), Ice Ages - ancient and modern. Geol.Jour.Spec. Issue, 6, 69-94.

Gerggren, W.A. & van Couvering, J. 1974. The Late Neogene: Biostratigraphy, geochronology and palaeoclimatology of the last 15 million years in marine and continental sequences. Palaeogeog. Palaeoclim. & Paleoecol., 16, 1-216.

Brindley, G.W. and Brown, G. 1980. Crystal structures of clay minerals and their x-ray identification. Mineralogical Society, London. Spottiswoode Ballantyne Ltd., 495pp.

Boulton, G.S. 1970. The deposition of subglacial and meltout tills at the margins of certain Svalbard glaciers, J.Glaciol. 9, 56, 231-45.

Boulton, G.S. 1971. Till genesis and fabric in Svalbard, Spitzbergen. In: Goldthwait, R.P. (ed.), Till, a symposium. Ohio State Univ. Press, 41-72.

Boulton, G.S. 1986. Push-moraines and glacier-contact fans in marine and terrestrial environments. Sed., 33, 5, 677-699.

Boulton, G.S. and Jones, A.S. 1979. Stability of temperate ice caps and ice sheets resting on beds of deformable sediment. Journ. Glaciol., 24 (90), 29-43.

Brown, A.G. 1985. Traditional and multivariate techniques in the interpretation of floodplain sediment grain size variations. Earth Surf.Proc. & Land., 10, 281-291.

Burbank, D.W. 1982. The chronologic and stratigraphic evolution of the Kashmir and Peshawar intermontane basins, northwestern Himalaya. Ph.D. thesis, Dartmouth college, N.Hamp., 291pp.

Burgisser, H.M., Gansser, A. and Pika, J. 1982. Late glacial lake sediments of the Indus Valley area, northwestern Himalayas, Ecol.Geol.Helv., 76, 51-63.

Conway, W.M. 1894. Climbing and exploration in the Karakoram Himalaya. London, T. Fisher Unwin, 709pp.

Cronin, V.S. 1982. Physical and magnetic polarity stratigraphy of the Skardu Basin, Baltistan, northern Pakistan. Unpublished M.A. thesis, Dartmouth College, N.Hamp., 226pp.

Cronin, V.S. 1988. Structural setting of the Skardu intermontane basin, Karakoram Himalaya, Pakistan. In: Malinconico, L.L. & Lillie, R.J. (ed.), Tectonics and Geophysics of the Western Himalayas, Geol.Soc.Am.Spec. paper.

Cronin, V.S. & Johnson, W.P. 1988. Chronostratigraphy of the Late Cenozoic Bunthang sequence and possible mechanism controlling base level in Skardu intermontane basin, Karakoram Himalaya, Pakistan. In: Malinconico, L.L. & Lillie R.J. (ed.), Tectonics and Geophysics of the Western Himalayas. Geol.Soc.Am.Spec. paper.

Croot, D.G. 1987. Glacio-tectonic structures: a mesoscale model of thin-skinned thrust sheets? Journ.Struc.Geol., 9, 7, 797-808.

Dainelli, G. 1922. Studi sul glaciale: Spedizone Italiana de Filippi nell' Himalaia, Caracorum e Turchestan/Cinese (1913-1914), Ser.II, 3, 658pp.

Derbyshire, E., Li Jijun, Perrott, F.A., Xu Shuying and Waters, R.S. 1984. Quaternary glacial history of the Hunza Valley, Karakoram Mountains, Pakistan. In: Miller, K.J. (ed.), International Karakoram Project volume 2, pp.456-495. Cambridge: Cambridge Univ. Press.

Dredge, L.A. and Grant D.R. 1986. Glacial deformation of bedrock and sediment, Magdalen Islands and Nova Scotia, Canada: Evidence for a regional grounded ice sheet. In: van der Meer, J.J.M. (ed.) Tills and Glaciotectonics, pp.183-197. Rotterdam: A.A.Balkema.

Drew, F. 1873. Alluvial and lacustrine deposits and glacial records of the Upper Indus basin. Quat.Jour.Geol.Soc. Lond., 29, 441-71.

Ferguson, R.I. 1984. Sediment load of the Hunza River. In: Miller, K.J. (ed.), International Karakoram Project volume 2, Cambridge: Cambridge Univ. Press.

Ferguson, R.I., Collins, D.N. & Whalley, W. B., 1984. Techniques for investigating meltwater runoff and erosion. In: Miller, K.J. (ed.), International Karakoram Project volume 2, pp456-495. Cambridge: Cambridge Univ. Press.

de Filippi, F. 1922. Karakoram and Western Tibert, 1909, London.

Folk, R.L. and Ward, W.C. 1956. A study in the significance of grainsize parameters, J.SEd.Pet., 27, 3-26.

Godwin-Austin, H.H. 1864. Geological notes of part of the northwestern Himalayas, Quat.Jour.Geol.Soc., 20, 383-387.

Harrison, P.W. 1957. A clay-till fabric - its character and origin, J.Geol., 65, 275-308.

Heller, F. & Lui Tungsheng. 1984. Magnetism of Chinese loess deposits (abstr.), Geophys. J.R., 77, 125-141.

Hewitt, K. 1967. Ice front sedimentation

and the seasonal effect: a Himalayan example. Trans.Inst. British Geographers, Publ.42, 93-106.

Johnson, W.P. 1986. The Physical and Magnetic Polarity Stratigraphy of the Bunthang Sequence, Skardu Basin, northern Pakistan, Unpublished M.Sc. thesis, Dartmouth College, N. Hamp.

Kusumger, S. 1980. Geochronology of the palaeoclimatic events of Late Cenozoic Period in the Kashmir Valley, Unpublished Ph.D. thesis, University of Bombay.

Li Jijun, Derbyshire, E. and Xu Shying, 1984. Glacial and paraglacial sediments of the Hunza valley north-west Pakistan: A preliminary analysis. In: Miller, K.J. (ed.), International Karakoram Project Volume Two, 496-535. Cambridge: Cambridge University Press.

Li Jijun, Wen Shixuan, Zhang Qingseng, Wang Fubao, Zhang Benxing and Li Bingyuan, 1979. A discussion on the period, amphitude and type of the uplift of the Qinghai-Xizang Plateau, Scientia Sinica, 22, 11, 1314-1327.

Lydekker, R. 1883. Geology of the Cashmir and Chamba Territories and British District of Khagar: Mem.Geol.Surv. India, 22, 186pp.

Moran, S.R. 1971. Glaciotectonic structures in drift. In: Goldthwait, R.P. (ed.), Till - a symposium. Columbus: Ohio State Univ. Press, 127-148.

Moran, S.R., Clayton, R., Hooke, LeB., Fenton, M.M. & Andriashek, 1980. Glacier-bed landforms of the prairie region of North-America. Journ.Glaciol. 27 (95), 457-476.

Morley, C.K. 1986. A classification of thrust fronts. Amer.Assoc.Pet.Geol.Bull., 70, 12-25.

Muller, A. 1947. Cited in Banham, 1975.

Ostreich, K. 1906. Die taler des nordwestlichen Himalaya. Petermann's Mittheilungen, Erganzungsheft, vol.155.

Owen, L.A. 1988a. Neotectonics and glacial deformation in the Karakoram Mountains, Western Himalaya. Tectonophysics Spec. Pub. "Neotectonics". in press.

Owen, L.A. 1988b. Terraces, uplift and climate, the Karakoram Mountains, northern Pakistan. Unpublished Ph.D. thesis, University of Leicester.

Virkkala, K. 1952. On the bed structure of till in eastern Finland. Bull.Com.Geol. Fin., 157, 97-109.

Vuichard, D. & Zimmermann, M. 1987. The 1985 catastrophic drainage of a moraine-dammed lake, Khumbu Himal, Nepal: cause and consequences. Mountain research and Development, 7, 2, 91-110.

Wateren, F.M. van der, 1985. A model of glacial tectonics, applied to the ice-pushed ridges in the Central Netherlands. Bull.Geol.Soc. Denmark 34: 55-74.

Glaciotectonics: Forms and Processes, Croot (ed.), © 1988 Balkema, Rotterdam. ISBN 90 6191 848 0

Glacially deformed diamictons in the Karakoram Mountains, northern Pakistan

Lewis A.Owen & Edward Derbyshire
Department of Geography, University of Leicester, Leicester, UK

ABSTRACT: The deformation of glacial and paraglacial sediments takes many forms and occurs at several scales owing to the very variable, high-energy glacial system and the large number of juxaposed sedimentary environments. The Karakoram glaciers and adjacent Pleistocene and Recent glacial sediments provide good examples of such variability. Two main types of glacial sedimentary environments are present:

1. Ghulkin type, consisting of ice-contact fan sedimentation dominated by glaciofluvial, debris flow and slide processes;
2. Pasu type, characterised by roches moutonnées hummocky moraines and glaciofluvial outwash plains dominated by processes of subglacial lodgement and meltout.

Glacial diamictons display structures and fabrics which arise from ice-push and subglacial shear. However, many of the structures within these diamictons cannot be explained as the result of ice-push. Rather, they relate to resedimentation by debris-slide and debris flow processes. Field study aided by laboratory work involving both optical and scanning electron microscopy of the microfabrics can help resolve the question of the mode of deposition or emplacement of these diamictons.

1 INTRODUCTION

The mountain glacial sedimentary environment is complex, being marked by subglacial and supraglacial meltout, lodgement, debris flow, sliding, glaciofluvial and glaciotectonic processes. Sediments deposited in such an environment are commonly coarse-grained and poorly sorted; they have a bimodal or multimodal grain size distribution and may be referred to as dimicts or diamictons, and as diamictites when lithified (Flint et al. 1960). Their sedimentary characteristics including texture, grain shape, stratification, pebble fabric, microfabric, mesostructures and macrostructures show differing degrees of variability. These are a function of the petrology, transport, deposition and post-depositional modification or resedimentation of the material. Many of these diamictons have structures such as folds, faults and irregular discontinuities that may be loosely referred to as "deformation structures". These are commonly attributed to glaciotectonic processes as the result of push and thrusting related to the movement of glacial ice over or adjacent to the sediment, such that the structures verge in the ice movement direction. However, close examination of many of the deformed diamictons suggests that such strcutures, although initially produced by deformation, are later modified by slope processes to such an extent that the slide-produced structures are often dominant. Deformation structures are present on all scales, examples having been observed from a few mm to several hundred metres in extent. In order to understand how they form, it is necessary to understand the glacial system giving rise to them.

Examples from the western end of the trans-Himalayan mountain belt are used here to demonstrate the variability of these diamicton structures and to develop a model of their formation based on the Karakoram mountain glacial depositional system.

2 KARAKORAM GLACIERS

The Karakoram Mountains contain some of the longest glaciers outside the polar regions (Fig. 1). These are glaciologically complex and of high activity type and can be

Figure 1. Map of the study area in northern Pakistan, showing the main glaciers (in solid black), rivers, mountains and settlements.

classified as alpine. They have high altitude source areas (above 4,500m) with mean annual precipitation totals in excess of 2000mm, while their snouts extend down to semi-arid valley floors (c. 2,700m) where precipitation is less than 100-200m/year, and summer temperatures frequently exceed 25°C (Zhang Xiangsong et al. 1980 and Goudie et al 1984). The glaciers are several hundred metres thick and are among the steepest in the world (e.g. Hasanabad glacier, 7.5°; Ghulkin glacier, 8.1°; and Pasu glacier 6.0°). A combination of high ablation rates in their lower reaches (e.g. 1841mm/year on bare ice, Batura glacier snout, and an ablation season lasting 315 days; Shi and Zhang, 1984) induces high mass flux values. Given their thickness, steepness and high activity indices, these glaciers have flow rates of between 100 and 1,000m/year; (e.g. Batura Glacier 517m/year, Batura Glacier Investigation Group, 1979; 1000m/year Shi and Wang, 1980; Pasu Glacier 157m/year; Pillewizer, 1957). Glacial surging has been recognised from

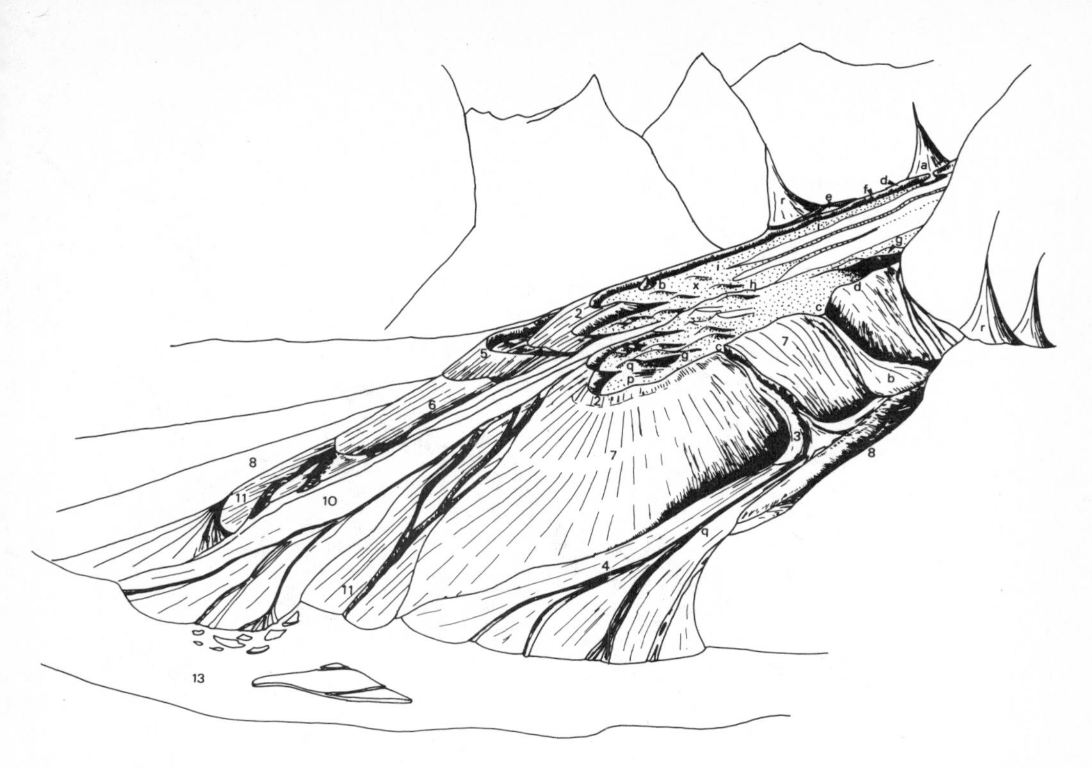

Figure 2. Landforms and sediment associations of ice-contact facies based on the Ghulkin Glacier.
a. debris flow; b. flowslide; c. gullied lateral moraine; d. lateral moraine; e. ablation valley lake; f. ablation valley; g. supraglacial lake; h. supraglacial stream; i. ice-contact terrace; j. lateral lodgement till moraine; k. roche moutonnée; l. fluted moraines; m. diffluence col; n. high level till remnant; o. diffluence col lake; p. fines washed from supraglacial debris; q. ice cored moraines; r. screes; s. hummocky moraines; t. dead ice; u. subglacial stream; w. kettle hole lakes; x. supraglacial debris; y. outwash fan; z. ancient roche moutonnée; 1. truncated scree; 2. latero-terminal dump moraine; 3. outwash channel drained laterally; 4. glaciofluvial outwash fan; 5. slide moraine; 6. slide-debris flow cones; 7. slide modified lateral moraine; 8. abandoned lateral outwash fan; 9. meltwater channel; 10. meltwater fan; 11. abandoned meltwater fan; 12. bare ice areas; 13. trunk valley river.

historical documentation (Hewitt, 1969) but does not appear to be a frequent and consistent feature of the Karakoram glaciers.

The glacier ice is of cold metamorphic type, only the lower reaches being isothermal with behaviour characteristic of temperate glaciers. As a result of the great thickness of most glaciers, however, basal sliding is important producing polished and moutonnée bedrock surfaces. The high sliding velocities and isothermal ice produced abundant englacial and subglacial meltwaters which leave the glacier by way of subglacial tunnels. The positions of these exits migrate owing to thermal erosion, ice-collapse and subglacial debris removal resulting in the shift of proglacial streams (Zhang Xiangsong et al. 1980). The steep unstable valley sides made up both of rock and other glacial deposits provide large volumes of debris, often mixed with snow which avalanches on to the glacier surface. Debris falling on to the glacier above the equilibrium line becomes interbedded with ice and enteres the net downward flowlines to be transported towards the zone of

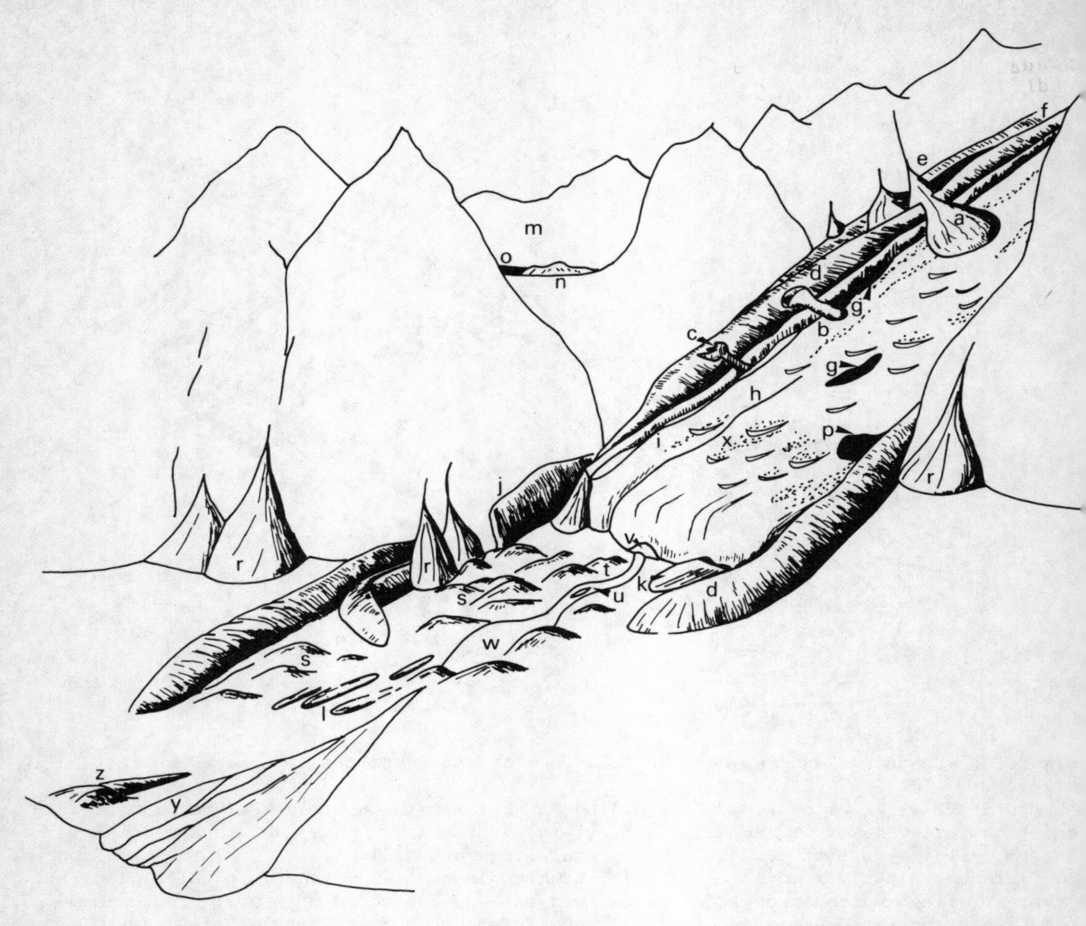

Figure 3. Perspective sketch of sediments and landforms near the Pasu Glacier snout, Captions as in figure 2.

traction, and similar material falling into crevasses may eventually be carried downwards. However, the steepness of these glaciers and the substantial supply of rock waste from valley sides to the glacier surface below the equilibrium line means that much of the englacial material may never enter the active traction zone. The maintenance of basal ice at pressure melting point in the lower reaches of the glacier results in a lodgement till, usually less than 2m thick, frequently being plastered against bedrock knolls. Basal sliding is demonstrated by the presence of fluted moraine and small push moraines. Supraglacial and englacial moraines dominate and are usually a composite of meltout, slide and push processes. The high proportion of debris on the surface of the glacier effectively insulates the ice and so reduces the rate of ablation. However, much surface water is present forming supraglacial streams or ponds on the ice surface.

The glaciers can be broadly divided into two main types on the basis of the sediment landform associations in the lower reaches and on the proglacial plains. These are shown in Figures 2 and 3, and are referred to as the Ghulkin and Pasu types.

2.1 Ghulkin type

The Ghulkin Glacier provides the best example of this type of glacier (Fig. 2 and 4). It is characterised by a cone of sediments surmounting its snout, comprising till glaciofluvial, and debris flow sediments. This may be referred to as an end moraine, although definitions can be vague (Gripp 1938, Flint 1971 and Embleton and King 1975), and Sugden and John (1982) suggest that when glaciofluvial processes are important in modifying end moraines, the term "kame moraine" is more appropriate. However, debris-flow and sliding processes are very important in the formation of these cones in the Karakoram, and it may be more accurate to refer to such landforms as ice-contact fans or cones (Owen, 1988c).

Ice-contact fans may reach several hundreds of metres in height, the present ice levels often being several tens of metres below the crest of the fan owing to recent ice wastage (Goudie et al. 1984). The dominant depositional process is the meltout of thick deposits of supraglacial till which forms an extensive and thick veneer over most of the lower reaches of the glacier. Individual moraines are not present in this region and the distribution of till is chaotic due to widespread seracs and crevassing resulting from glacial tension and shear. Although these thick mantles of supraglacial till insulate the ice surface, large discharges of englacial meltwater from the clean ice exposed by thrusting occur during the summer half of the year (Goudie et al. 1984 and Ferguson 1984; Fig. 5). This may be discharged along with meltwater exiting from englacial meltwater caves via meltwater channels with steep floors (15-18°), accompanied by frequent shifting of the position of the englacial meltwater exits, leaving ancient abandoned channel courses and creating new ones (Zhang Xiangsong et al. 1980). This may also produce episodic meltwater fans, consisting of bouldery gravels with slopes of up to 25°. Jones et al. (1983) described examples of these forming off the snout of the Ghulkin glacier.

The large quantities of water present and the melting of ice cores on the margins of these steep slopes produces large debris

Figure 4. Oblique aerial view of the Ghulkin Glacier showing the ice-contact fan.

Figure 5. Oblique aerial view looking up the Ghulkin glacier. Note the large quantities of supraglacial seiment compared to the relatively debris free ice (dark patches) exposed in the foreground. Note also the sliding of debris which is one of the most dominant processes in this environment.

flows and till slides. These processes give rise to massive, bouldery diamictons, with thick bedding (one to tens of metres) and depositional slopes as steep as 25°. Impressive examples can be seen near the snouts of the Pasu, Minapin, Ghulkin and Ghulmit glaciers (Fig. 6). A common feature is the progressive destruction or modification of older ice-contact fans by fluvial processes or by resedimentation as debris flow and slide processes take over upon glacial retreat.

2.2 Pasu type

Ice-contact fans are not present adjacent to this glacier type (e.g. Pasu, Rakhiot, Buldar, Hinarche and Mani Glaciers). Rather, these glaciers are characterised by hummocky moraines and a glaciofluvial outwash plain (Fig. 3 and 7). The Pasu, as the type example, is characterised by small end moraines consisting essentially of till in the form of hummocks and small parallel ridges. Thin (<2m thick subglacial lodgement tills are present and form on the lee sides of bedrock knolls, or roche moutonnées. These tills are generally overconsolidated and show well-developed shear systems and strongly orientated pebble fabrics.

Supraglacial debris is less common on the Pasu glacier than on those glaciers characterised by contact fans. It thus appears that the development of large ice-contact fans of the Karakoram type is dependent on the presence of large amounts of supraglacial debris.

3 DEFORMATION STRUCTURES IN DIAMICTONS PRODUCED BY SUBGLACIAL SHEARING AND PUSH

Deformation structures may occur in glacial, paraglacial and proglacial sediments. The glaciotectonic deformation of proglacial sediments has been described by Owen (1988a) in the Skardu Basin, Baltistan. These structures comprise thrusts and folds and at some locations, up to five duplexes of stacked sediments with sheath-fold geometries occur in essentially fine-grained sediments. Structures are present in similar sediments throughout the Karakoram Mountains and relate to the movement of glacial ice over or adjacent to the sediment (Owen 1988a). However, we are concerned here with describing deformation structures in diamictons of glacial and paraglacial origin. The deformation of these sediments is controlled by the many

Figure 6. Sediments of ice-contact fan facies near the Ghulkin Glacier.

processes present within the glacial environment and the structures produced vary from microscopic scale to several tens of metres in length.

3.1 Macrostructures

Large-scale thrusting is very common and displacements of over one hundred metres can be seen in the Karakoram valleys. Figure 8 shows an example of a till which overthrusts glaciofluvial sediments and verges down the Indus valley. Owen (1988b and c) attributed this till to the main Indus valley ice; it is overconsolidated and may be referred to as a tillite. Folding is commonly associated with thrusting, but it is often difficult to identify in the field in an essentially homogeneous sediment. Usually it is only when other sediments are present as intercalations that folding can be readily identified. Figure 9 shows an exception to this general statement in the form of a major sheath fold produced by the Indus valley ice movement, the structure being picked out by differential erosion.

An example of a number of impressive sections of glacially deformed diamictons and associated sediments to be seen in this region is one on the north side of the Skardu Basin. This was considered by Cronin and Johnson (1988) in their palaeomagnetic study, but they did not describe any of the evidence of glaciotectonics. Figure 10 shows a field sketch of the section. At the base of the exposure is a folded basal till interbedded with contorted silts. This is overlain by lenticular beds of sands and bouldery gravels intercalated with planar bedded silts. Tills are thrust over these and form an imbricated thrust stack. The thrust planes dip towards the NE and are lined with dm-thick layers of silt which facilitate their recognition. The thrust planes can be traced for several tens of metres to a well-defined basal décollement. The thrust planes are spaced between 10 and 15m apart. Unconformably onlapping the tills are dm- to m-thick planar bedded sands and bouldery gravels. These dip 5° SW (downvalley) and are several hundred metres thick in total. The sands and gravels have been cut by southward-trending channels near the top of the section and then infilled with similar materials. These are overlain by more than 50m of planar bedded, laminated silts. The terrace is capped by deposits of bouldery diamictons which form a fan-like deposit radiating from the tributary valley which is tangential to the trend of the section.

Figure 11 shows a series of simplified schematic sections illustrating the

Figure 7. View of the snouth of the Pasu Glacier, with hummocky moraines (middle ground) and a fluted moraine (foreground), August 1980.

the sedimentary environment.

A further ice advance produced a series of push moraines which were thrust over the glaciofluvial sediments but did not deform them (Fig. 11C and D). The silts that line the thrust planes probably represent ice marginal ponds formed within the till hummocks before they were deformed. With subsequent glacial retreat, additional glaciofluvial sediments were deposited, onlapping the second suite of tills (Fig. 11E). Ice was probably present in the side valleys adjacent to the Bunthang Sequence and, with further climatic amelioration, it began to melt and retreat. This resulted in increased meltwater discharges from the side valleys and produced southward-trending channels which cut into the glaciofluvial sediments in the trunk valley. These channels were subsequently filled by sediment derived from the melting tributary tributary glaciers (Fig. 11F). Later advances of tributary valley glaciers may have blocked the trunk valley locally, allowing lakes and ponds to form in which lacustrine silts were deposited (Fig. 11G). Above these, ice-contact fans developed by the resedimentation of till off the snouts of the tributary glaciers (Fig. 11H). It appears that, as the tributary valley glaciers diminished and became encased within high valley-side rock walls, so the proportion of supraglacial debris increased and there was a change in the glacier-sediment association, i.e. a shift from Pasu to Ghulkin type.

development of the Bunthang terraces. The sediments were deposited within a palaeovalley which was probably deepened by ice which extended down the Shigar valley and into the Skardu Basin. The basal till was deposited and subsequently deformed along with the adjacent proglacial sediments owing to fluctuations in the glacier's position as it advanced down the Bunthang palaeovalley. These sediments are considered to be a push moraine (Fig. 11A). Glaciofluvial sediments were then deposited over the proximal flanks of the moraine as the glacier retreated (Fig. 11B). It is likely that the glacier was of the Pasu type during this stage as there is little evidence for extensive ice-contact sediment, a large outwash plain dominating

3.2 Mesostructures

These include thrusts, normal faults, folds and joints. Thrusts are present as discrete planes or anastomosing systems of shears with an alignment of clays and pebbles along the zone of deformation (Fig. 12). Folding may accompany the thrusting along with associated normal faulting, in zones of extension on the fold nose. The thrust planes or shear

Figure 8. A sequence of glacially-thrust glaciofluvial sands and gravels in the middle Indus valley. The dark bouldery unit is a tillite. The petrol tanker near the right edge of the photograph indicates scale. Ice movement was right to left.

Figure 9. Glacially-folded till with a sheathfold geometry in the middle Indus valley. The length of the exposure is about 120m.

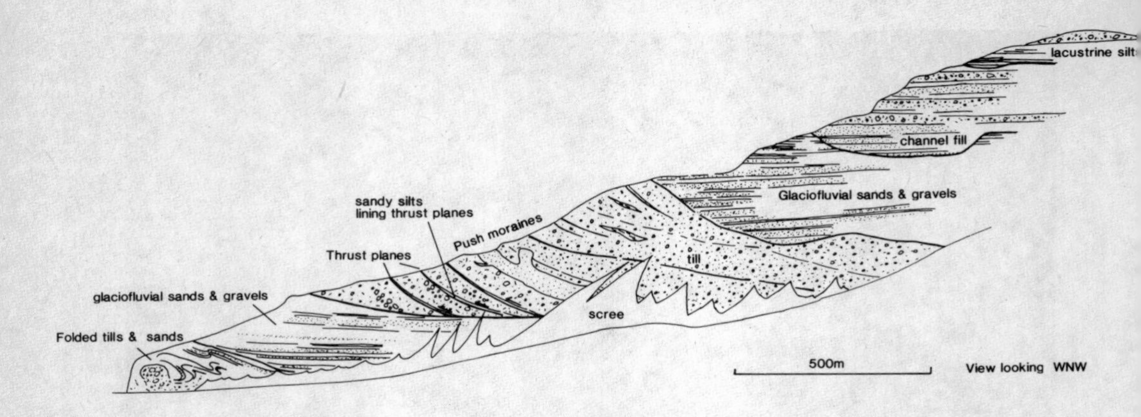

Figure 10. Field sketch of the Bunthang terraces.

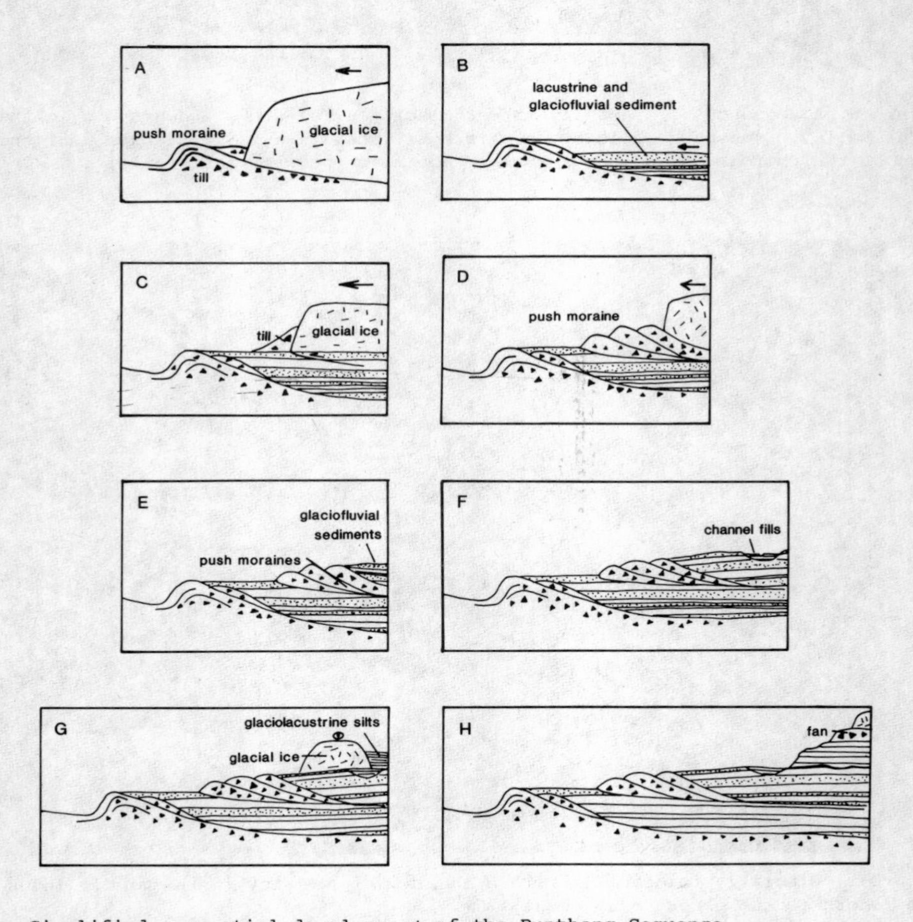

Figure 11. Simplified sequential development of the Bunthang Sequence.

Figure 12. Glacially sheared till. Strain hardening along the shear makes it more resistant to weathering and gives it a positive relief.

planes may be slickensided, and strain hardening often strengthens the zones so that it has a higher relief than the weathered till adjacent to it.

Several varieties of joints are recognised. Subhorizontal joints with openings from 1 to 40mm wide, for example, may be traced for several tens of metres to a few mm (Fig. 13: Owen 1988c). They frequently have a weak subhorizontal fissility developed parallel to the joint plane and usually occur in overconsolidated tills. Such features result from ice loading which frequently overconsolidates a till, and the process may involve re-orientation of platy minerals with their a-b planes disposed normal to the stress direction i.e. horizontally aligned. Unloading releases the stress and the dilation occurs producing the joints. Subvertical joints also occur with openings varying from 1 to 100mm with decimetre to metre spacing. Although these, too, can be seen in overconsolidated tills, no fissility is developed parallel to the joint surface. They may be primary features caused by the vertical stresses induced by an ice overburden, this process being aided by hydraulic fracturing similar to that described in the experimental work of Price (1981). Some of these joints may also be produced by pressure release in the vertical plane, as the till is excavated by river erosion or fails in mass movements. In some heavily overconsolidated tills, a stockwork system of irregular joints with cm spacing and mm wide openings may result from pressure release.

3.3 Microstructures

A large variety of microstructures have been observed in tills, using both optical (van der Meer 1985 and 1987, Sitler and Chapman 1955, Evenson 1970 and Ostry and Deane 1963, Derbyshire et al. 1988, and Owen 1988c) and scanning electron microscopy (Derbyshire 1978, Love and

Figure 13. Horizontal dilation joints in till in the middle Indus valley.

Derbyshire 1985, McGown and Derbyshire 1974, and Owen 1988c).

3.3.1 Stratification

Subtle grain size variations can be observed within some of the tills, and may include the following.

1. Laminations in silt and sand may result from subglacial meltwater deposition in cavities within the ice and become incorporated within the till during its deposition.
2. Grain size differentiation is usually transitional from areas dominated by large clasts to matrix-rich areas. Such contrasts in grain size occur in patches or may be aligned subhorizontally or dip steeply (45-30°), and are considered to be a result either of differential comminution or the translocation of clay minerals during shearing and dewatering.
3. Mud balls have also been described by van de Meer (1987) as soft zones of till deformed within the matrix. Two types of mud ball exist in the Karakoram tills; those forming clasts made up of material different from the matrix, and often rimmed by fine clays, and those made up of the same material as the bulk till, being irregular in shape with voids around the margins and often associated with water escape structures (Fig. 14). The first type probably represents small mud balls of till formed subglacially and added to another till during deposition, i.e. they are interclasts. The second type may also be explained in this way, but they may also represent areas of differential consolidation and dewatering.
4. Water escape structures are injections of fine matrix and may vary from mm to cm widths and represent dewatering of the till. Van der Meer (1987) described similar features from ice-cored moraines in the Alps (Fig. 15 and 16).

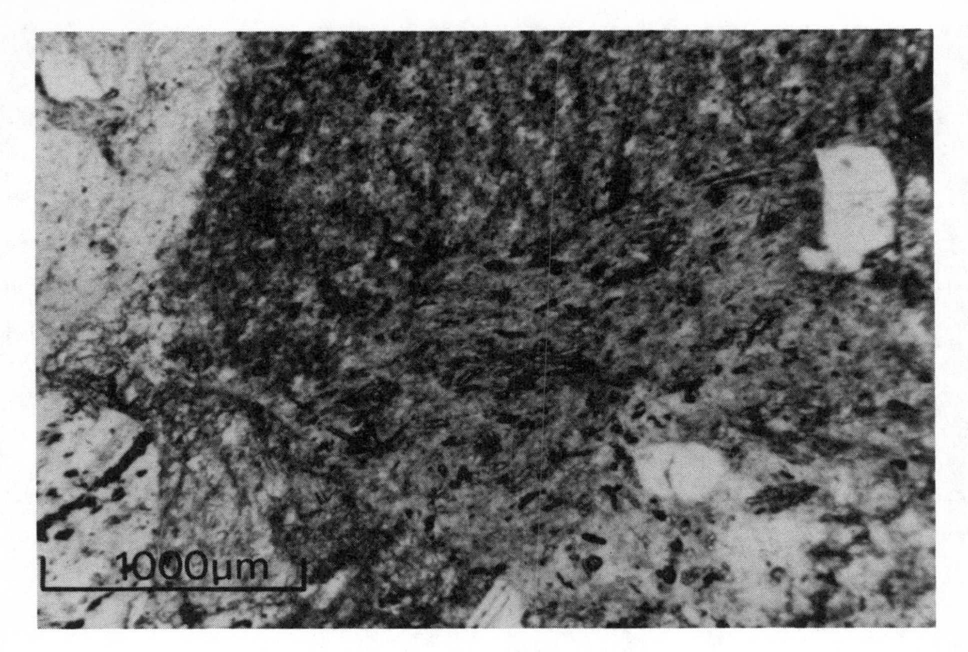

Figure 14. Photomicrograph of a thin section of subglacial till showing zones of differing matrix (plane polarised light: vertical section; top = top of page).

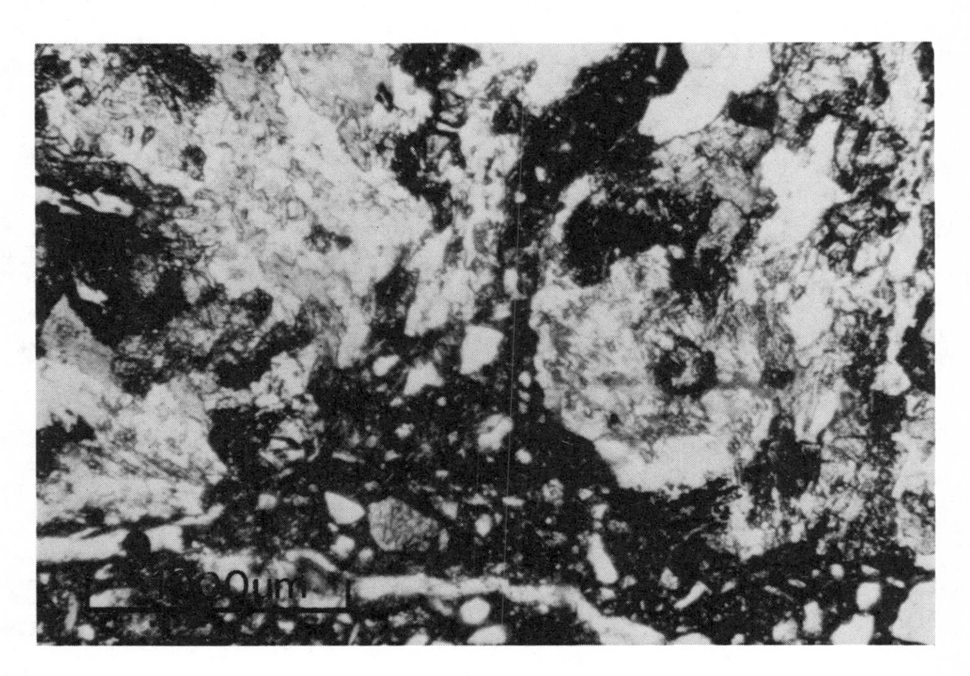

Figure 15. Photomicrograph of a thin section of subglacial till showing the squeezing of till matrix into areas of differential consolidation (plane polarised light: vertical section; top = top of page).

Figure 16. Sample showing a dyke of till with a composition different to the main till mass. Note also that the till is now overconsolidated and an irregular stockwork system of joints has developed as a result of stress release.

3.3.2. Shearing

Shear systems within tills have been recognised by many workers (e.g. Boulton 1970, Moran 1971, Derbyshire 1981, 1983, 1984a and b, van der Meer 1987, and Owen 1988c), as resulting from subglacial shearing and internal deformation within the till (Boulton 1970, 1971 and 1987) or from sliding and flowage defined by discrete zones. Shearing on the microscopic scale is represented by a strong alignment of clays dipping against the ice movement direction and frequently forming shear bands and anastomosing shear systems similar to those frequently described in the deformation of metamorphic rocks (e.g. Lister and Snoke 1980). These are most characteristic of subglacial tills deposited by Pasu type glaciers. Shearing varies from:

1. grain boundary sliding where sand and silt grains in tightly compacted zones slide over one another and may produce zones of dilation;
2. discrete shear zones with sub-parallel to parallel alignments of clays and silts and with anastomosing and bifurcating geometries (Fig. 17A & B);
3. discrete shears within larger shear zones (Fig. 18);
4. shear zones with large rotated quartz grains in a clay matrix (Fig. 19);
5. shear zones between two large grains which, in moving, have aligned the material lying between them;
6. shearing of individual grains along cleavage planes in micas (Fig. 20A) and as fractures in quartz, feldspars and lithoclasts, and in which the fractured clast may be streaked out parallel to the shear direction (Fig. 21); and
7. stacking of grains within a shear zone (Fig. 20, 22A and 22B).

Figure 23 summarises the variety of microstructures observed in tills from the Karakoram Mountains (Owen 1988b and c). This variation of shear types is not fully understood but is probably a function of the following.

1. Mineralogy, and hence grain shape and grain size distribution. The micaceous, slaty tills are frequently highly sheared with shear zones and anatomosing systems, because of the platy shape of most of the particles. In constrast, quartz- and feldspar-rich till matrices deform by grain boundary sliding and may have large zones of dilation, producing a fabric similar to that produced in sand box experiments, since the grains are generally coarse and equant in form. Limestone-rich tills show little evidence of shearing, because the grains and clasts are of the equidimensional shape. They have areas of compact and open frameworks which may relate to shear. Frequent carbonate cements may overprint shear structures.
2. Recent work (Maltman 1977 and Arch et al. 1986) has shown that water content may be the most important control on shear system geometries. However, no experimental work has yet taken into account bleeding of water from the system and the effect of multimodal grain sizes.
3. Shear and stress conditions are a function of ice overburden, pore water

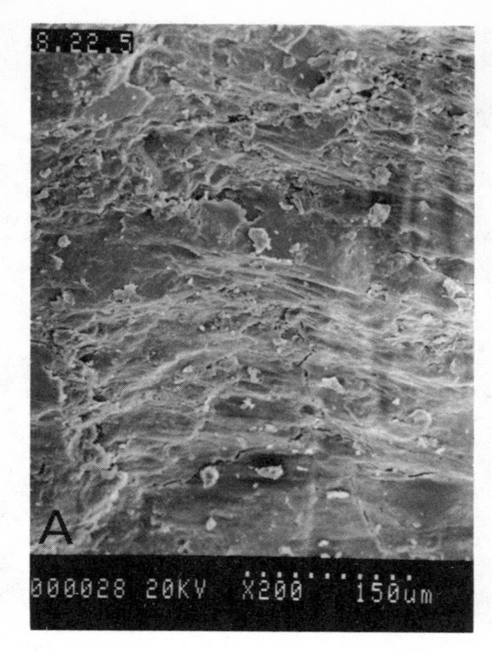

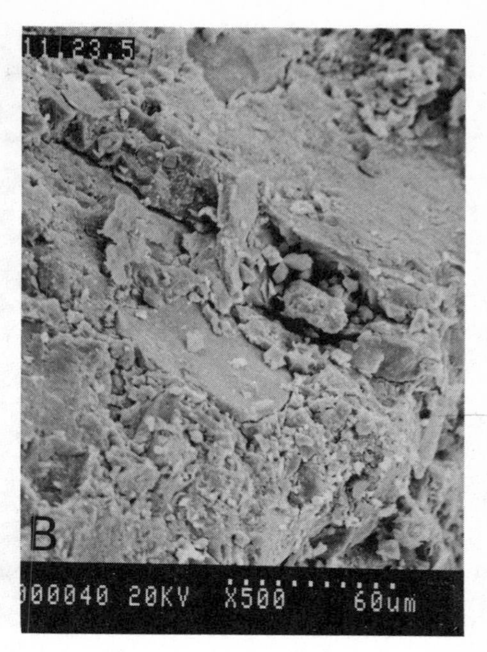

Figure 17. A) Scanning electron photomicrograph of a subglacial till showing a discrete shear zone with a tight packing of clays aligned parallel to the shear direction. The shear direction is left to right and the surface is subhorizontal. b) Scanning electron photomicrograph of vertical face of a subglacial till showing a shear surface. Ice movement direction was right to left. (Vertical sections; top = top of page.)

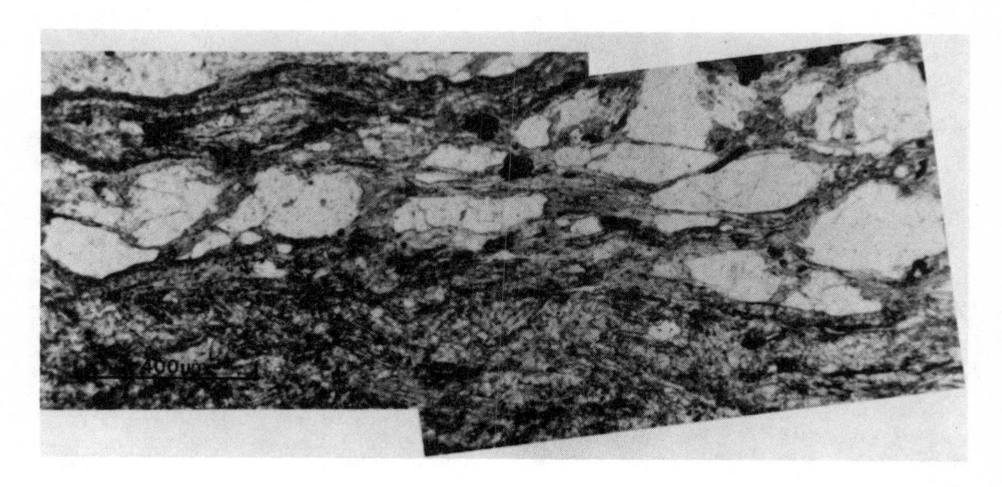

Figure 18. Photomicrograph of a thin section of a vertical face of a subglacial till showing a complex shear zone (plane polarised light: vertical section; top = top of page). Ice movement direction was right to left.

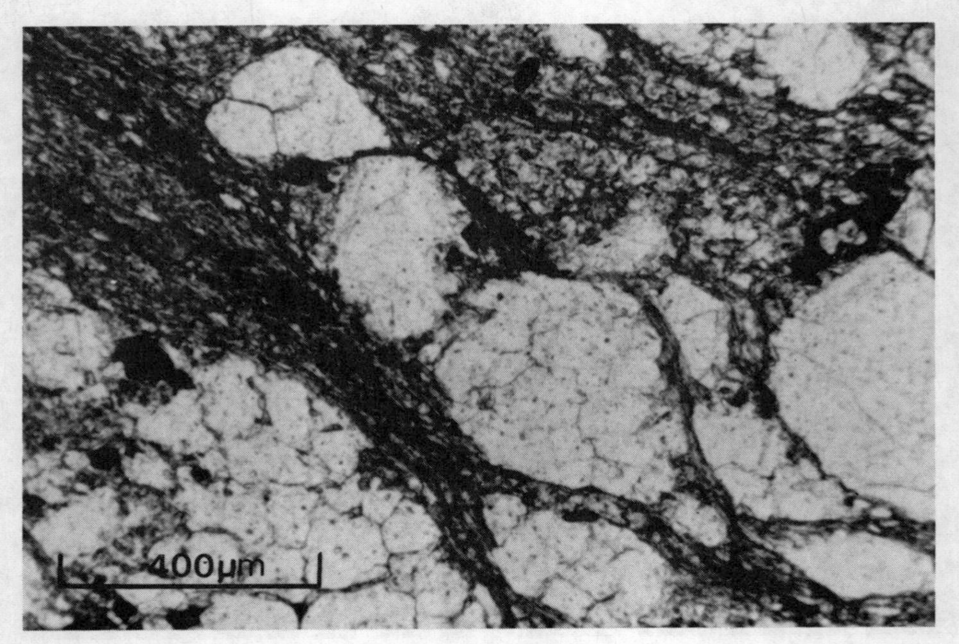

Figure 19. Photomicrograph of a thin section of a vertical face of a subglacial till showing the partial rotation of clasts within a shear zone (plane polarised light: vertical section; top = top of page).

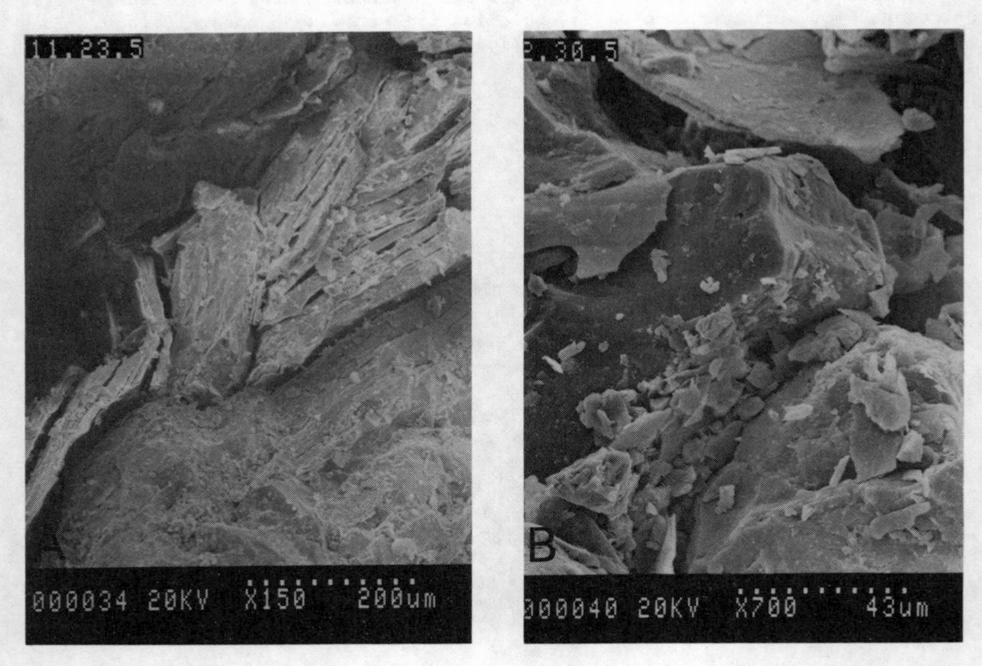

Figure 20. A) Scanning electron photomicrograph of a subglacial till showing the splitting of mica grains along their cleavage planes. B) Scanning electron photomicrograph of a subglacial till showing the grinding of grains due to subglacial shear. Vertical face, ice movement left to right.

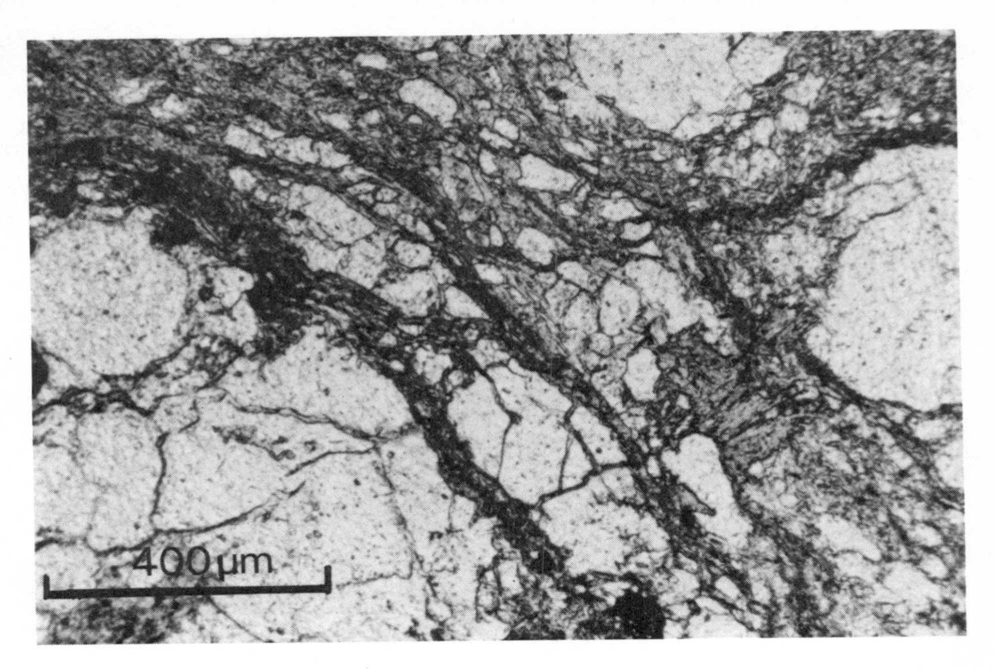

Figure 21. Photomicrograph of a thin section of a vertical face of a subglacial till showing the fracturing of clasts within a shear zone (plane polarised light, vertical sections).

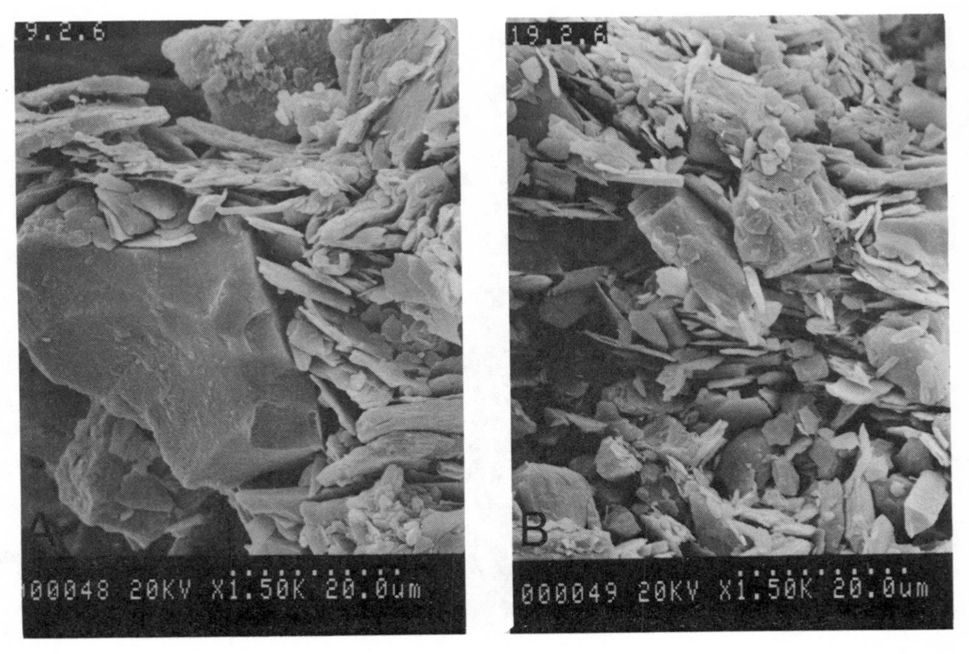

Figure 22. A & B) Scanning electron photomicrographs of a subglacial till from the end moraine at the mouth of the Bagrot valley (Vertical sections; top = top of page).

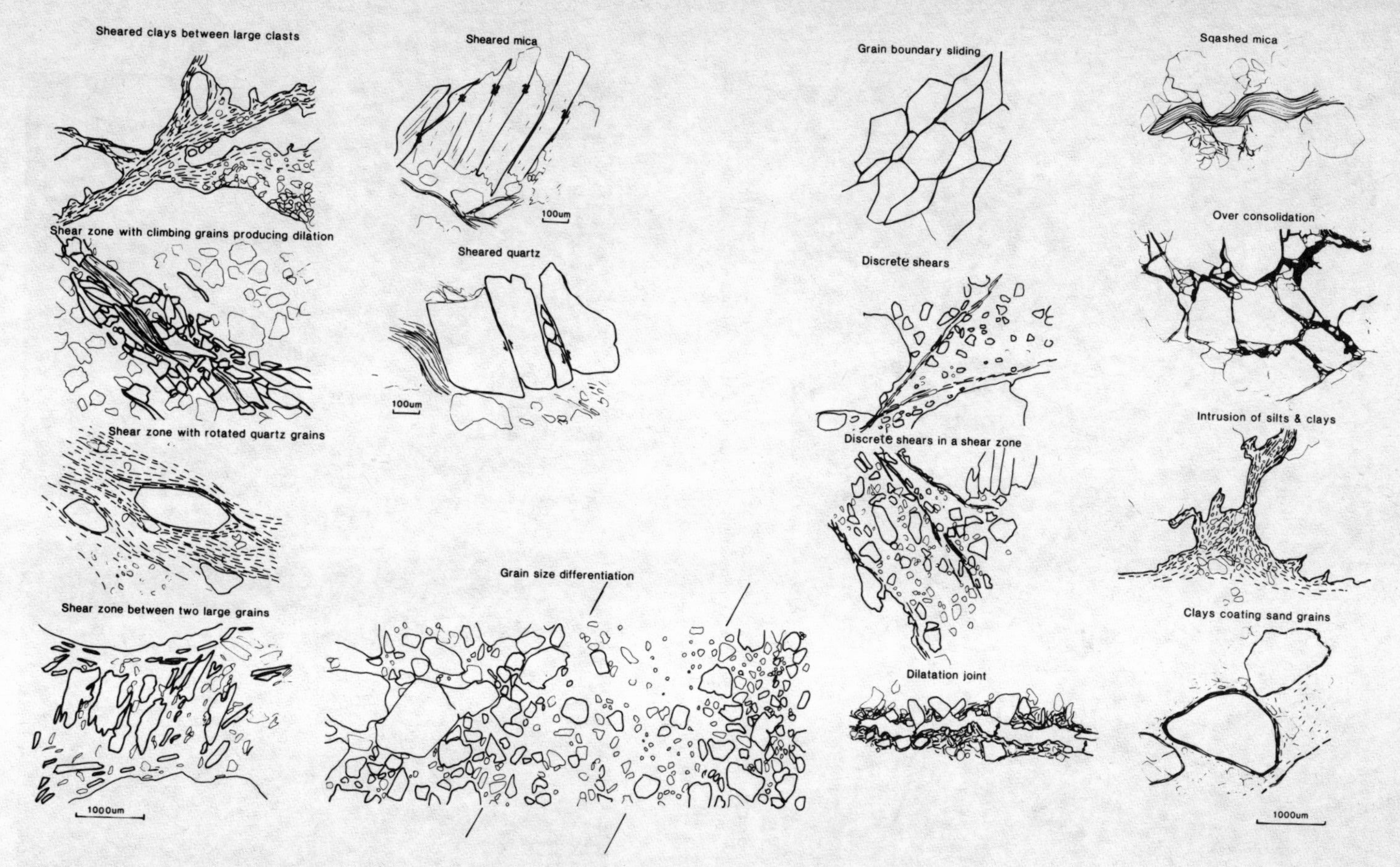

Figure 23. Diagram summarising variation in the microstructures found in subglacial tills.

pressure, the cohesion of the sediment, and local variations in the bed-ice interface. All these functions are highly variable and such variations may explain the observed complexity of matrix shear systems.

Most of the structures present in the glacial sediments result from the process of subglacial deposition. However, tills deposited from the margins of the glacier are also deformed by processes of slide and flow (cf. Lawson, 1981).

4 DEFORMATION STRUCTURES IN DIAMICTONS RESULTING FROM DEBRIS FLOW AND SLIDING PROCESSES

Sliding of glacial debris off the margins of the glaciers is an important and common process especially in the Ghulkin type glaciers which have steep fronts. Resedimentation of till may also occur after glacial retreat as till remnants are often left perched on steep slopes and so are highly vulnerable to reworking by gravity and running water in particular. Structures formed by these processes of resedimentation are often similar to structures produced by glacial push. However, when the orientation of the structures, their facies associations and their microfabrics are considered together, it becomes apparent that they are the product of slide and mass flow processes.

4.1 Resedimented tills

Good examples of resedimented tills can be seen in the Hindi Embayment (Li Jijun et al. 1984) and around the Ghulkin glacier (Fig. 6). These sediments have accumulated to thicknesses of up to 90m, making extensive fans up to 4km in diameter. These sediments are remarkably uniform, being made up of metre-thick, matrix-supported diamictons interbedded with thin waterlain sediments ranging from gravels to silts. Palaeoslopes may be as steep as 20°. The textural properties of supra-

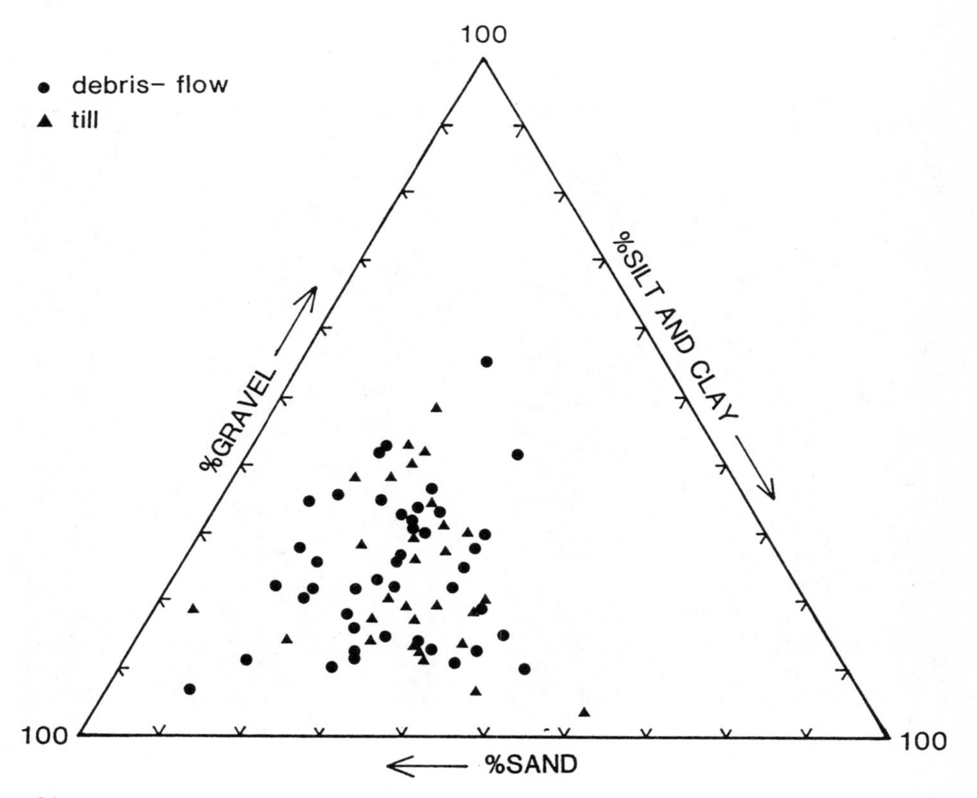

Figure 24. Ternary plot showing the similarity between supraglacial tills and debris flow sediments (many of which are resedimented tills).

glacial till are very similar to those of suprglacial till which has been reworked by slope processes, their grain size distributions being similar but somewhat finer-grained (Fig. 24). However, fabrics are more random than found in most tills, with a larger percentage of steeply-orientated clasts; fabrics are more diffuse with high voids ratios and a low micro-fabric anistropy (Fig. 25A). Bulk densities are low (Table 1), but overconsolidation associated with desiccation has affected the surface zones of many of these deposits. A translocation of clays is often apparent in some of the samples. Microshears are present usually dipping $<20^{\circ}$ and parallel to the slope direction (Fig. 25B). These deposits can be differentiated from glacial diamictons formed by glacial processes involving push, although at some sites the distinction is not so clear-cut. This is the case near the Batura Glacier and in the lower Gilgit valley. Here sliding of glacial debris has occurred but not large-scale resedimentation by debris flow and flow slide processes (Owen 1988c), and particular examples will be discussed below.

Table 1. Bulk densities of tills and tills resedimented by mass flow processes.

Lodgement till	2.45gcm^{-3}
Lodgement till	2.41
Debris slide	2.39
Debris slide	2.38
Massive till (lateral moraine)	2.22
Lodgement till	2.17
Debris flow	2.17
Debris flow	2.08
Hummocky moraine	2.08
Debris flow	2.04
Hummocky moraine	1.97

4.2 Slide structures in tills

4.2.1 Deformed tills in the Pasu-Batura Glacier area

Figure 26 is a map of the area near the Batura glacier and the village of Pasu. A number of glacial advances have been recognised in the Quaternary record, including two major advances of the Batura

Figure 25. A) Scanning electron photomicrograph of a debris flow sediment. Note the diffuse fabric and open framework, B) Scanning electron photomicrograph of vertical face of a debris flow sediment showing a discrete shear plane.

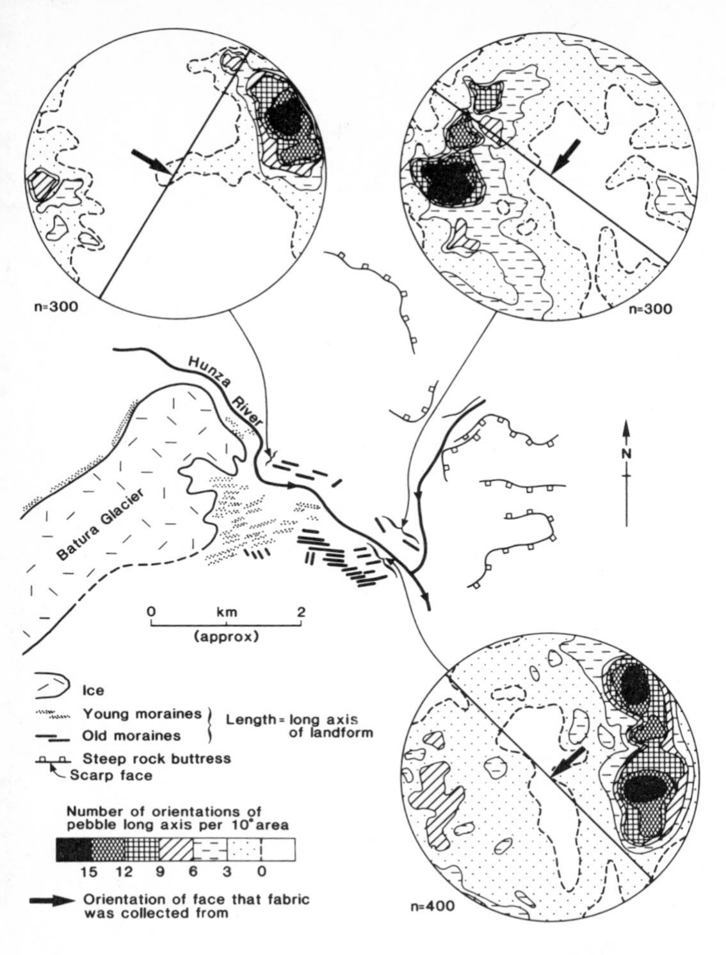

Figure 26. Map of tills near the Batura Glacier and their pebble fabrics (lower hemisphere equal area plots).

glacier in the Pleistocene (the Yunz and Borit Jheel) and, during the Holocene, an expanded foot stage (Ghulkin I) and four minor advances (Ghulkin II, Batura, Pasu I, Pasu II) as well as some recent fluctuations of the ice front (Derbyshire et al. 1984c, Li Jijun et al. 1984). The oldest glacial sediments derive from the Borit Jheel stage and are overlain by glaciofluvial and glaciolacustrine sediments which, in turn, are overlain by Ghulkin I tills. The Ghulkin I till is referred to on Figure 26 as old moraines. These form a series of inlier ridges within the sediment fans on the north side of the Hunza river and are made up of a series of subdued hummocks with very dark and weathered varnished boulders. The tills of the "young moraines" include those with strong, dark varnish (Ghulkin II stage) on surface boulders, brown varnish (Batura stage) and the yellow varnish of the Pasu stage. These form distinct, arcuate ridges which are readily correlated using morphostratigraphy and weathering criteria.

These lateral moraine accumulations of the Batura till plateau and the lateral moraines on the east side of the Hunza valley result from fluctuations in the position of the Batura ice front, a common feature in the Upper Hunza area (Batura investigation Group 1979, Shi and Zhang 1984). Boulton and Eyles (1979) describe the development of lateral and laterofrontal dump moraines by mountain glacier fluctuations. A similar mechanism was envisaged by Osborn (1978) in explanation of the lateral moraines near the Bethartoli Glacier, Garwal Himalaya. The surface morphology of the tills on the Batura till plateau north of Pasu resembles that of a series of lateral moraines resulting from the advance of the Batura glacier (Fig. 27). Listric discontinuities can be identified along the river section, which cuts through the entire deposit normal to the ridge trend. The thrusted and folded silts are consistent with ice movement directions determined from several lines of evidence, the surface expression of this glacial push being parallel series of ridges. However, detailed pebble fabrics along the discontinuities show mean orientations which are consistent with sliding rather than push, and incorporated within one moraine are several metres of silt. This has large en echelon tension gashes which are also consistent with sliding.

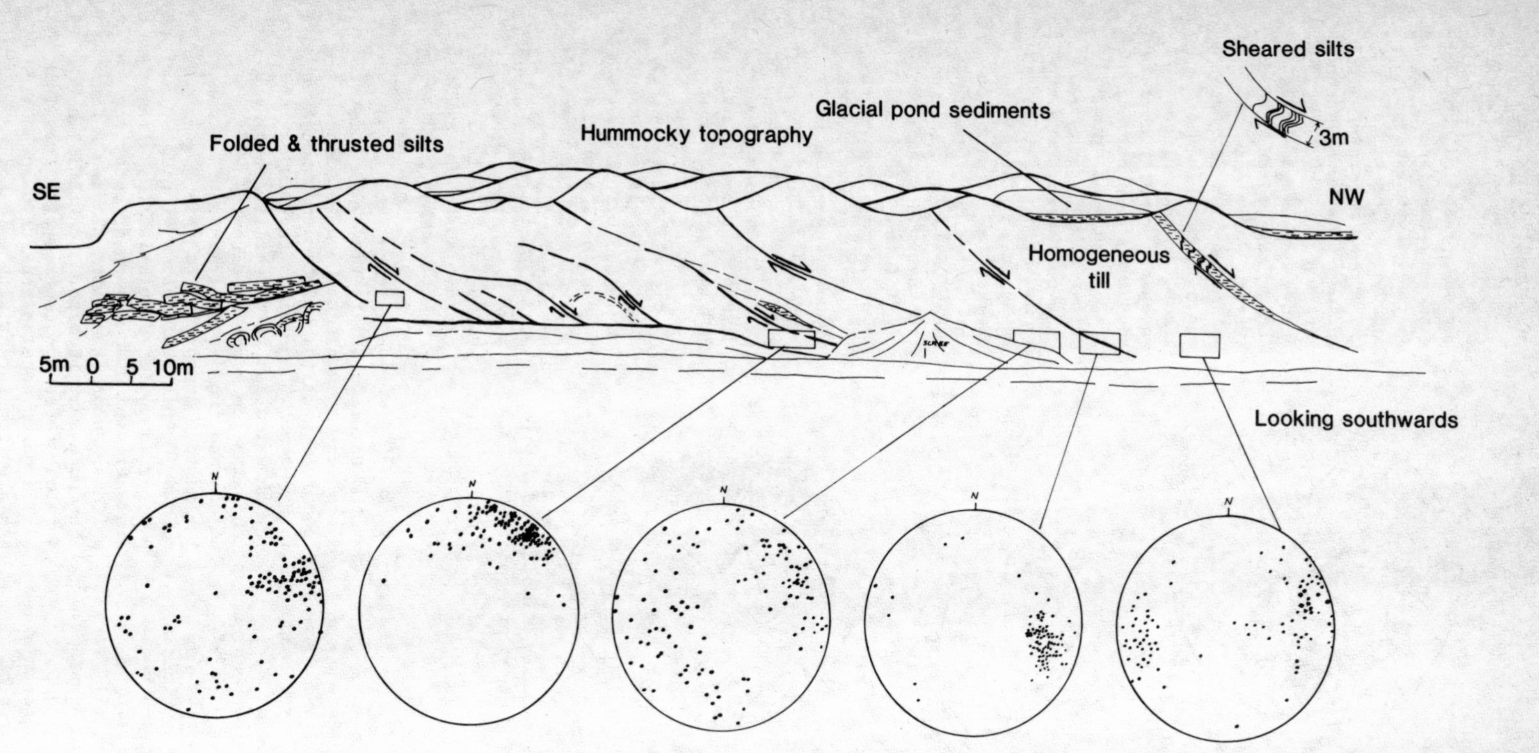

Figure 27. Section through the Batura Till Plateau, showing structural discontinuities and pebble fabrics (lower hemisphere equal area plots).

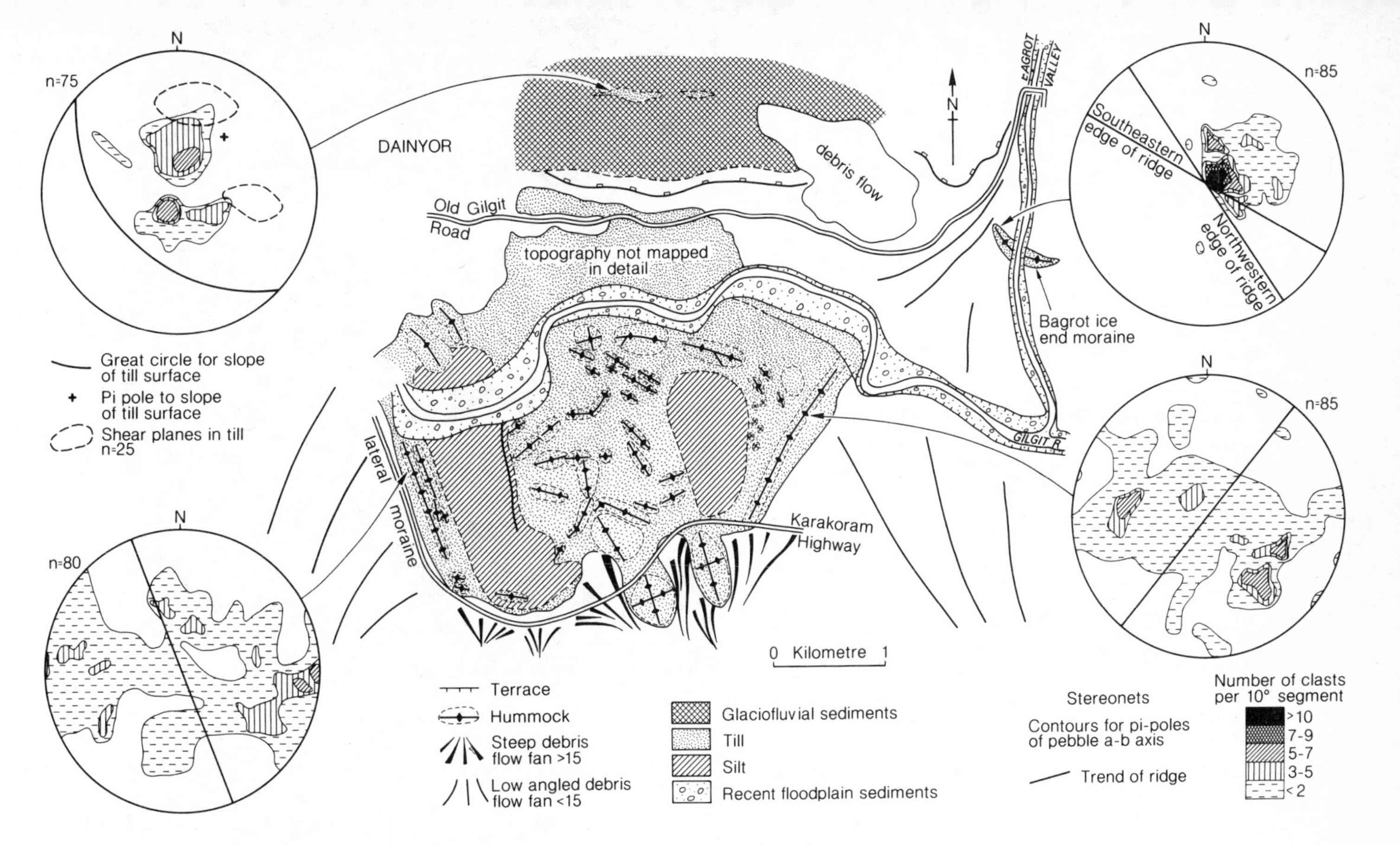

Figure 28. Dak Choki moraine in the middle Gilgit valley, showing the landforms and till pebble fabrics (lower hemisphere equal area plots).

Therefore, it seems likely that, following accumulation of moraines by glacial push, sliding often occurs along discontinuities which originated during lodgement.

4.2.2 Structures within the Dak Choki and Dainyor moraine, southeast of Gilgit

Desio and Orombelli (1971) described a series of moraines southeast of Gilgit which they referred to as the Dak Choki moraine. However, they gave little attention to the sedimentology of the constituent deposits. Figure 28 is a map showing the surface sediments in this area, including the moraines which Owen (1988c) and Owen and Derbyshire (1988) attribute to a trunk valley glacier which retreated up the Gilgit and Hunza valleys during the Borit Jheel glaciation. The southeastern margin of the morainic area is bounded by a large linear ridge trending northeastwards and its westernmost edge is delimited by a northward-trending ridge. These represent recessional and lateral moraines respectively. The moraines confined to the area bounded by these ridges are hummocky and made up of polymictic sediments with poorly sorted, fine-skewed to nearly symmetrical grain size distributions, the mean grain size being in the medium to coarse sand grade. The tills include chaotically arranged clasts from metre size boulders to pebbles. Irregular discontinuities are present within the tills defined by silty horizons, and shearing, although rare, can be recognised. Many of the discontinuities may be shears resulting from push or sliding of glacial ice before ice wastage, followed by slumping. Coarse silts and silty sands infill depressions, onlap diamictons and fill spaces between boulders, the bedding being horizontal and planar, becoming rippled up-section. There are cm- to dm-wide scours filled with sandy gravels. These deposits were laid down in ponds which filled depressions between the hummocks. However, the silts are often deformed and slumped, indicating that they were probably deposited before the ice cores melted. Closely associated are discontinuities within the tills: these originated as the tills slumped and slid during ice melting and collapse of ice blocks (Fig. 29).

Figure 29. Section through the Dak Choki hummocky moraine. The bed of fine-grained sediments in the middle of the section is slumped and represents a small pond filled with silt before meltout was complete. Discontinuities within the till represent sliding and slumping of the till during the meltout process. The height of the section is about 10m.

The lateral moraine defining the western margin of the Dak Choki moraine has a-b plane fabrics parallelling the slope of the ridge. These are consistent with sliding of debris rather than glacial movement, and are similar to the fabrics produced by sliding of material from the margin of the Batura Glacier (see above).

The recessional moraine forming the southeastern ridge of the Dak Choki moraine is a polymictic, massive and poorly sorted sediment. It has a bimodal clast fabric with the a-b planes of clasts lying parallel to the frontal and lee slopes, indicating that they are probably a product of sliding rather than glacial push.

Of course, it is not suggested that all recessional or end moraines are produced by the sliding of material off the ice front. A case in point is the small end moraine attributed to an advance of ice down the Bagrot valley (Fig. 28). Although the fabric of this polymictic, poorly sorted sediment appears diffuse in the field (Fig. 30), it has a strongly orientated pebble fabric dipping up-valley and the highly ordered microfabric shows a very strong anisotropy (Fig. 22A and B). These characteristics bespeak lodgement deformation by drag or push rather than by sliding, and this moraine is considered to have been formed at and beneath the snout of an active ice front.

East of the Dak Choki moraine a 700m-thick sequence of sediments infills a palaeovalley, the Dainyor Palaeovalley (Wiche 1959, Desio and Orombelli 1971, Owen and Derbyshire 1988, Derbyshire and Owen 1988, Owen 1988c). This sequence contains glaciofluvial, debris flow and till sediments of ice-contact fan facies. Near the top, tills are intimately associated with folded glaciofluvial sediments. These tills probably formed after the retreat of the main valley ice and mark minor ice advances over the sediment. The folds are of the recumbent and sheath types plunging in a direction between 100° and 270°N. They are overlain by over-consolidated, sheared, poorly sorted, positively skewed diamictons with edge-rounded and striated clasts. Two sets of shears are recognised; a westward dipping set at 30-45° and a southward dipping set at 25-45°. The pebble fabrics measured also fall into two sets (southerly and north-northwesterly), the shears dipping more steeply than the pebbles. The folded glaciofluvial sediments indicate that the ice advanced over fan sediments in a south to southwesterly direction, and probably

Figure 30. View of the end moraine at the mouth of the Bagrot valley. The till appears to have a diffuse fabric, but detailed pebble measurements show highly consistent orientations and the microfabric is markedly anisotropic (see plate 17A & B).

deposited the tills during the advance. However the pebble fabrics and shear orientations are not consistent with this ice movement; rather, they appear to have been produced by downslope sliding of till in a westerly direction. It is difficult to invoke true debris flow processes to explain the structures in this till because the till is overconsolidated, a feature we have rarely noted in the debris flow sediments of this region. Sliding within a closely packed fabric seems more likely, dewatering inducing large surface tensions between particles and so helping to induce a degree of overconsolidation.

5 CONCLUSION

The mountain glacial environment in the Karakoram mountains is complex, till deposition dominated by supraglacial processes with subglacial deposition being less important. The very high and steep valley-side slopes on which the tills have been deposited, together with the abundant meltwaters resulting from the high ablation rates in the summer half year, leads to widespread resedimentation of till; observed structures and fabrics in till may thus be explained as much by mass flow and slide, and by meltout processes as by subglacial shearing and ice-push. The observed structures form a continuum ranging from microscopic scale to several hundreds of metres in length. It is important to relate the three-dimensional orientation of the structures and fabrics of these sediments at all scales (Derbyshire, McGown and Radwan 1976) to what is known or can be concluded about the ice-movement direction in order to assign processes to the structures present within the tills.

Clearly, fabrics are highly variable and complex resulting from both the depositional mechanisms and syngenetic or later deformation. Broad models can be created to explain the characteristics of the Karakoram tills and, within these models, the following several phases of glacial deposition and deformation of subglacial till can be recognised.

1. Deposition of till by ice: the till may be internally deformed by a variety of shears and folds (cf. Boulton 1987). The till may be thrust on to previously deposited sediments which are frequently glaciofluvial sands.

2. Ice may override previously deposited sediments, deforming them and introducing till into the sequence by thrusting process.

3. Varying degrees of overconsolidation may occur as a result of variations in the drainage of subglacial till as a function of the overburden weight of the ice and the permeability of the sediment and its substrate.

4. Desiccation following ice retreat may produce within the till masses with different viscosities, the interaction between them giving rise to grain size differentiation and dewatering structures. Desiccation also produces overconsolidation (Terzaghi and Peck 1967).

5. Dilation occurs after desiccation and the release of overburden pressure during ice wastage, resulting in sub-horizontal pressure release joints or sub-vertical joints on steep exposed faces.

6. Slumping and sliding of the till may occur during (1) to (5) and may occur after deposition.

The sequence for supraglacial till is simple:

1. Deposition by meltout, or by mass flow and sliding from the margins of the glacier.

2. Dewatering of ice cores, producing further slides and flows plus dewatering structures and translocation of clays. Desiccation may produce weak consolidation and crusting of the sediment.

3. Later modification by mass flow processes.

The importance of later modification of glacial and paraglacial diamictons cannot be over-emphasised, and resedimentation appearst to be the norm, especially in the case of Ghulkin type glaciers, Owen (1988c), Owen and Derbyshire (1988) and Derbyshire and Owen (1988) have shown that the resedimentation of tills occurs quite soon after the end of a glaciation and constitutes a major component of the valley fills made up essentially of landforms which they refer to as "sediment fans". Tills and associated sediments deformed by glaciotectonic processes frequently form an important component as inliers within these sediment fans.

ACKNOWLEDGEMENTS

The authors would like to thank S. Haywood for type setting the manuscript and K. E. Moore and R. E. Pollington for drafting some of the diagrams. This research was supported by NERC grant GT4/85/95/46 as part of a Ph.D. programme (LAO). The assistance of the helicopter group of the Pakistan army is gratefully acknowledged (ED).

REFERENCES

Arch, J. Maltman, A.J. & Knipe, R.J. 1986. Shear zones geometries in experimentally deformed sediments. Abstracts Tectonic Studies Group, University of Hull.

Batura Glacier Investigation Group 1979. The Batura Glacier in the Karakoram Mountains and its variations. Scientica Sinica, 22, 8.

Boulton, G.S. 1970. The deposition of subglacial and meltout tills at the margins of certain Svalbard glaciers. J. Glaciol. 9, 56, 231-45.

Boulton, G.S. 1971. Till genesis and fabric in Svalbard, Spitzbergen. In: Goldthwait, R.P. (ed.), Till, a symposium, p.41-72. Ohio State Univ. Press.

Boulton, G.S. 1987. A theory of drumlin formation by subglacial sediment deformation. In: Menzies, J. & Rose, J. (eds.) Drumlin Symposium, p.25-81. Balkema, Rotterdam.

Boulton, G.S. & Eyles, N.I. 1979. Sedimentation by valley glaciers: a model and genetic classification. In: Schluchter, C.H. (ed.), Moraines and Varves, A.A., p.11-25. Balkema, Rotterdam.

Cronin, V.S. & Johnson, W.P. 1988. Chronostratigraphy of the Late Cenozoic Bunthang Sequence and the possible mechanisms controlling base level in Skardu intermontane Basin, Karakoram Himalaya, Pakistan. In: Malinconico, B. B. & Billi, R.J. (eds.), Tectonics and Geophysics of the Western Himalayas, Geol. Soc. Amer. Spec. Paper. In press.

Derbyshire, E. 1978. A pilot study of till microfabric using the scanning electron microscope. In: Whalley, W.B. (ed.), Scanning electron microscopy in the study of sediments, p.41-61. Geo Abstracts, Norwich.

Derbyshire, E. 1981. Glacier regime and glacial sediment facies: a hypothetical framework for the Qinghai-Xizang Plateau. In: Proceedings of symposium on Qinghai-Xizang (Tibet) Plateau, Beijing, China. Geological and Ecological Studies of Qinghai-Xizang Plateau, Vol.2, p.1649-1656. Science Press, Beijing.

Derbyshire, E. 1983. The Lushan Dilemma: Pleistocene glaciation south of the Chang Jiang (Yangtze River). Z. Geomorph. N.F. 27, 445-471.

Derbyshire, E. 1984a. Sedimentological analysis of glacial and proglacial debris: a framework for the study of Karakoram glaciers. In Miller, K. (ed.), International Karakoram Project, p. Cambridge University Press, Cambridge.

Derbyshire, E. 1984b. Till properties + Glacier regime in parts of High Asia: Karakoram and Tian Shan: The Evolution of the East Asian Environment, vol.1. Geology and Palaeoclimatology. Whyte, R. O. (ed.), p.84-110. Centre for Asian Studies, University of Hong Kong.

Derbyshire, E., Billard, A., Van-Vliet-Lanoe, Lautridou, J.-P. & Cremaschi, M. 1988. Loess and Palaeoenvironment: some results of a European joint programme of research. J. Quat. Sci., in press.

Derbyshire, E., Li Jijun, Perrot, F.A., Xu Shuying & Waters, R.S. 1984. Quaternary glacial history of the Hunza Valley, Karakoram Mountains, Pakistan. In: Miller, K. (ed.), International Karakoram Project, p.456-495. Cambridge University Press, Cambridge.

Derbyshire, E., McGown, A. & Radwan, A. 1976. 'Total' fabric of some till landforms. Earth Surface Processes, 1:17-26.

Derbyshire, E. & Owen, L.A. 1988. Quaternary sediment fans in the Karakoram Mountains. In: Chuch, M. (ed.), Alluvial fans - a field approach. J. Wiley and Sons, Chichester, in press.

Desio, A. & Orombelli, G. 1971. Preliminary note on the presence of a large valley glacier in the middle Indus Valley (Pakistan) during the Pleistocene, Atti della Accademia Nazionale dei lincei, 51, 387-392 (in Italian).

Embleton, C. & King, C.A.M. 1975. Glacial and Periglacial Geomorphology, Arnold, London.

Evenson, E.B. 1970. A method for 3-dimensional microfabric analysis of tills obtained from exposures or cores. J. Sed. Petr. 40, 762-764.

Ferguson, R.J. 1984. Sediment load of the Hunza River. In: Miller, K. (ed.), International Karakoram Project, p.374-382. Cambridge University Press, Cambridge.

Flint, R.F. 1971. Glacial and Quaternary Geology. Wiley, New York.

Flint, R.F., Sanders, J.E. & Rodgers, J. 1960. Diamictite, a substitute term for symmictite. Geol. Soc. Am. Bull., 71, 1809.

Goudie, A.S., Brunsden, D., Collins, D.N., Derbyshire, E., Ferguson, R.I., Hashnet, Z., Jones, D.K.C., Perrott, F.A., Said, M., Waters, R.S., & Whalley, W.B. 1984. The geomorphology of the Hunza Valley, Karakoram Mountains, Pakistan. In: Miller, K. (ed.), International Karakoram Project, p.359-411, Cambridge University Press, Cambridge.

Gripp, K. 1938. Endmoranen. Int. Geogr. Congr. Abstracts IIA.

Hewitt, K. 1969. Glacier surges in the Karakoram Himalaya (Central Asia)

Canad. J. Earth Sci. 6, 1009-1018.
Jones, D.K.C., Brunsden, D. & Goudie, A.S. 1983. A preliminary geomorphological assessment of part of the Karakoram Highway. Quat. J. Eng. Geol. Lond., 16, 331-355.
Lawson, D.E. 1981. Distinguishing characteristics of diamictons at the margin of the Matanuska Glacier, Alaska. Annual Glac., 2, 78-84.
Li Jijun, Derbyshire, E. & Xu Shuying 1984. Glacial and paraglacial sediments of the Hunza valley North-West Karakoram, Pakistan: A preliminary analysis. In: Miller, K. (ed.), International Karakoram Project, p.496-535, Cambridge University Press, Cambridge.
Lister, G.S. & Snoke, A.W. 1984. S-C Mylonites. J. Struct. Geol. 6, 6, 617-638.
Love, M.A. & Derbyshire, E. 1985. Microfabric of glacial soils and its quantitive measurement. In: Forde, M.C. (ed.), Glacial Tills 85, Proc. Int. Conf. on glacial tills and boulder clays, p.129-135. Edinburgh, Engineering Technics Press.
McGown, A. & Derbyshire, E. 1974. Technical developments in the study of particulate matter in glacial tills. J. Geol., 82, 225-235.
McGown, A. & Derbyshire, E. 1977. Genetic influences on the properties of tills. Quart. J. Eng. Geol. 10, 389-436.
van der Meer, J.J.M. 1985. Sedimentology and genesis of glacial deposits in Goudsberg, Central Netherlands. Meded. Rijks Geol. Dienst. 39-2, 29.
van der Meer, J.J.M. 1987. Micromorphology of glacial sediments as a tool in distinguishing genetic varieties of till. In: Kujansuu, R. & Saarnisto, M. (eds.), INQUA Till Symposium. Geol. Surv. Finland, p.77-89, Spec. Paper 3.
Moran, S.R. 1971. Glaciotectonic structures in drift. In: Goldthwait, R.P. (ed.), Till- a symposium, p.127-148. Ohio State Univ. Press, Columbus.
Osborn, G.D. 1978. Fabric and origin of lateral moraines, Bethartoli Glacier, Garhwal Himalaya, India. J. Glac., 20, 84, 547-553.
Ostry, R.C. & Deane, R.E. 1963. Microfabric analysis of till. Bull. Geol. Soc. Amer. 74, 165-168.
Owen, L.A. 1988a. Wet sediment deformation of Quaternary and recent sediments in the Skardu Basin, Karakoram Mountains, Pakistan. In: Croot, D. (ed.), Glaciotectonics, A.A. Balkema, Rotterdam, this volume.
Owen, L.A. 1988b. Neotectonics and glacial deformation in the Karakoram Mountains, Northern Pakistan. Tectonophysics, Spec. Suppl. Neotectonics and palaeoseismicity, in press.
Owen, L.A. 1988c. Terrace, uplift and climate in the Karakoram Mountains, Northern Pakistan, Unpubl. Ph.D. thesis, Univ. of Leicester.
Owen, L.A. & Derbyshire, E. 1988. Quaternary sediment fans in a high active mountain belt, Karakoram Mountains, Northern Pakistan. J. Geol. Soc. Lond., in press.
Pillewizer, W. 1957. Bewegungstudien an Karakorum-Gletschem. Geomorph. Studien. Machatschek, Festschr. Erg. Petermanns, Mitt. Hft., 262, 53-60.
Price, N.J. 1981. Fault and joint development in brittle and semi-brittle rock. Pergamon Press, Oxford.
Shi Yafeng and Wang Jingtai 1981. The fluctuations of climate, glaciers and sea level since late Pleistocene in China. In: Sea level, Ice and Climatic Change, I.A.H.S. Publ. 131, 281-293.
Shi Yafeng & Zhang Xiangsong 1984. Some studies of the Batura glacier in the Karakoram mountains. In: Miller, K. (ed.), International Karakoram Project, p.51-62. Cambridge Univ. Press, Cambridge.
Sitler, R.F. & Chapman, C.A. 1955. Microfabrics of till from Ohio and Pennsylvania, J. Sed. Pet. 25, 262-269.
Sugden, D.E. & John, B.S. 1982. Glaciers and Landscape. Edward Arnold, London.
Terzaghi, K. & Peck, R.B. 1967. Soil Mechanics in Engineering Practice. John Wiley and Sons Inc., London.
Wiche, K. 1959. Klimamorphologische Untersuchungen im westlichen Karakorum, Verhandl. dtsch. Geographentages 32, 190-203.
Zhang Xiangsong, Shi Yafeng & Cai Xiangxing 1980. The migrating sub-glacial channel of the Batura Glacier and the tendency of the new channel. In: Shi Yafeng (ed.), op. cit., p.153-165, (in Chinese).

Glaciotectonics: Forms and Processes, Croot (ed.), © 1988 Balkema, Rotterdam. ISBN 90 6191 848 0

Observations on glaciodynamic structures at the Main Stationary Line in western Jutland, Denmark

S.A.S.Pedersen, K.S.Petersen & L.A.Rasmussen
Geological Survey of Denmark, Copenhagen, Denmark

ABSTRACT: The Bovbjerg profile is the most impressive cross section through the Weichselian Main Stationary Line. The following division of the structural stockverk is observed: Holsteinian marine clay probably forms the basis of the dark Saalian till. Above the Saalian till a sandur deposit grades into a grey sandy deposit with lignite. In the dislocated section of the Bovbjerg profile the lower stockwerk includes also the lowermost Weichselian tillbed. The deformation related to this lower stockwerk consists of thrust fault imbrications dipping to the north. In the Upper structural Stockwerk 2 tillbeds are found but outside to the south of the dislocated section the Weichselian ice advance can be separated into three minor tillbeds representing one kinetostratigrafical unit.

Geological mapping of areas in eastern and central parts of Jutland supports the evidence of a continuous upper kinetostratigraphical unit related to the Late Middle Weichselian iceadvance.

1. INTRODUCTION

The concept of the Main Stationary Line (MSL) and the description of this geological and geomorphological feature as representative of the maximum extension of the Weichselian glaciation were originally given by Ussing (1903 & 1907). The MSL is the most impressive glaciodynamic and sedimentological feature in the Quaternary of Denmark, and the Line can be traced from Bovbjerg at the West Coast to Viborg in the Central part of Jutland, and from the Central part southwards through Jutland to Germany, where it bends towards the east, the C line on fig. 1. (Milthers 1948).

Locally, the MSL is prominent through the geomorphological contrast between the Weichselian till deposits and the braided sandur outwash deposits. Sections through the MSL are, however, rare. The best exposed section is the Bovbjerg profile at the West coast of Jutland. Here in a north-south trending cross-section the succession related to a glaciodynamic zone can be studied. Although being of great importance in the Quaternary history of Denmark it has only been described very briefly in publications by Ussing (1903, 1907) and Nørregaard (1912).

This paper presents observations on the glaciodynamic structures of the Bovbjerg profile. The obtained data are related to a kinetostratigraphical conceptual Model, and the glaciodynamic model is discussed in relation to geological mapping of areas in eastern and central Jutland.

2. THE DESCRIPTION OF THE CLIFF SECTION AT BOVBJERG.

The Bovbjerg profile represents a coast eroded section through the western part of a ground

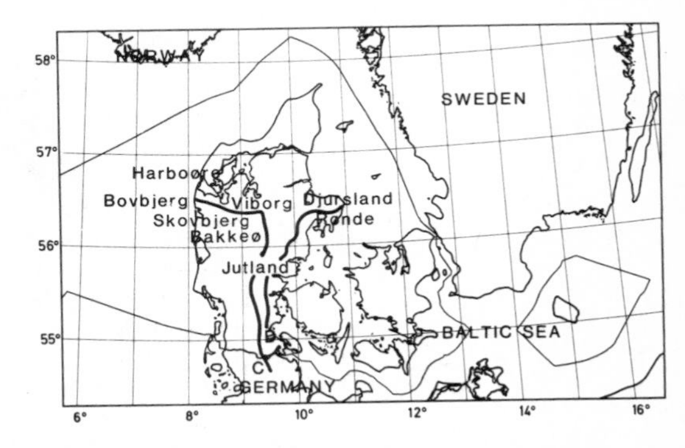

Fig.1. Location map with the Main Stationary Line indicated by C and East Jutlandic Borderzone by D.

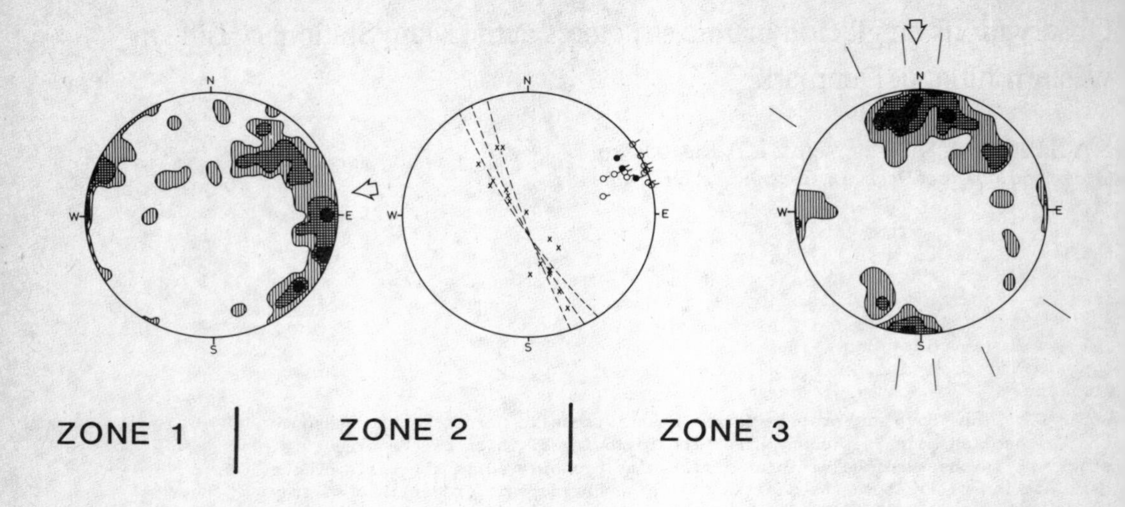

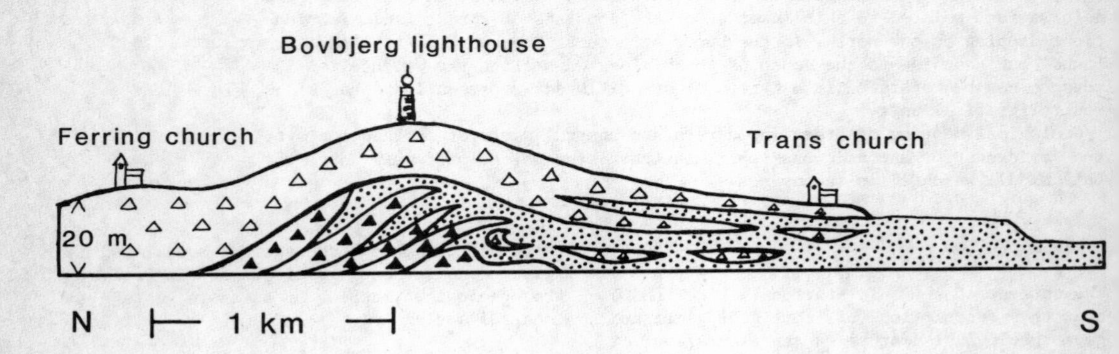

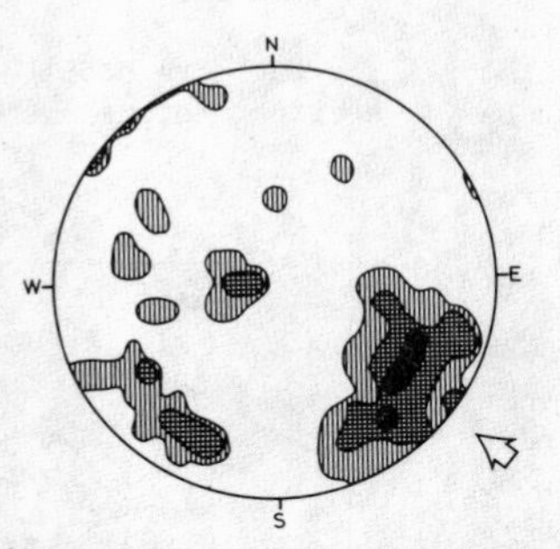

Fig.2. Simplified profile of the Bovbjerg coastal cliff from Ferring in the north to Trans in the south. Open triangles are Weichselian tills, solide triangles are Saalian tills and dotted areas are sand and sandur deposits. Note the vertical distance is not to scale. The lower diagram represents the fabric in the Saalian till. The diagram in zone 3 shows the icemovement direction of the dislocating till unit of Late Middle Weichselian age. The diagram in zone 2 shows the directions of fold axes from the dislocated strata. The diagram in zone 1 shows the direction of the final icemovement within the area.

moraine landscape which has been mapped in the forties by the Geological Survey of Denmark (Hansen, 1948). South of this area the transitionzone to the sandur reveals an example of kettlehole landscape between Trans church and the village Fjaltring. The section to be described runs over a distance of 3 km from the village Ferring to Trans church in the south.

In the following the references are made to the breakwaters with letter symbols seen on the two topographic maps, scale 1:25.000, 1115 IV NW and 1115 IV SW published by Geodætisk Institut, København 1964. The distance between the breakwaters is approximately 400 m.

To the north at Ferring till is seen in the topmost part of the cliff section with an elevation of 20 m.a.s.l. situated at breakwater B. Towards the south the cliff section rises slowly to 22 m at the breakwater C where the cliff is exposed from 4 m above sea level showing a sandy grey till with Norwegian indicator boulders at the base (Larvikite and Rombeporfyrite). The topmost part is weathered and no dynamic features have been observed. Between breakwatersC and D fabric analysis was carried out half way up the profile at about 10 m a.s.l. The diagram (fig. 2) shows an a-fabric with a plunge in the eastern direction. Consequently, the ice movement was from east to west.

Between breakwater D and E about 100 m north of E the first and northernmost dislocations can be observed. The central part of the dislocated strata can be studied further to the south in the profile between breakwater E and F where the height of the cliff culminates at 40 m above sea level at the lighthouse. The interpretation of the structures in the profile suggests a lower unit consisting of a dark grey clayey massive till with a thickness of more than 10 m. This till is overlain by finegrained, well sorted, grey sand with lignite. The thickness of this sand unit exceeds 5 m but is highly influenced by erosion and deformation. Superjacent to this sand unit a 1 to 2 m thick till is found. This complex of units was dislocated in forming overturned folds with a foldaxis direction trending between 64° to 73° and with a plunge of about 10°. An undisturbed complex of two tills rests unconformably on top of the formations mentioned above within the dislocated section of the cliff. Fabric analysis has been carried out in the lowermost dislocated unit - the dark grey clayey till. The diagram (fig. 2) is adjusted for the strike and dip according to the surrounding dislocated strata (72°-34°N). In the diagram it is seen that the majority of the pebbles plunge in a southeastern direction, representing an a-fabric. Consequently, the ice movement has been from southeast to northwest.

Within the dislocated sequence along the coast a landslide south of breakwater E obscures the structures for nearly 300 m. However, in the topmost part of the cliff section above the slide material it is seen that floes of till exist. It is suggested that the landslide was activated due to the occurence of marine clay in the lowermost part of the section. Marine clay is recorded from wells to the north of the Bovbjerg profile (see later), and this clay would have typical mechanical properties for initiation of landslides. However, marine strata has not been observed during the investigation of this part of the cliff section and hopefully, future investigations will provide the evidence for this speculation. The southern part of the dislocated units continue to south of breakwater F where the structures are well exposed and offer the possibility to measure both the orientation of bedding planes and the foldaxis directions within the drag folds. The construction of foldaxes and observed axes show a very consistant diagram (fig. 2).

South of breakwater F the dislocations come to an end. Between breakwater G and H the intercallations of till beds and sandur deposits can be studied in a section about 30 m high. The observed stratigraphy is as follows: The lowermost unit consists of the grey finegrained sand with clasts of lignite with a thickness exceeding 2 m, 6 m above sealevel. The sand is overlain by a 1-2 m thick till unit which has also been observed in the dislocated strata further to the north which are the uppermost part of the dislocated sequence. Superjacent to this till about 7 m of sandur deposits with fining upward sediments are found. The coarser material is dominated by flint (chert). From about 16 m a.s.l. and to the top of the cliff section two till units can be seen separated by 1 m of gravelly sand. Fabric analyses have been carried out in the central part of each of the three till units. The diagram from the lowermost till unit shows a pronounced maximum in the north-north-western direction (a-fabric), thus indicating an ice movement from north-north-west. Striations on the boulders confirm the fabric diagram. The fabric diagram from the midmost till unit shows an a-fabric with a plunge in the northern direction indicating an ice movement from north to south. Again an accordance is found between observed striations on boulders and the fabric analysis confirming the ice movement. Likewise, the topmost till unit reveals an a-fabric showing an ice movement from the north. However, in this case the striations on the boulders display a north-north-east, south-south-west direction, see fig. 3.

Observations made further to the south are depicted in fig. 3 based on six logs to around 150 m south of Trans church. The two youngest till units can be followed nearly as far as Trans church where the transition to the ice

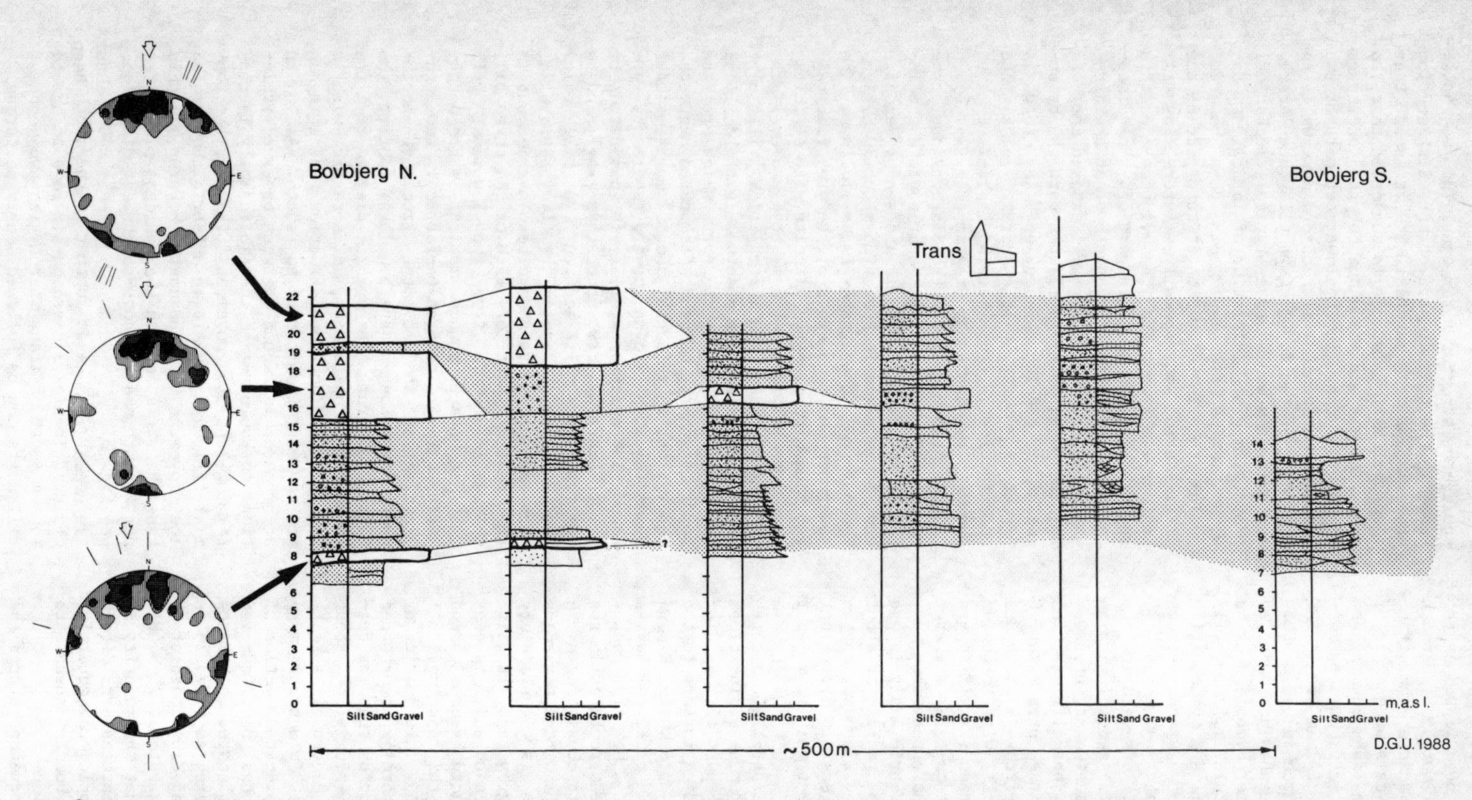

Fig.3. Diagrammatic section through the southern part of the Bovbjerg profile showing the interfingering of three Weichselian till units and the Sandur which represents the upper Kinetostratigraphical drift unit related to the Late Middle Weichselian ice advance. Sandur deposits shown by screen tint and till beds in open triangles. Fabric diagrams given for each till bed.

contact sediments is found in the proximal part of the sandur exposed in the central part of the cliff section. At this place the cliff has a height of 23 m, and outcrops of the lowermost till unit have not been observed.

3. GEOLOGICAL INTERPRETATION OF THE BOVBJERG CLIFF SECTION.

The oldest unit observed is the dark grey clayey till occurring in the Central part of the cliff section. The fabric shows that this till unit has been laid down by a glacier moving from southeast towards northwest viz. of a Baltic origin. Such directions of movement are traced south of the Main Stationary Line which may be regarded as being a part of a Middle Pleistocene landscape area, the northwestern part of the hilly island, Skovbjerg Bakkeø. The maximum age of this till might be found by studying the underlying strata in the Bovbjerg cliff section which are not exposed at present. However, as mentioned during the description of the cliff profile, Holsteinian marine deposits are found north of Bovbjerg at Harboøre (Knudsen 1987 fig. 1). Marine clay of equal age can be expected to occur within the section where the landslide has taken place south of breakwater E. Besides, the lowermost till unit which only occurs within the dislocated strata, at most three younger till beds can be seen in the central and southern part of the section. The older till bed of these three youngest tills takes part in the dislocations, fig. 4.

The Bovbjerg cliff section demonstrates a change of ice movement direction from north-north-west in the older part of the three till beds to the north in the youngest till bed as observed in the fabric diagrams in the southernmost part of the profile (Fig. 3). A consistent foldaxis direction trending east-north-east with a small dip is found within the dislocated part of the section. This implies that the folding of the formations was from north-north-west considering the overturned folds. In the northernmost part of the Bovbjerg profile where a massive till unit is found within the entire section the fabric analysis indicates an ice advance from the east.

The extension of the lowermost of the three youngest till beds is only to be followed as far as breakwater I north of Trans church. However, information from the geological map shows a landscape with kettleholes between Trans church and Fjaltring further to the south which might be taken as a proof of the maximum extension of the ice cap in the north to Fjaltring in the south. Such an extension might be read from the unpublished excursion guide by N.V. Ussing in 1907. In the guide he regarded the till found at Fjaltring as older than the period of stagnation. This is in accordance with his naming of the Bovbjerg section as a part of the Main Stationary Line. The build-up of the sandur during occillations of the ice cap as can be studied in the Bovbjerg section,took place before the disintergration of the ice from the maximum extention which shows that the whole sequence was formed within a rather short timespan.

4. DISCUSSION OF THE BOVBJERG PROFILE IN RELATION TO THE AREAS IN CENTRAL AND EASTERN JUTLAND.

Considering the melting stages, as worked out by Ussing 1903-10 (Milthers 1935), on the basis of the changing of the paleo current of the meltwater certain stages have been demonstrated, indicating that the wasting icecap had a northeastern origin. This leads to the idea that from the westernmost part of the Main Stationary Line at Bovbjerg the impact of the Northeast ice might be traced.

In the Viborg area an analogous situation of change in ice movements has been demonstrated. The dislocation found within the Viborg map sheet (Rasmussen & Petersen 1986) refered to the icemovement directions during the Late Middle Weichselian. A till related to this ice advance is the Store Klinthøj Formation. Till fabric analyses at St. Klinthøj show an ice advance form north to south dated to be younger than 22,000 yrs. This mappable till unit is related to the ice advance extending to the Main Stationary line, indicating that the St. Klinthøj Formation can be correlated to the upper till unit in the Bovbjerg section. Within the Viborg area the Melbjerg Hoved locality demonstrates older events including ice-movements from the south - as is also seen in the Bovbjerg profile.

According to the kinetostratigrapic principle the St. Klinthøj Formation has been correlated to the Djursland region, Eastern Jutland, with the application of fabric analyses and glaciomorphological features such as drumlins. The kinematic pattern has demonstrated a dynamically continuous transition from the north towards northeast. These ice movements belong to the Main Stationary Line advance.

In the eastern part of Jutland within the Djursland region the dynamically continuous transition can be followed to its final movement from southeast represented by the Young Baltic iceadvance. This advance dislocated the strata of the St. Klinthøj Formation and formed the moraine ridges of the so-called East Jutlandic Borderzone (Harder 1908). Sandur deposits related to this ice margin has been dated by inbedded mammoth remains to the latest Middle Weichselian cfr. Mangerud et al 1974, between 13,000 and 14,000 BP. (Petersen and Rasmussen 1987)

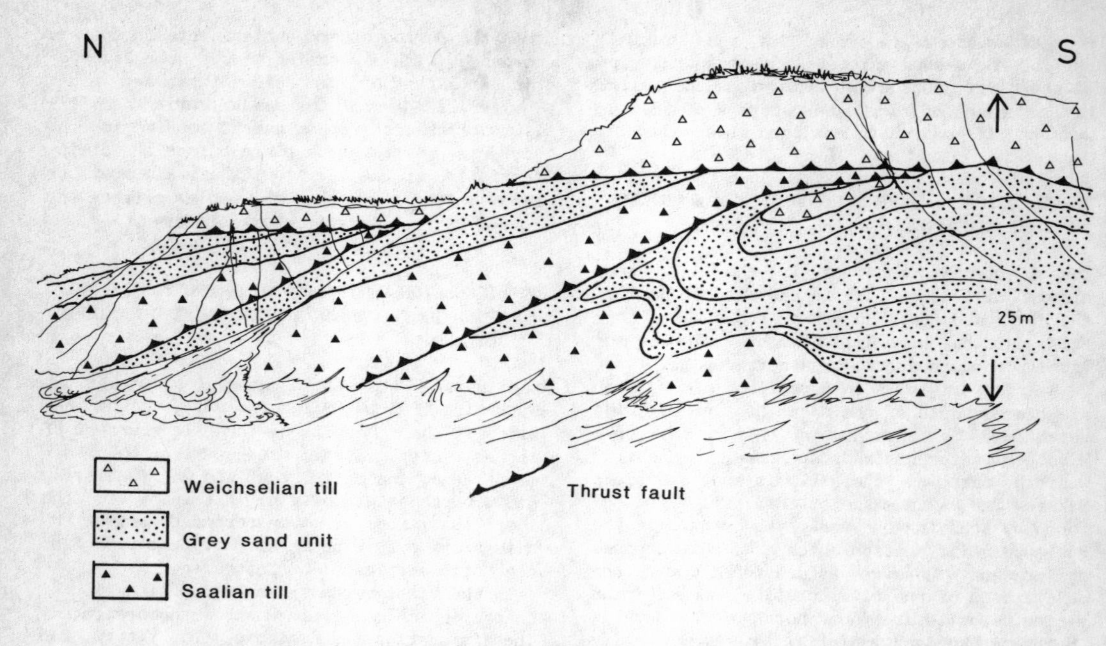

Fig.4. Segment of the coastal cliff section of Bovbjerg within the dislocated zone. In the lower stockwerk steeply north dipping thrust sheets of the Saalian till are seen discordantly superpost by the main Weichselian till units. In the foot wall syncline the lowermost Weichselian till is trapped beneath the southern steeply dipping thrust fault.
The profile is seen in the direction of the foldaxis.

An older till complex is also demonstrated within the Rønde map sheet on Djursland. Fabric analysis indicates an icemovement form south-south-east to north-north-west.

Among Skandinavian geologists there are contrasting views on the glaciation pattern during the Weichselian. Regarding the above concept of a dynamically continuous transition during the Late Middle Weichselian Ringberg (1988 fig. 11) speaks in favour of an ice succession initiating with a Norwegian Ice reaching the northern and central Danmark and, subsequently, a so-called Old Baltic advance again followed by the Northeast Ice extending to the MSL. However, Houmark-Nielsen (1987) supposes that the Old Baltic Ice advance antedates the Norwegian Ice but also in Late Middle Weichselian around 22 ka BP. In the investigated area dealt with in this paper the Old Baltic icemovement has not been demonstrated and it is considered to belong to the Early Middle Weichselian events (Petersen 1985). However, the Norwegian ice advance appears to be part of the demonstrated Late Middle Weichselian dynamically continuous transition leading to the final icemovements from Southeast represented by the younger Baltic iceadvance.

The last ice movement within Central Jutland seems to indicate a transition from west to east of the area of origin for the glaciers, and that this transition only in the outermost part of the glaciated area was differentiated into several till beds. In the Bovbjerg cliff section the initial stage of icemovements to the Main Stationary Line is represented by the Norwegian ice (from NNW). This ice advance is also responsible for the dislocations observed in the Bovbjerg profile. However, in the northernmost part of the cliff section the icemovement from ENE represents the impact of the Northeast ice. This implies that the whole sequence represents a dynamic transition from NNW to ENE observed within one till unit. This till unit can be followed from the western part of Jutland over the Viborg area where it is named the St. Klinthøj Formation to the eastern part of Jutland. Among the older ice movements the glaciers of a Baltic (southern) origin are dominating extending far up towards the northwestern part of Denmark and, by far, more extensive than the Younger Baltic ice lobe forming the East Jutlandic Border Zone. The morphological features of the older ice streams cannot be followed in the landscape but the movements of such icecaps can be reconstructed applicating the fabric analyses on the preserved till units.

5. CONCLUSION

The Bovbjerg profile is a classical profile for the study of a cross-section through a sedimentary sequence, deformed and deposited at a main Stationary ice border zone. The glaciodynamical development related to Late-Middle Weichselian ice advance may be summarized as follows:

An initial advance form north to south reached the south of the Bovbjerg profile. A continuous ice advance form the north dislocated the sandur sequence including the early till unit and an older Saalian till and sand sequence. The glacitectonical deformation formed a thrust fault imbrication complex.

Three important relationships are indicated form the till fabric analysis:

1) Till fabric in the till-deposits of the Norwegian Ice coincides with directions of glacitectonical deformations.
2) Till fabric observed in the northern part of the profile indicates a late stage ice movement form ENE.
3) In relation to the Bovbjerg locality the regional till fabric pattern within the central and eastern part Jutland demonstrates a clockwise turn in the movement of the Late-Middle Weichselian ice sheet.

6. ACKNOWLEDGEMENTS

Our thanks are directed to Inge Martine-Legene, Torben Friis Jensen and Kirsten Sloth who provided technical accistance and Janne Høybye for correcting the English.

REFERENCES

Berthelsen, A. 1973: Weichselian ice advances and drift successions in Denmark. Bull. Geol. Inst. Univ. Upps., 5, 21-29.

Berthelsen, A. 1978: The methodology of kinetostratigraphy as applied to glacial geology. Bull. geol. Soc. Denmark, vol. 27, Special Issue, 25-38.

Hansen, S. 1948: Landskabets geologiske Udformning. In: Mathiassen, Th. Studier over Vestjyllands Oldtidsbebyggelse. Nationalmuseets Skrifter Arkæologisk-Historisk Række II.

Hansen. S. 1978: Sidste nedisnings maksimumsudbredelse i syd- & Midtjylland Danm. geol. Unders. Årbog 1976, pp. 139-152.

Harder, P. 1908: En østjydsk Israndslinie og dens Indflydelse paa Vandløbene. Danm. geol. Unders. II, No. 19, 227 p.

Houmark-Nielsen, M. 1987: Pleistocene stratigraphy and glacial history of the central part of Denmark. Bull. geol. Soc. Denmark, vol. 36 pp. 1-189.

Knudsen, Karen Luise 1987: Elsterian-Holsteinian faraminiferal stratigraphy in the North Jutland and Kattegat areas, Denmark. Boreas, Vol. 16, pp. 359-368.

Mangerud, J., Andersen, S.T., Berglund, B.E. and Donner, J.J., Norden, a proposal for terminology and classification. Boreas, Vol. 3, pp. 109-128.

Milthers, K. 1935: Landskabets Udformning mellem Alheden og Limfjorden. Danm. geol. Unders. II. Række. Nr. 56, 36 p.

Milthers, V. 1948: Det danske Istidslandskabs Terrænformer og deres Opstaaen. Danm. geol. Unders. III. Række No. 28, 234 p.

Nørregaard, E.M. 1912: Bovbjerg-Profilet. Medd. dansk geol. Foren. B. 4. pp. 46-54.

Petersen, K. Strand 1985: The Late Quaternary Historie of Danmark. Journal of Danish Archaeology 4, 7-22.

Petersen, K. Strand and Rasmussen L. Aabo 1987: A geological concept of the map sheet Rønde based on dynamic structures. DGU series C, No. 8, 58 pp.

Rasmussen, L. Aabo and Petersen, K. Strand 1986: Geological map of Denmark Map sheet 1215 IV Viborg. Geological Survey of Denmark Mapseries No. 1.

Ringberg, B. 1988: Late Weichselian geology of southernmost Sweden Boreas, Vol. 17 pp. 243-263.

Ussing, N.V. 1903: Om Jyllands Hedesletter og Teorierne for deres dannelse. Overs. K. danske Videnskb. Selsk. Forh., 1903, No. 2, 99-165.

Ussing. N.V. 1907: Om Floddale og Randmoræner i Jylland. Overs. K. danske Vidensk. Selsk. Forh., 1907, No. 4, 161-213.

Glaciotectonics: Forms and Processes, Croot (ed.), © 1988 Balkema, Rotterdam. ISBN 90 6191 848 0

Sand-filled frost wedges in glaciotectonically deformed mo-clay on the island of Fur, Denmark

S.A.S.Pedersen & K.S.Petersen
Geological Survey of Denmark, Copenhagen, Denmark

ABSTRACT: The occurences of sand-wedges in the glacitectonically deformed Tertiary mo-clay in the Limfjorden Region in Northern Jutland have been investigated. The sand wedges are regarded as pollution of the mo-clay rawmaterial, and the present investigation was carried out to explain and map the occurence of this unusual formation.

The investigation resulted in the suggested model for the glaciodynamic development of the glaciogeological setting of Fur.

1. INTRODUCTION

On the island of Fur (fig. 1) deposits of a Palaeogene diatomite, the mo-clay, has been exploited for nearly a hundred years and used in the industrial production of insulation bricks. The diatomite is very clean with no primary sedimentary intercalations of sand and gravel. The occurences of the mo-clay rawmaterial are structurally related to folded and thrust-faulted sheets of the early Tertiary strata overlaid by Pleistocene drift, and the mo-clay pits are excavated along strike following the direction of the foldaxis of the anticlines. However, an alarming occurrence of small pockets and wedges of sand and gravel in the mo-clay was reported from one of the pits in the summer 1985. The authors were asked to investigate and map these sand pockets and wedges, and a report on the subject was finished in November 1985.

The aim of this paper is to sum up the results of this applied glacitectonic investigation. The implication of the near-surface occurence of sand wedges is discussed, and a description of the relationship between glacial deformation and the syn-tectonical erosion and sedimentation is given.

2. GEOLOGICAL SETTING

The outcrops of mo-clay on the island of Fur belong to the mo-clay setting well known from the Limfjorden region in the northwestern part of Jutland, Denmark (Bøggild 1918; Gry 1940, 1979; Pedersen 1981). The mo-clay is a diatomite which forms the main constitution of the Fur Formation (Pedersen & Surlyk 1983). The formation is 60 m thick and is divided into a lower member with only few intercalations of non-diatomitic lithology, and an upper member characterized by more than a hundred thinly bedded volcanic ash layers. The Fur Formation was deposited in the Tertiary Danish-Norwegian - North Sea basin and interfingers with the Ølst Formation and the London Clay. The latest data on the age of the formation indicates that the Palaeocene-Eocene boundary is placed near the middle of the formation (Knox 1984, Heilman-Clausen 1985).

During the Pleistocene the Tertiary sequences overlying the mo-clay were eroded away, and except for a few pockets of overlying plastic clay preserved in structural traps, sediments of Pleistocene drift discordantly overly the deposits of the Fur Formation. The last great ice advance in the Late-Middle Weichselian pressed thrust sheets of mo-clay up into a number of thrust fault and fold complexes. Geomorphologically, these glacitectonic deformed complexes are preserved in the landscape as ice-push-ridges forming a rough hilly topography. The northern part of the island of Fur constitutes a very characteristic crescent-shaped glacitectonic complex. In the post-glacial time this hilly complex formed an island in the archipelago related to the Littorina Sea. During the Holocene isostatic elevation of the near shore deposits around the glacitectonic complex was added to form the present shape of the Fur island.

The investigated area is a mo-clay pit si-

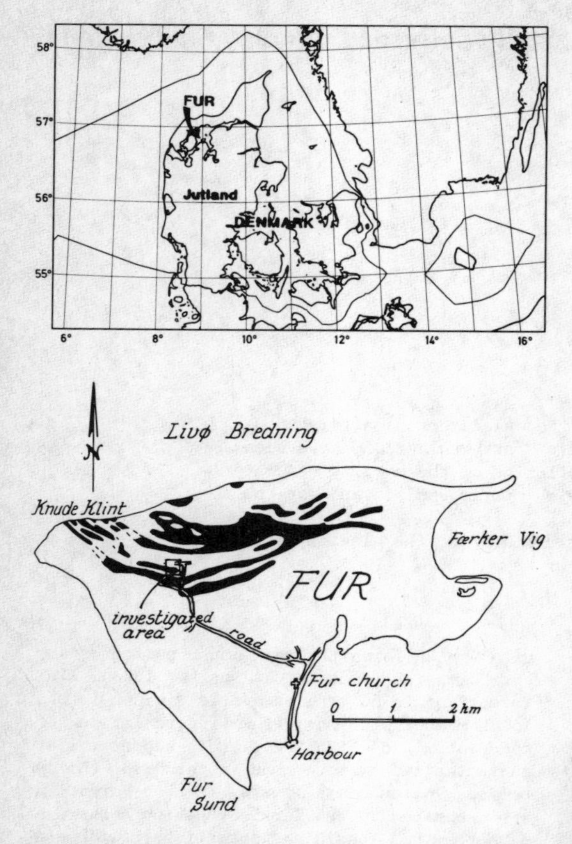

Fig.1. Location map of the position of Fur, and a detailed map of the island of Fur with the mo-clay strikes given in black.

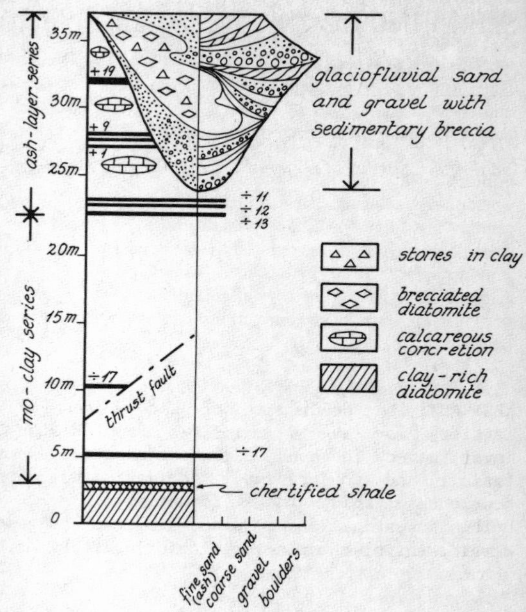

Fig.2. Composit Lithostratigraphical section of the rock types investigated.

Fig.3. Sand wedge structure in northern flank of synclinal structure with meltwaterchannel.

tuated in the western part of the crescent-shaped glacitectonic complex about 30-50 m above sea level (Fig. 1).

3. LITHOLOGY

In the investigated area the following rock units have been differentiated:
1) The Tertiary sequence consisting of three units related to the lower and middle part of the Fur Formation, and
2) the glacigene sequence which consists of glaciofluvial deposits and various sedimentary breccias.

3.1. The Tertiary sequence

For description of rawmaterial and the glacio-dynamic features in this mo-clay pit three informal units of the Fur Formation have been applied:
1) The shale series,
2) the mo-clay series, and
3) the ash layer series. (Fig 2.).

The shale series. The shale series are situated in the lower part of the Fur Formatior and consist of dark clay-rich diatomite. A 30 cm thick brown-weathering chertyfied shale forms the upper limitation of this unit. In the mo-clay pit the excavation has not proceeded below the chertified shale. The mo-clay series. This series constitutes the essential exploitation unit. It overlies the shale series and is 15 m thick with approximately 7 thin volcanic ash layers of which the one numbered - 17 reaches a thickness of 4 cm. The mo-clay series are overlaid by the ash layer series. The ash layer series. The ash layer series are characterized by more than 100 volcanic ash layers interbedded the diatomite. In the present investigation this unit is limited to the approx. 15 m thick part of the Fur Formation situated between ash layer - 13 and + 19. Except for the latter which is a 20 cm thick grey dacitic ash layer the main composition of the approx. 1 - 8 cm thick black fine-sand ash layers is tholeiitic. Additionally, a number of calcareous concreations appear in this unit.

3.2. The glacigene sequence

The glacigene sediments are mainly glaciofluvial sand and gravel. Moreover, glaciolacustrine clay and flow-till like syntectonically breccias occur (Fig. 2). The glaciofluvial deposits. Sand, gravel and boulders related to this unit occur in channel-like structures which roughly follows the trend of the synclines of the folded ash layer series. Within the channel structure large scale through cross-bedding is found (Fig. 2). Along the sides of the channels erosion cavity structures filled by coarse sand occur in the flanks of the synclines (see Fig. 4 stages 2). The sedimentary breccias. In the channels irregular bodies of clay or mo-clay with erratic stones and boulders appear. These muddy breccias are closely related to the structural development of folds guiding the trends of the glaciofluvial deposition. They may be regarded as flow-till intercalations or simply as debris-flows derived from the popping up anticlines (Fig. 4 stages 3).

3.3. The sand wedges

The sand wedges occur in the ash layer and mo-clay series in a zone down to 10 m below the surfaces between the glacigene and the Tertiary sequences. The sand wedges are steeply dipping (meaning they were intially formed orthogonally to the bedding), and the width of the wedges at the top is about 30 cm narrowing towards the depth, where they disappear in discrete fractures (Fig. 3).

The wedges are filled with coarse sand, gravel and stones up to 10 cm in size. The sand is quartzitic and the main composition of stones is cherts or limestone with or without chert. Besides, variegated crystalline pebbles derrived from the Scandinavian Shield form a minor constituent of the sediment.

The best preserved sand wedges display a characteristic symmetrical zonation in a vertical cross-section (Fig. 5). Thus the sand wedges are interpreted as ice wedges or frost wedge structures formed in frozen ground sediments during the Glacial Age. This is supported by studies of recent ice wedge structures from Jameson Land, East Greenland (Fig. 6).

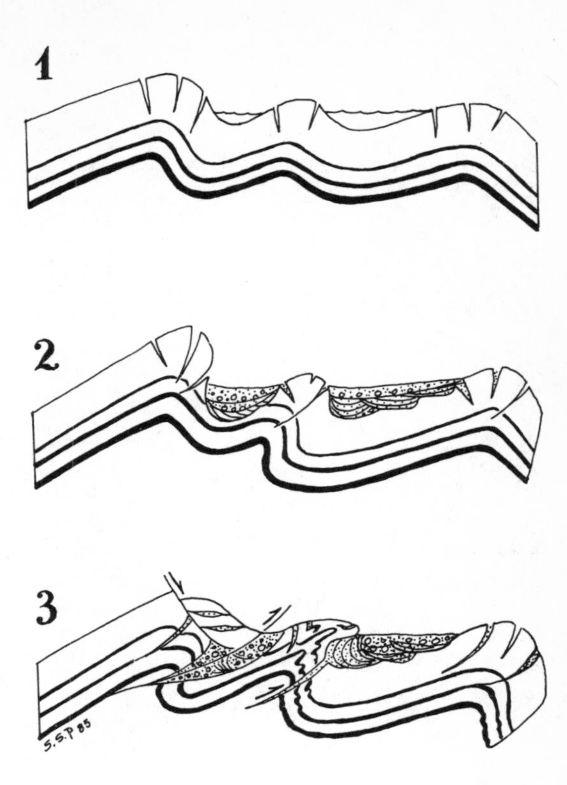

Fig.4. Three main stages of the glaciodynamic development. 1) Initial folding with cracking. Formation of meltwater streams takes place in the synclinal structures. 2) Continuous folding, cracking, deposition of glacial sand and gravel in the synclinal channels. 3) Sliding of the mo-clay anticlines into the meltwater-channels, faulting and thrusting of the anticlines and shearing out of sand wedge structures. Note that the sand wedge structures are worked into the moclay series by weakly inclined thrusts.

4. STRUCTURES

The glacitectonical structures are instructively outlined by the black ash layers in the mo-clay (Fig. 7). The folds are close to tight and overturned to the south, and fold axes are directed WNW-ESE (Fig. 8). The thrust faults are dipping steeply to the north and implicate a separation of the mo-clay into several thrust sheets. The structures indicate glaciodynamic deformation from N-NNE to S-SSW.

Fig.5. Detail of sand wedges in the mo-clay pit of Fur.

Fig.6. Recent ice wedge from Jameson land, Greenland.

In the synclines filled with sand and gravel tilting of the lowermost trough cross-bedding has been observed. This corresponds to the syn-tectonical sedimentation of glaciofluvial sand in glacitectonical imbricate complexes as described from the northernmost part of Jutland by Pedersen (1987).

The sand wedges show a marked bent appearence. This is partly due to the reorientation caused by the buckle-type folding, but in the deeper parts the wedges are displaced by small thrust faults dipping slightly to the north. These thrust faults appear in "massive" mo-clay and separate the sand wedges into small lense-shaped pockets.

5. DISCUSSION

The glaciogeological features described here are very complex. The different types of geology involved in an investigation of this type are incalculable and no mono-geological type of study can be applied. The fascinating aspect in the geological investigation is the dense accumulation in time of highly different geological events. This is exemplified in the presented investigation.

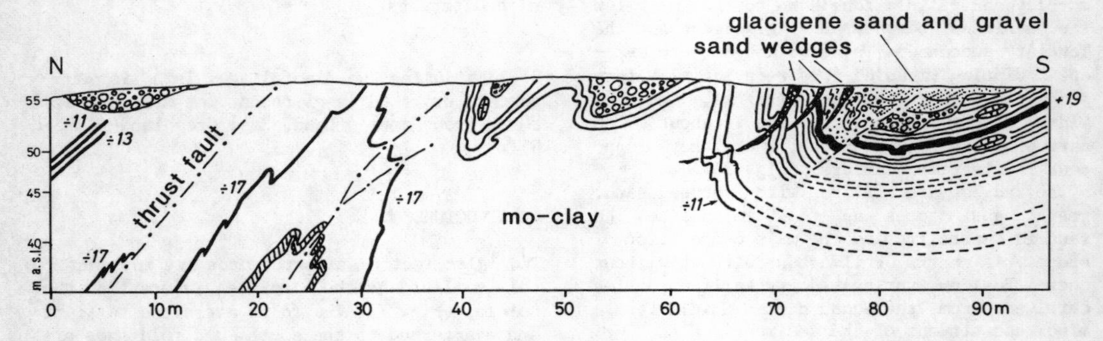

Fig.7. Cross-section of the mo-clay pit drawn perpendicular to the direction of the main fold axis.

Note that the glacigene sequence overlies the folded mo-clay on a erosive unconformity. Numbers refer to ash layers according to the nomenclator of Bøggild (1918).

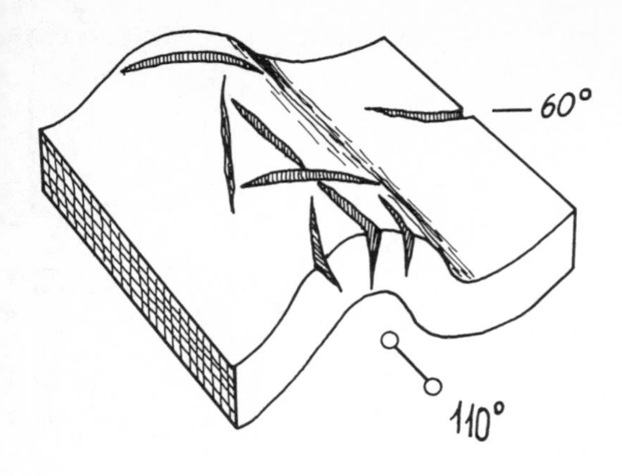

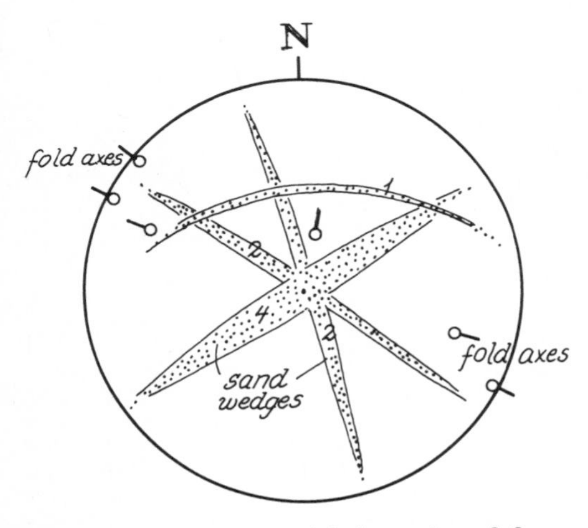

Fig.8. Fold structure with formation of fractures, the predominant extensional fractures being the 60o direction. On the Wulfnet diagram below the constructed fold axes of different areas in the pit are shown. Furthermore, the dotted areas state the sand wedge directions measured on near vertical wedges segments. Numbers on the wedges (dotted) gives the number of measurements.

The Tertiary sequence forms a weakly compacted soft sedimentary substratum for the glaciodynamic deformation. Above this sequence strata of late Eocene, Oligocene and probably Miocene have been eroded away. This material is possibly forming the in-basinal redeposition during the Pleistocene further to the south in Jutland.

The thrust faults and folds in the mo-clay were formed during the Late-Middle Weichselian glaciation, and the deformation is correlated to the ice-stream derived from Norway (Gry 1940, 1979). The style of deformation is related to a pro-glacial thrust fault imbrication (Pedersen 1987), and the first question to be put foreward is, what kind of environment did the ice-front transgress.

The interpretation of the sand wedges indicates that the mo-clay was permafrozen, and the wedges possibly formed pattern ground. Thus the ice advanced over a terrestrial surface or possibly a surface that during initial uplift due to glacitectonics became terrestrial.

The next question concerns the explanation of the sedimentary breccia. The coincidence of synclines and melt-water streams indicates an ice-border foreland with ridges and elongated troughs. The ridges were subjected to erosion in the crest by permafrost activity and in the flanks by melt-water stream erosion respectively. In this environment it is easy to imagin irregularly parts of the anticlines to drop or slide into the troughs. Due to the considerable rate of deposition the mudslides, partly frozen, were preserved due to the fast burial.

Finally, the investigation was asked to give an explanation for the sandpockets. It was demonstrated that the lower, narrow parts of the wedges were affected by small-scale thrusting. This thrusting is a late phase in the progressive deformation as it is seen to displace the reorientated sand wedges i.e. rotated due to folding.

The glacitectonic complex of Fur was transgressed by the Norwegian ice without building up a sandur in the foreland on the southern part of Fur (Gry 1979). Thus it is implied that the ice flattened the formed structures, and the small-scale thrusting developed in relation to sub-glacial shearing characteristically located to the homogeneous lithology, namely the mo-clay.

6. CONCLUSION

Investigation of sand wedges in the glacitectonically deformed Tertiary mo-clay on the island of Fur has resulted in a geological-glaciodynamical model for the formation of a glacitectonic complex in the Limfjorden Region in Denmark. The area was affected by the Norwegian iceadvance in the Late-Middle Weichselian time, and the following dynamic phases are distinguished:

1) Initial folding elevated the area and the surface was subjected to permafrost conditions resulting in formation of frost-ice wedges.
2) Contemporaneous with the folding,melt-water streams developed in the synclines of the folded mo-clay.
3) Thrust faulting and folding progressed proglacially. The frost wedges were filled with sand, and erosion of the anticlinal

ridges caused the appearence of sedimentary breccia in the glaciofluvial throughs or channels.

4) The structures and depositional features were transgressed by the ice, and small-scale thrusting developed due to shearing in the sub-stratum beneath the sole of the ice.
5) The sand-pockets have caused the production of mo-clay bricks great troubles because the quartz sand and cherts are impurities in the rawmaterial, but the sand-pockets are only recognized and located with the application of a glaciodynamic - glacitectonic concept.

7. ACKNOWLEDGEMENTS

Our thanks are directed to Kirsten Sloth who provided technical accistance and Janne Høybye for correcting the English.

REFERENCES

Bøggild, O.B. 1918: Den vulkanske Aske i Moleret. DGU Rk. II, Nr. 34, 84.

Gry, H. 1940: De istektoniske forhold i Molerområdet. Med bemærkninger om vore dislocerede Klinters dannelse. Medd. dansk geol. Foren. 9, 586-627.

Gry, H. 1979: Beskrivelse til Geologisk Kort over Danmark, Kortbladet Løgstør. Danm. geol. Unders. I. Række, 26, 58.

Heilmann-Clausen, C. 1985: Dinoflagellate stratigraphy of the uppermost Danian to Ypresian in the Viborg 1 borehole, central Jylland, Danmark. Danm. geol. Unders. Serie A, 2, 69.

Knox, R.W.O´B. 1984: Nannoplankton zonation and the Paleocene/Eocene boundary beds of NW Europe: an indirect correlation by means of volcanic ash layers. J. Geol. Soc. London 141, 993-999.

Pedersen, G.K. 1981: Anoxic events during sedimentation of a Paleogene diatomite in Denmark. Sedimentology 28; 487-504.

Pedersen, G.K. & Surlyk, F. 1983: The Fur Formation, a late Paleocene ash-bearing diatomite from northern Denmark. Bull. geol. Soc. Danmark, Vol. 32, 43-65.

Pedersen, S.A.S. 1987: Comparative Studies of graity tectonics in Quaternary sediments and sedimentary rocks related to fold belts. In Jones, M.E. & Preston, R.M. (eds.): Deformation of Sediments and Sedimentary Rocks. Geol. Soc. Spec. Publ. No. 29, 165-180.

Glaciotectonics: Forms and Processes, Croot (ed.), © 1988 Balkema, Rotterdam. ISBN 90 6191 848 0

Glaciotectonics and its relationship to other glaciogenic processes

Hanna Ruszczyńska-Szenajch
Department of Geology, Warsaw University, Warsaw, Poland

ABSTRACT: Glaciotectonics is defined as the mechanical action of glacial ice on the substratum. The author seeks to differentiate more clearly between glaciotectonic processes and other glacial processes, particularly erosion, and deposition by lodgement. It is, however, concluded that since the processes naturally coexist, one should not always expect to be able to separate out their geological products.

This paper is a brief contribution to the rising discussion concerning the nature of glaciotectonic processes and their relation to the other processes arising from glacial activity. The problem has been considerably confused, even in the last several years.

Firstly: glacial tectonics has previously been defined by some researchers as "incomplete glacial erosion" (Dreimanis 1976), so it was, in a way, identified with glacial erosion.

Secondly: despite the fact that glaciotectonically deformed sediments are usually shifted from their primary bed position, there is usually a lack of clear distinction between "deformation" and "transport" and consequently a lack of specific study of glaciotectonic transport (which has been commonly studied within general "glacial transport").

Thirdly: the most confused question concerns the processes of glaciotectonics associated with the deposition of lodgement tills. A lack of clear distinction of the former resulted in the creation and widespread use of the term "deformation till", with deformation identified by some researchers with the process of glacial (sensu stricto) depositior (Boulton 1982).

The author's study of glaciotectonic processes has been closely connected with her longstanding study of glacial rafts. These studies inclined her to define glaciotectonics as a "mechanical action of glacial ice upon its substratum" (Ruszczyńska-Szenajch 1980).

Since the origin of glacial rafts involves raft detachment, transport and subsequent deposition (Ruszczyńska-Szenajch 1987), the author has separately analysed these three groups of processes, and this has also helped her to clarify relationships among the whole set of glaciogenic processes. These relationships are shown in Table 1.

The main processes - i.e. erosion, transport and deposition - are given the general name "glaciogenic" on the Table, thus including: glaciogenic erosion, glaciogenic transport and glaciogenic deposition. Each may cover three different groups of processes i.e. glacial sensu stricto, glaciotectonic and glacioaqueous ones.

The glacioaqueous processes - melt water erosion, transport and deposition - have already been distinguished from the two former ones without considerable controversy (though the effects of their interpreted action, especially with regard to some larger-scale subsurface features in lowland areas, are sometimes not sufficiently well-documented to answer questions concerning their origin: glacioaqueous, glaciotectonic or glacial sensu stricto?).
Aqueous processes in non-glacial environments have also been thoroughly studied thus facilitating close comparisons with melt water activity. Hence the glacioaqueous group need not be discussed here.

However, the distinction between glacial sensu stricto and glaciotectonic processes

Table 1. Interrelationships between glaciogenic processes.

GLACIOGENIC PROCESSES

	GLACIOGENIC EROSION	GLACIOGENIC TRANSPORT	GLACIOGENIC DEPOSITION
GLACIAL SENSU STR.	freezing-on	englacial (incl. basal) supraglacial	lodgement meltout flow
GLACIOTECTONIC	tectonic detachment →	squeezing pushing dragging	tectonic re-deposition
GLACIOAQUEOUS	melt water erosion	melt water transport	melt water deposition

has developed into a significant problem, as mentioned above, and will be briefly outlined with reference to publications concerning the more detailed studies of these processes, which have supplied a basis for the present concise discussion.

The question of differentiation between glacial erosion sensu stricto and glaciotectonics was orally reported by the author during the INQUA Congress in 1977, and subsequently published (Ruszczyńska-Szenajch 1980). A major conclusion of this work is a proposal to attribute "the term glacial tectonics to the mechanical action of glacial ice on its substratum, and the term glacial erosion to freezing-on processes of substratum material to the glacier sole. In such an approach glacial tectonics and glacial erosion are essentially different groups of processes" (Ruszczyńska-Szenajch 1980 p. 75). This view has been confirmed by subsequent works by the author (Ruszczyńska-Szenajch 1983, 1987) and is reflected in Table 1 within this paper. According to these works and the Table, **glaciotectonic erosion** is identified with (glacio) tectonic detachment of substratum material, i.e. with squeezing, pushing and dragging. The main mechanisms involved in this action are: shearing and plastic forcing out. The above view will hopefully continue the discussion, which it had already provoked (Aber 1985).

Mechanical (tectonic) detachment of substratum material is usually followed by a mechanical displacement of the detached sediments without incorporation into the ice (Ruszczyńska-Szenajch 1987). The detachment and the immediate **glaciotectonic transport** form a continuous process and are so closely connected with each other that they are usually given common names such as already mentioned squeezing, pushing and dragging (Table 1). The processes of such transport have been discussed in more detail, based on Pleistocene evidence (Ruszczyńska-Szenajch 1983, 1987) and with reference to the observations of recent glaciers and their substrata (Boulton 1979). It becomes clear from such research that mechanical (glaciotectonic) transport of substratum material is essentially different in nature from englacial or supraglacial (i.e. glacial sensu stricto) transport, because the material is not carried within the ice body or upon it but is shifted subglacially and/or proglacially through (successive) shearing or plastic "flow", caused by ice pressure and movement. The distance of glaciotectonic transport is usually short when compared with that of glacial sensu stricto transport, and it probably does not exceed several kilometres. The evidence examined by the writer points to common distances of tens to hundreds of metres (for large-scale glaciotectonic features). However, it must be added that the recognition of these two types of transport (glaciotectonic and glacial sensu stricto) is sometimes very difficult (compare Ruszczyńska-Szenajch 1987, with Fenton 1983 and Moran et al. 1980).

The most complicated problems have been encountered, as already mentioned, in the studies of glacial deposition associated with glacial tectonics.

With reference to glacial tectonics the writer underlined, in the Table, **glaciotectonic re-deposition** because glaciotectonics - like any tectonic process - causes deformation and displacement of deposits, but it does not create a new sediment, as glacial sensu stricto and glacioaqueous deposition do. This is obvious when large portions of substratum are glaciotectonically displaced en masse eg. when they form glaciotectonic end moraines. However, glaciotectonism is also commonly recorded within thin layer-like zones composed of substratum sediments, often containing inclusions of till, which are overlain by subglacial tills. The existence of such "layers" brought about the introduction of a term "deformation till", with deformation suggesting a process of deposition - as already mentioned above. The author's study of lodgement tills (Ruszczyńska-Szenajch 1983) supplied further evidence indicating that regular lodgement processes may be synchronous with glaciotectonic action. This synchroneity results in the formation of layer-like zones comprising redeposited substratum material and till inclusions. The inclusions are created by regular lodgement processes but substratum sediments only change their position, they do not change their origin (eg. lacustrine, fluvial, marine etc). Consequently glacial deposition may be accompanied by glaciotectonic deformation, but deformation itself cannot be regarded as a process of deposition (which by definition results in creation of a unique type of sediment).

An even more complicated question is posed by the process of lodgement itself, especially "hard" lodgement dominated by pressure and friction, which results in strong compaction and consistent orientation of particles within the resulting till. It seems most probable that this process is transitional between glacial sensu stricto deposition and glaciotectonic action (as for example the formation of flow till, which is transitional between glacial sensu stricto deposition and glacioaqueous processes). Once again it is evident that nature favours transitional boundaries rather than sharp limits.

In spite of transitional boundaries between some of the processes discussed, the different nature of glaciotectonic and glacial sensu stricto processes becomes more and more clear. It must be stressed, however, that both these groups of processes frequently operate together: at the same time and in the same place. They simply often coexist. Hence the examination of the whole set of their effects - recorded in exposures or in other geological evidence - is necessary for characterising the corresponding subglacial environment, as well as for the recognition of individual processes.

ACKNOWLEDGEMENTS. The author is obliged and grateful to Dr. D. G. Croot for critical reading of the manuscript and for his valuable suggestions and improvements of the text.

REFERENCES

Aber, J.S. 1985. The character of glaciotectonism. Geologie en Mijnbouw 64:389-395.

Boulton, G.S. 1979. Processes of glacier erosion on different substrata. Journal of Glaciology 23: 15-38.

Boulton, G.S. 1982. Subglacial processes and the development of glacial bedforms. In: Davidson-Arnott,R., Nickling, W., Fahey, B.D. (eds): Research in Glacial, Glacio-fluvial and Glacio-lacustrine Systems: 1-31. Geo Books Norwich.

Dreimanis, A. 1976. Tills: their origin and properties. In: Leggett R.F. (ed). Glacial Till: 11-49. Royal Society of Canada and National Research Council of Canada. Ottawa.

Fenton, M.M. 1983. Deformation terrain Mid-Continent Region: properties, subdivision, recognition. Annual meeting of the Geological Society of America, Abstracts 15:250. Madison, Wisconsin.

Moran, S.R., Clayton L., Hooke, R.L., Fenton, M.M. and Andriasheck, L.D. 1980. Glacier-bed landforms of the Prairie Region of North America. Journal of Glaciology 25:457-476.

Ruszczyńska-Szenajch, H. 1980. Glacial erosion in contradistinction to glacial tectonics. In: Stankowski, W. (ed): Tills and Gliogene Deposits: 71-76. Poznań.

Ruszczyńska-Szenajch, H. 1983. Lodgement tills and syndepositional glaciotectonic processes related to subglacial thermal and hydrologic conditions. In: Evenson, E.B. Schlüchter, Ch. and Rabassa J. (eds): Tills and Related Deposits: 113-117. A.A. Balkema, Rotterdam.

Ruszczyńska-Szenajch, H. 1987: The origin of glacial rafts: detachment, transport, deposition. Boreas 16: 101-112.

Glaciotectonics: Forms and Processes, Croot (ed.), © 1988 Balkema, Rotterdam. ISBN 90 6191 848 0

Bibliography of glaciotectonic references

James S.Aber
Earth Science Department, Emporia State University, Emporia, Kans., USA

ABSTRACT: This bibliography contains nearly 300 references to published works on the subject of glaciotectonic structures, landforms, and other features created by glacially induced deformation of pre-existing bedrock and sediment strata. The bibliography is a project of INQUA's Work Group on Glacial Tectonics to which many members contributed. References are listed in standard format, followed by a geographic index.

1. INTRODUCTION

This bibliography includes nearly 300 references to published articles, maps or books, in which glaciotectonics is the major subject or a significant secondary aspect of the reference. The bibliography is meant to be representative and selective, not comprehensive. Therefore, abstracts, theses, and other informal or unpublished works are not included. Works in which glaciotectonics are mentioned only briefly or casually (less than one paragraph) with little description or analysis are likewise not included. The two oldest references contained in the bibliography are by Puggaard (1851) and Lyell (1863). References through early 1988 are included.

The term glaciotectonic refers to structures and landforms created by glacially induced deformation of pre-existing substratum bedrock and sediment. Shallow (<200 m deep) crustal features are normally included, but crustal depression and rebound are usually excluded. On this basis, several related glacial phenomena are not considered in this bibliography: primary or penecontemporaneous features (i.e. till fabric, till lodgement, dead-ice collapse, soft-sediment deformation, etc.). Deformations within glaciers and ice sheets are likewise omitted.

This bibliography represents a project of the INQUA Work Group on Glacial Tectonics. J. S. Aber served as the primary compiler, and many members of the work group contributed information concerning references. The references are arranged in alphabetical order according to the author's last name. Scandinavian letters ae, ø and â are listed after z; other letters with accent marks are listed in normal position. Standard reference format is used with names of periodicals and publishing organizations spelled out as fully as possible.

The bibliography is followed by a geographic index. For each geographic category, references are classified in two broad groups. Group A includes brief to detailed local investigations of limited areas or reconnaissance studies of features over larger areas. Group B includes regional syntheses involving large areas, demonstrations of methods, or theoretical analyses. In addition to the geographic categories, a "general" category contains references that are not related to certain geographic regions. References are listed in several languages: Danish, Dutch, English, French, German, Polish and Swedish; however, no Russian references are included.

2. BIBLIOGRAPHY

Aber, J.S. 1979. Kineto-stratigraphy at Hvideklint, Møn, Denmark and its regional significance. Geological Society Denmark, Bulletin 28:81-93.

Aber, J.S. 1982a. Two-ice-lobe model for Kansan glaciation. Transactions Nebraska Academy Sciences 10:25-29.

Aber, J.S. 1982b. Model for glaciotectonism. Geological Society Denmark, Bulletin 30:79-90.

Aber, J.S. 1985a. The character of glaciotectonism. Geologie en Mijnbouw 64:389-395.

Aber, J.S. 1985b. Definition and model for Kansan glaciation. Ter-Qua Symposium Series, vol. 1:53-60.

Aber, J.S. 1988a. Structural geology exercises with glaciotectonic examples. Hunter Textbooks, Winston-Salem, North Carolina, 140 p.

Aber, J.S. 1988b. Spectrum of constructional glaciotectonic landforms. In Goldthwait, R.P. and Matsch, C.L. (eds.), Genetic classification of glacigenic deposits, in press. A.A. Balkema, Rotterdam.

Aber, J.S. 1988c. West Atchison drift section. Geological Society America, Centennial Field Guide 4:5-10.

Aber, J.S. and Aarseth, I. 1988. Glaciotectonic structure and genesis of the Herdla Moraines, western Norway. Norsk Geologisk Tidsskrift 68 (in press).

Adrielsson, L. 1984. Weichselian lithostratigraphy and glacial environments in the Ven-Gumslöv area, southern Sweden. Lundqua Thesis 14, Lund University, Sweden, 120 p.

Andersen, S.A. 1944. Det Danske Landskabs Historie. I. Bind, Undergrunden. Populaer-Videnskabeligt Forlag, Copenhagen, 480 p.

Andrews, D.E. 1980. Glacially thrust bed rock--an indication of late Wisconsin climate in western New York State. Geology 8:97-101.

Babcock, E.A., Fenton, M.M. and Andriashek, L.D. 1978. Shear phenomena in ice-thrust gravels, central Alberta. Canadian Journal Earth Sciences 15:277-283.

Banham, P.H. 1975. Glacitectonic structures: a general discussion with particular reference to the contorted drift of Norfolk. In Wright, A.E. and Moseley, F. (eds.), Ice ages: ancient and modern. Geological Journal, Special Issue 6:69-94.

Banham, P.H. 1977. Glacitectonics in till stratigraphy. Boreas 6:101-105.

Barbour, E.H. 1913. A minor phenomenon of the glacial drift in Nebraska. Nebraska Geological Survey, vol. 4, part 9:161-164.

Ber, A. 1987. Glaciotectonic deformation of glacial landforms and deposits in the Suwalki Lakeland (NE Poland). In Meer, J.J.M. van der (ed.), Tills and glaciotectonics, p. 135-143. A.A. Balkema, Rotterdam.

Berg, M.W. van den and Beets, D.J. 1987. Saalian glacial deposits and morphology in the Netherlands. In Meer, J.J.M. van der (ed.), Tills and glaciotectonics, p. 235-251. A.A. Balkema, Rotterdam.

Berthelsen, A. 1971. Fotogeologiske og feltgeologiske undersøgelser i NV-Sjaelland. Dansk Geologisk Forening, Arsskrift for 1970:64-69.

Berthelsen, A. 1973a. Skriften på vaeggen. Varv, Nr. 1:3-5.

Berthelsen, A. 1973b. Weichselian ice advances and drift successions in Denmark. Bulletin Geol. Instit. Univ. Uppsala, n.s. 5:21-29.

Berthelsen, A. 1974. Nogle forekomster af intrusivt moraeneler i NØ-Sjaelland. Dansk Geologisk Forening, Arsskrift for 1973:118-131.

Berthelsen, A. 1975. Geologi på Røsnaes. Varv, Ekskursionsfører 3.

Berthelsen, A. 1978. The methodology of kineto-stratigraphy as applied to glacial geology. Geological Society Denmark, Bulletin 27, Special Issue, p. 25-38.

Berthelsen, A. 1979a. Recumbent folds and boudinage structures formed by sub-glacial shear: An example of gravity tectonics. In Linden, W.J.M. van der (ed.), Van Bemmelen and his search for harmony. Geologie en Mijnbouw 58:253-260.

Berthelsen, A. 1979b. Contrasting views on the Weichselian glaciation and deglaciation of Denmark. Boreas 8:125-132.

Berthelsen, A., Konradi, P. and Petersen, K.S. 1977. Kvartaere lagfølger og strukturer i Vestmøns klinter. Dansk Geologisk Forening, Arsskrift for 1976:93-99.

Billings, M.P. 1972. Structural Geology, 3rd edition. Prentice-Hall, New Jersey, 606 p.

Bluemle, J.P. 1966. Ice thrust bedrock in northwest Cavalier County, North Dakota. North Dakota Academy Science 20:112-118 (North Dakota Geological Survey, Miscellaneous Series 33).

Bluemle, J.P. 1971. Geology of McLean County, North Dakota. North Dakota Geological

Survey, Bulletin 60, part 1.

Bluemle, J.P. 1973. Geology of Nelson and Walsh Counties, North Dakota. North Dakota Geological Survey, Bulletin 57, part 1.

Bluemle, J.P. 1975. Geology of Griggs and Steele Counties. North Dakota Geological Survey, Bulletin 64, part 1.

Bluemle, J.P. 1981. Geology of Sheridan County, North Dakota. North Dakota Geological Survey, Bulletin 75, part 1.

Bluemle, J.P. and Clayton, L. 1984. Large-scale glacial thrusting and related processes in North Dakota. Boreas 13:279-299.

Brodzikowski, K. and Loon, A.J. van 1985. Inventory of deformational structures as a tool for unravelling the Quaternary geology of glaciated areas. Boreas 14:175-188.

Brykczynska, E. and Brykczynski, M. 1974. Geologia przekopu Trasy Lazienkowskiej na tle problematyki zaburzen osadow trzeciorzedu i czwartorzedu w Warszawie. (The geological section in the Lazienkowska Route and the problems of the deformations of the Tertiary and Quaternary sediments in Warsaw.) Prace Muz. Ziemi 22:199-218. Warszawa.

Brykczynski, M. 1982. Glacitektonika krawedziowa w Kotlinie Warszawskiej i Kotlinie Płockiej. (Valley-side glacitectonics in the Warsaw Basin and the Płock Basin.) Prace Muz. Zieme 35:3-68. Warszawa.

Byers, A.R. 1959. Deformation of the Whitemud and Eastend Formations near Claybank, Saskatchewan. Transactions Royal Society Canada 53, series 3, sect. 4:1-11.

Carle, W. 1938. Das innere Gefüge der Stauch-Endmoränen und seine Bedeutung für die Gliederung des Altmoränengebietes. Geol. Rundschau 29:27.

Carlson, C.G. and Freers, T.F. 1975. Geology of Benson and Pierce Counties, North Dakota. North Dakota Geological Survey, Bulletin 59, part 1.

Christiansen, E.A. 1956. Glacial geology of the Moose Mountain area Saskatchewan. Saskatchewan Department Mineral Resources, Report 21.

Christiansen, E.A. 1959. Glacial geology of the Swift Current Area, Saskatchewan. Saskatchewan Department Mineral Resources, Report 32.

Christiansen, E.A. 1961. Geology and ground-water resources of the Regina Area Saskatchewan. Saskatchewan Research Council, Geology Division, Report 2.

Christiansen, E.A. 1971a. Tills in southern Saskatchewan, Canada. In Goldthwait, R.P. (ed.), Till/a symposium, p. 167-183. Ohio State University Press, Ohio.

Christiansen, E. A. 1971b. Geology and groundwater resources of the Melville Area (62K, L) Saskatchewan. Saskatchewan Research Council, Geology Division, Map No. 12.

Christiansen, E.A. 1979. The Wisconsinan deglaciation of southern Saskatchewan and adjacent areas. Canadian Journal Earth Sciences 16:913-938.

Christiansen, E.A. and Whitaker, S.H. 1976. Glacial thrusting of drift and bedrock. In Legget, R.F. (ed.), Glacial till. Royal Society Canada, Special Publ. 12:121-130.

Clayton, L. and Moran, S.R. 1974. A glacial process-form model. In Coates, D.R. (ed.), Glacial geomorphology, p. 89-119. SUNY-Binghamton Publications in Geomorphology, Binghamton, New York.

Clayton, L., Moran, S.R. and Bluemle, J.P. 1980. Explanatory text to accompany the Geologic Map of North Dakota. North Dakota Geological Survey, Report of Investigation No. 69.

Clayton, L., Teller, J.T. and Attig, J.W. 1985. Surging of the southwestern part of the Laurentide Ice Sheet. Boreas 14:235-241.

Crommelin, R.D. and Maarleveld, G.C. 1949. Een nieuwe geologische kaartering van de zuidelijke Veluwe. Tijdschr. Koninkl. Ned. Aardrijkskundig Genoot. 66:41.

Croot, D.G. 1987. Glacio-tectonic structures: A mesoscale model of thin-skinned thrust sheets? Journal Structural Geology 9:797-808.

Dahl, K. 1978. Kort over Danmark--Tekst og detailkort over fredede områder, 3rd ed. Danmarks Naturfredningsforening, Copenhagen, 184 p.

Dellwig, L.F. and Baldwin, A.D. 1965. Ice-push deformation in northeastern Kansas.

Kansas Geological Survey, Bulletin 175, part 2, 16 p.

Dredge, L.A. and Grant, D.R. 1987. Glacial deformation of bedrock and sediment, Magdalen Islands and Nova Scotia, Canada: Evidence for a regional grounded ice sheet. In Meer, J.J.M. van der (ed.), Tills and glaciotectonics, p. 183-195. A.A. Balkema, Rotterdam.

Dreimanis, A. 1988. Genetic classification of tills, the history of its development, and terminology in 21 languages. In Goldthwait, R.P. and Matsch, C.L. (eds.), Genetic classification of glacigenic deposits, in press. A.A. Balkema, Rotterdam.

Dreimanis, A., Hamilton, J.P. and Kelley, P.E. 1987. Complex subglacial sedimentation of Catfish Creek Till at Bradtville, Ontario, Canada. In Meer, J.J.M. van der (ed.), Tills and glaciotectonics, p. 73-87. A.A. Balkema, Rotterdam.

Dreimanis, A. and Lundqvist, J. 1984. What should be called till? Striae 20:5-10.

Dreimanis, A. and Schlüchter, C. 1985. Field criteria for the recognition of till or tillite. Palaeogeography, Palaeoclimatology, Palaeoecology 51:7-14.

Drewry, D. 1986. Glacial geologic processes. Edward Arnold, London, 276 p.

Drozdowski, E. 1981. Pre-Eemian push-moraines in the lower Vistula region, northern Poland. In Ehlers, J. and Zandstra, J.G. (eds.), Glacigenic deposits in the southwest parts of the Scandinavian Icesheet. Mededelingen Rijks Geologische Dienst 34-1/11:57-61.

Dyke, A.S. and Prest, V.K. 1987. Late Wisconsinan and Holocene history of the Laurentide Ice Sheet. Geographie physique et Quaternaire XLI:237-263.

Dylik, J. 1961. The Lodz region. VI INQUA Congress, Guidebook of excursion, Lodz.

Dylik, J. 1967. The main elements of Upper Pleistocene paleogeography in central Poland. Biuletyn Peryglacjalyn 16:85-115.

Edwards, M.B. 1978. Glacial environments. In Reading, H.G. (ed.), Sedimentary environments and facies, p. 416-438. Elsevier, New York.

Ehlers, J. 1982. Different till types in North Germany and their origin. In Evenson, E.B., Schlüchter, C. and Rabassa, J. (eds.), Tills and related deposits, p. 61-80. A.A. Balkema, Rotterdam.

Ehlers, J. (editor) 1983a. Glacial deposits in North-west Europe. A.A. Balkema, Rotterdam, 470 p.

Ehlers, J. 1983b. The glacial history of North-west Germany. In Ehlers, J. (ed.), Glacial deposits in North-west Europe, p. 229-238. A.A. Balkema, Rotterdam.

Ehlers, J., Gibbard, P.L. and Whiteman, C.A. 1987. Recent investigations of the Marly Drift of northwest Norfolk, England. In Meer, J.J.M. van der (ed.), Tills and glaciotectonics, p. 39-54. A.A. Balkema, Rotterdam.

Erdmann, E. 1873. Iaktagelser öfver moränbildningar och deraf betäckta skiktade jordarter i Skåne. Geologiska Föreningen i Stockholm, Förhandlingar 2:13-24.

Eybergen, F.A. 1987. Glacier snout dynamics and contemporary push moraine formation at the Turtmannglacier, Wallis, Switzerland. In Meer, J.J.M van der (ed.), Tills and glaciotectonics, p. 217-231. A.A. Balkema, Rotterdam.

Feldmann, R.M. 1964. Preliminary investigation of the paleontology and structure of Sibley Buttes, central North Dakota. The Compass 41:129-135.

Fenton, M.M. and Andriashek, L.D. 1983. Surficial geology Sand River area, Alberta (NTS 73L). Alberta Geological Survey, map scale = 1:250,000.

Figge, K. 1983. Morainic deposits in the German Bight area of the North Sea. In Ehlers, J. (ed.), Glacial deposits in North-west Europe, p. 299-303. A.A. Balkema, Rotterdam.

Flint, R.F. 1971. Glacial and Quaternary geology. J. Wiley, New York, 892 p.

Fraser, F.J., McLearn, F.H., Russell, L.S., Warren, P.S. and Wickenden, R.T.D. 1935. Geology of Saskatchewan. Geological Survey Canada, Memoir 176, 137 p.

Frederiksen, J. 1976. Hvad Sønderjyske klinter fortaeller. Varv, Nr. 2:35-45.

Fuller, M.L. 1914. The geology of Long Island, New York. United States Geological Survey, Professioal Paper 82.

Funder, S. and Petersen, K.S. 1979. Glacitectonic deformations in East Greenland.

Geological Society Denmark, Bulletin 28:115-122.

Galon, R. 1961. North Poland, area of the last glaciation. VI INQUA Congress, Guidebook of excursion. Lodz.

Gans, W. de, Groot, T. de and Zwaan, H. 1987. The Amsterdam basin, a case study of a glacial basin in the Netherlands. In Meer, J.J.M. van der (ed.), Tills and glaciotectonics, p. 205-216. A.A. Balkema, Rotterdam.

Gijssel, K. van 1987. A lithostratigraphic and glaciotectonic reconstruction of the Lamstedt Moraine, Lower Saxony (FRG). In Meer, J.J.M. van der (ed.), Tills and glaciotectonics, p. 145-155. A.A. Balkema, Rotterdam.

Gravenor, C.P. 1985. Glacial tectonic and flow structures in glaciogenic deposits: A cautionary note. Geological Society Denmark, Bulletin 34:3-11.

Gravenor, C.P., Green, R. and Godfrey, J.D. 1960. Air photographs of Alberta. Alberta Research Council, Bulletin 5.

Green, R. and Copeland, F.L. 1972. Geological Map of Alberta. Alberta Geological Survey, map scale = 1:1,267,000.

Gripp, K. 1929. Glaciologische und geologische Ergebnisse der Hamburgischen Spitzbergen-Expedition 1927. Abh. naturwiss. Ver. Hamburg XXII:147-249.

Grube, F. and Vollmer, T. 1985. Der geologische Bau pleistozaner Inlandgletschersedimente Norddeutschlands. Geological Society Denmark, Bulletin 34:13-25.

Gry, H. 1940. De istektoniske forhold i moleromraadet. Meddelelser Dansk Geologisk Forening 9:586-627.

Gry, H. 1965. Furs geologi. Fur Museum, Dansk Natur-Dansk Skole, Arsskrift for 1964, 55 p.

Gry, H. 1979. Beskrivelse til geologisk kort over Danmark, Kortbladet Løgstor, Kvartaere aflejringer. Danmarks Geologiske Undersøgelse, I Raekke, Nr. 26.

Haarsted, V. 1956. De kvartaergeologiske og geomorfologiske forhold på Møn. Meddelelser Dansk Geologisk Forening 13:124-126.

Hansen, D.E. 1967. Geology and ground water resources Divide County, North Dakota. North Dakota Geological Survey, Bulletin 45, part 1 Geology.

Hansen, S. 1965. The Quaternary of Denmark. In Rankama, K. (ed.), The Quaternary, vol. 1, p. 1-90. J. Wiley, New York.

Heginbottom, J.A. 1985. Glacially deformed ground-ice bodies in a wall of the community ice cellar, Tuktoyaktuk, western Arctic coast, Canada (July 28, 1983). Geology 13:312.

Hicock, S.R. and Dreimanis, A. 1985. Glaciotectonic structures as useful ice-movement indicators in glacial deposits: four Canadian case studies. Canadian Journal Earth Sciences 22:339-346.

Hillefors, A. 1973. The stratigraphy and genesis of stoss- and lee-side moraines. Bulletin Geol. Inst. Univ. Uppsala 5:139-154.

Hillefors, A. 1985. Deep-weathered rock in western Sweden. Fennia 163:2, p. 293-301.

Hintze, V. 1937. Moens Klints geologi. C.A. Reitzels, Copenhagen, 410 p. (edited posthumously by E.L. Mertz and V. Nordmann).

Höfle, H.-C. and Lade, U. 1983. The stratigraphic position of the Lamstedter Moraine within the Younger Drenthe substage (Middle Saalian). In Ehlers, J. (ed.), Glacial deposits in North-west Europe, p. 343-346. A.A. Balkema, Rotterdam.

Hollick, A. 1894. Dislocations in certain portions of the Atlantic coastal plain strata and their probable causes. Transactions New York Academy Sciences 14:8-20.

Holst, N.O. 1903. Om skriftkritan i tullstorpstrakten och de båda moräner, in hvilka den är inbäddad. Sveriges Geologiska Undersökning, C194.

Holst, N.O. 1911. Beskrifning till kartbladet Börringekloster. Sveriges Geologiska Undersökning, Aa138.

Hopkins, O.B. 1923. Some structural features of the plains area of Alberta caused by Pleistocene glaciation. Geological Society America, Bulletin 34:419-430.

Houmark-Nielsen, M. 1976. En glacialstratigrafisk oversigt fra Nordsamsø og Tunø. Dansk Geologisk Forening, Arsskrift for 1975:11-13.

Houmark-Nielsen, M. 1981. Glacialstratigrafi i Danmark øst for Hovedopholdslinien. Dansk Geologisk Forening, Arsskrift for 1980:61-76.

Houmark-Nielsen, M. 1983. Glacial stratigraphy and morphology of the northern Baelthav region. In Ehlers, J. (ed.), Glacial deposits in North-west Europe, p. 211-217. A.A. Balkema, Rotterdam.

Houmark-Nielsen, M. 1987. Pleistocene stratigraphy and glacial history of the central part of Denmark. Geological Society Denmark, Bulletin 36:1-189.

Houmark-Nielsen, M. and Berthelsen, A. 1981. Kineto-stratigraphic evaluation and presentation of glacial-stratigraphic data, with examples from northern Samsø, Denmark. Boreas 10:411-422.

Houmark-Nielsen, M. and Kolstrup, E. 1981. A radiocarbon dated Weichselian sequence from Sejerø, Denmark. Geologiska Föreningens i Stockholm, Förhandlingar 103:73-78.

Howe, W.B. 1968. Guidebook to Pleistocene and Pennsylvanian formations in the St. Joseph area, Missouri. Missouri Geological Survey and Water Resources, Association Missouri Geologists, 15th annual field trip and meeting, 45 p.

Hubbert, M.K. and Rubey, W.W. 1959. Role of fluid pressure in mechanics of overthrust faulting. Geological Society America, Bulletin 70:115-166.

Humlum, O. 1978. A large till wedge in Denmark: implications for the subglacial thermal regime. Geological Society Denmark, Bulletin 27:63-71.

Humlum, O. 1983. Dannelsen af en disloceret randmoraene ved en avancerende isrand, Höfdabrekkujölkull, Island. Dansk Geologisk Forening, Arsskrift for 1982:11-26.

Hyyppa, E. 1955. On the Pleistocene geology of southeastern New England. Acta Geographica 14:155-225.

Håkansson, E. and Sjørring, S. 1982. Et molerprofil i kystklinten ved Salger Høj, Mors. Dansk Geologisk Forening, Arsskrift for 1981:131-134.

Ingolfsson, O. 1985. Late Weichselian glacial geology of the lower Borgarfjördur region, western Iceland: A preliminary report. Arctic 38:210-213.

Ingolfsson, O. 1987. Investigation of the late Weichselian glacial history of the lower Borgarfjördur region, western Iceland. Lundqua Thesis 19, Lund University, Sweden.

Jacobsen, E.M. 1976. En moraenestratigrafisk undersøgelse af klinterne pa Omø. Dansk Geologisk Forening, Arrskrift for 1975:15-17.

Jahn, A. 1950. Nowe dane o połozeniu kry jurajskiej w Lukowie (New facts concerning the ice transported blocks of the Jurassic at Lukow). Annales Societatis Geologorum Poloniae 19:372-385, Krakow.

Jahn, A. 1956. Wyzyna Lubelska, rzezba i czwartorzed. (Geomorphology and Quaternary history of Lublin Plateau.) Panstw. Wyd. Nauk., Warszawa, 453 p.

Jelgersma, S. and Breeuwer, J.B. 1975. Toelichting bij de kaart glaciale verschijnselen gedurende het Saalian, 1:600,000. In Zagwijn, W.H. and Staalduinen, C.J. (eds.), Toelichting bij geologische overzichtskaarten van Nederland, p. 93-103. Rijks Geologische Dienst, Haarlem.

Jensen, J.B. and Knudsen, K.L. 1984. Kvartaerstratigrafiske undersøgelser ved Gyldendal og Kås Hoved i det vestlige Limfjordsområde. Dansk Geologisk Forening, Arsskrift for 1983:35-54.

Jessen, A. 1930. Klinten ved Halkhoved. Danmarks Geologiske Undersøgelse, IV Raekke, vol. 2:8, 26 p.

Jessen, A. 1931. Lønstrup Klint. Danmarks Geologiske Undersøgelse, II Raekke 49, 142 p.

Johnstrup, F. 1874. Ueber die Lagerungsverhaltnisse und die Hebungs-phänomene in den Kreidefelsen auf Moen und Rügen. Zeit. deutsch. geol. Ges. 1874:533-585.

Jong, J.D. de 1952. On the structure of the pre-glacial Pleistocene of the Archemerberg (Prov. of Overijsel, Netherlands). Geologie en Mijnbouw 14:86.

Jong, J.D. de 1955. Geologische onderzoekingen in de stuwwallen van oostelijk Nederland. I, Archemerberg en Nijverdal. Mededel. Geol. Sticht. N.S. 8:33.

Jong, J.D. de 1967. The Quaternary of the Netherlands. In Rankama, K. (ed.), The Quaternary, vol. 2, p. 301-426. J. Wiley, New York.

Kabel-Windloff, C. 1987. Petrographical and structural investigations of the Brodtener

Ufer cliff. In Meer, J.J.M. van der (ed.), Tills and glaciotectonics, p. 89-96. A.A. Balkema, Rotterdam.

Kalin, M. 1971. The active push moraine of the Thompson glacier. Axel Heiberg Island Research Reports, no. 4, McGill University, Montreal.

Kaye, C.A. 1964a. Outline of Pleistocene geology of Martha's Vineyard, Massachusetts. United States Geological Survey, Professional Paper 501-C:134-139.

Kaye, C.A. 1964b. Illinoian and early Wisconsin moraines of Martha's Vineyard, Massachusetts. United States Geological Survey, Professional Paper 501-C:140-143.

Kaye, C.A. 1980. Geologic profile of Gay Head Cliff, Martha's Vineyard, Massachusetts. United States Geological Survey, Open-file Report 80-148.

Klassen, R.A. 1982. Glaciotectonic thrust plates, Bylot Island, District of Franklin. Geological Survey Canada, Current Research, part A, Paper 82-1A:369-373.

Klatkowa, H. 1972. Paleogeografia Wyzyny Lodzkiej i obszarow sasiednich podczas zlodowacenia warcianskiego. (Paleogeographie du Plateau de Lodz et de terrain avoisinant pendant la glaciation de Warta.) Acta Geogr. Lodz. 28, Lodz, 220 p.

Königsson, L.K. and Linde, L.A. 1977. Glaciotectonically disturbed sediments at Rönnerrum on the island of Oland. Geologiska Föreningens i Stockholm, Förhandlingar 99:68-72.

Konradi, P.B. 1973. Foraminiferas in some Danish glacial deposits. Bulletin Geol. Instit. Univ. Uppsala, n.s. 5:173-175.

Krüger, J. 1983. Glacial morphology and deposits in Denmark. In Ehlers, J. (ed.), Glacial deposits in North-west Europe, p. 181-191. A.A. Balkema, Rotterdam.

Krüger, J. 1987. Traek af et glaciallandskabs udvikling ved nordranden af Myrdalsjökull, Island. Dansk Geologisk Forening, Arsskrift for 1986:49-65.

Krüger, J. and Humlum, O. 1980. Deformations- og erosionsstrukturer i bundmoraenelandskabet ved Myrdalsjökull, Island. Dansk Geologisk Forening, Arsskrift for 1979:31-39.

Kupsch, W.O. 1955. Drumlins with jointed boulders near Dollard, Saskatchewan. Geological Society America, Bulletin 66:327-338.

Kupsch, W.O. 1962. Ice-thrust ridges in western Canada. Journal Geology 70:582-594.

Kupsch, W.O. 1965. Jointing of boulders caused by flowing ice. Alberta Society Petroleum Geologists, 15th Annual Field Conference Guidebook, part 1:112-115.

Lagerlund, E. 1987. An alternative Weichselian glaciation model, with special reference to the glacial history of Skåne, South Sweden. Boreas 16:433-459.

Lammerson, P.R. and Dellwig, L.F. 1957. Deformation by ice push of lithified sediments in south-central Iowa. Journal Geology 65:546-550.

Landvik, J. and Hamborg, M. 1987. Weichselian glacial episodes in outer Sunnmore, western Norway. Norsk Geologisk Tidsskrift 67:107-123.

Lange, W., Menke, B. and Picard, K.-E. 1979. Die Deutung glazigener sedimente in Schleswig-Holstein. In Ehlers, J. and Grube, F. (eds.), Glacigenic deposits in the southwest parts of the Scandinavian Icesheet. Verh. naturwiss. Ver. Hamburg, (NF)23:51-68.

Larsen, G., Jørgensen, F.H. and Priisholm, S. 1977. The stratigraphy, structure and origin of glacial deposits in the Randers area, eastern Jylland. Danmarks Geologiske Undersøgelse, II Raekke, nr. 111, 36 p.

Lavrushin, Y.A. 1971. Dynamische Fazies und Subfazies der Grundmoräne. Zeit. Angew. Geol. 17:337-343.

Lavrushin, Y.A. 1978. Texturen, Fazies und stoffliche Zusammensetzung der Grundmoränen. In, 100 Jahre Glazialtheorie im Gebiet der skandinavischen Vereisungen. Schriftenreihe für Geologische Wissenschaften 9:161:177.

Lewinski, J. and Rozycki, S.Z. 1929. Dwa profile geologiczne przez Warszawe. Sprawozd. Tow. Nauk. Warsz. 22:30-50. Warszawa.

Lundqvist, J. 1967. Submoräna sediment i Jämtlands Län. Sveriges Geologiska Undersökning, C618, 267 p.

Lundqvist, J. 1985. Glaciations and till or tillite genesis: Examples from Pleistocene glacial drift in central Sweden. Palaeogeography, Palaeoclimatology, Palaeoecology 51:389-395.

Lundqvist, J. 1987. Beskrivning til Jordartskarta över Västernorrlands Län och Förutvarande Fjällsjö K:N. Sveriges Geologiska Undersökning, Ca55, 270 p.

Lundqvist, J. and Lagerbäck, R. 1976. The Pärve Fault: A late-glacial fault in the Precambrian of Swedish Lapland. Geologiska Föreningens i Stockholm, Förhandlingar 98:45-51.

Lyell, C. 1863. The geological evidences of the antiquity of man, 3rd ed. John Murray, London.

Lykke-Andersen, A.-L. 1981. En ny C-14 datering fra AEldre Yoldia Ler i Hirtshals Kystklint. Dansk Geologisk Forening, Arsskrift for 1980:1-5.

Maarleveld, G.C. 1953. Standen van het landijs in Nederland. Boor en Spade 4:95-105.

Maarleveld, G.C. 1981. The sequence of ice-pushing in the central Netherlands. In Ehlers, J. and Zandstra, J.G. (eds.), Glacigenic deposits in the southwest parts of the Scandinavian Icesheet. Mededelingen Rijks Geologische Dienst 34-1/11:2-6.

Maarleveld, G.C. 1983. Ice-pushed ridges in the central Netherlands. In Ehlers, J. (ed.), Glacial deposits in North-west Europe, p. 393-397. A.A. Balkema, Rotterdam.

Mackay, J.R. 1959. Glacier ice-thrust features of the Yukon coast. Geographical Bulletin 13:5-21.

Mackay, J.R. and Mathews, W.H. 1964. The role of permafrost in ice-thrusting. Journal Geology 72:378-380.

Mackay, J.R., Rampton, V.N. and Fyles, J.G. 1972. Relic Pleistocene permafrost, western Arctic, Canada. Science 176:1321-1323.

Mackay, J.R. and Stager, J.K. 1966. Thick tilted beds of segregated ice, Mackenzie delta area, N.W.T. Biuletyn Peryglacjalny 15:39-43.

Madsen, V. 1916. Ristinge Klint. Danmarks Geologiske Undersøgelse, IV Raekke, vol. 1, nr. 2, 32 p.

Madsen, V., Nordmann, V. and Hartz, N. 1908. Eem-zonerne. Studier over Cyprinaleret og andre Eem-aflejringer i Danmark, Nord-Tyskland og Holland. Danmarks Geologisk Undersøgelse, II Raekke, nr. 17, 302 p.

Mangerud, J., Sønstegaard, E., Sejrup, H.-P. and Haldorsen, S. 1981. A continuous Eemian-early Weichselian sequence containing pollen and marine fossils at Fjøsanger, western Norway. Boreas 10:137-208.

Mathews, W.H. and MacKay, J.R. 1960. Deformation of soils by glacier ice and the influence of pore pressures and permafrost. Transactions Royal Society Canada 54, series 3, section 4:27-36.

Meer, J.J.M. van der (editor) 1987. Tills and glaciotectonics. A.A. Balkema, Rotterdam, 270 p.

Merrill, F.J.H. 1886a. On the geology of Long Island. Annals New York Academy Sciences 3:341-364.

Merrill, F.J.H. 1886b. On some dynamic effects of the ice-sheet. Proceedings American Association Advancement Science 35:228-229.

Meyer, K.-D. 1983. Saalian end moraines in Lower Saxony. In Ehlers, J. (ed.), Glacial deposits in North-west Europe, p. 335-342. A.A. Balkema, Rotterdam.

Meyer, K.-D. 1987. Ground and end moraines in Lower Saxony. In Meer, J.J.M. van der (ed.), Tills and glaciotectonics, p. 197-204. A.A. Balkema, Rotterdam.

Mickelson, D.M., Clayton, L., Fullerton, D.S. and Borns, H.W. Jr. 1982. The late Wisconsin glacial record of the Laurentide Ice Sheet in the United States. In Wright, H.E. Jr. (ed.), Late Quaternary environments of the United States, vol. 1, The late Pleistocene (Porter, S.C., ed.), p. 3-37. University of Minnesota Press, Minneapolis.

Mills, H.C. and Wells, P.D. 1974. Ice-shove deformation and glacial stratigraphy of Port Washington, Long Island, New York. Geological Society America, Bulletin 85:357-364.

Mojski, J.E. 1979. Zarys stratygrafii plejstocenu i budowy jego podłoza w regionie gdanskim. (Outline of the stratigraphy of the Pleistocene and the structure of its basement in the Gdansk region.) Biuletyn Inst. Geol. 317:50-60. Warszawa.

Moran, S.R. 1971. Glaciotectonic structures in drift. In Goldthwait, R.P. (ed.), Till/a symposium, p. 127-148. Ohio State University Press, Ohio.

Moran, S.R., Clayton, L., Hooke, R.LeB., Fenton, M.M. and Andriashek, L.D. 1980. Glacier-bed landforms of the Prairie region of North America. Journal Glaciology 25:457-476.

Muller, E.H. 1983. Till genesis and the glacier sole. In Evenson, E.B., Schlüchter, C. and Rabassa, J. (eds.), Tills and related deposits, p. 19-22. A.A. Balkema, Rotterdam.

Nielsen, A.V. 1967. Landskabets tilblivelse. In Norrevang, O. and Meyer, T.J. (eds.), Danmarks Natur, Bd 1, Landskabernes Opståen, p. 251-344. Politikens Forlag, Copenhagen.

Nielsen, J.B. 1987. Kvartaerstratigrafiske observationer langs østsiden af Roskilde Fjord. Dansk Geologisk Forening, Arsskrift for 1986:41-47.

Nielsen, P.E. 1980. Kvartaergeologiske undersøgelser i Korsør-området. Dansk Geologisk Forening, Arsskrift for 1979:55-62.

Nielsen, P.E. 1982. Till fabric reoriented by subglacial shear. Geologiska Föreningens i Stockholm, Förhandlingar 103:383-387.

Nielsen, P.E. and Jensen, L.B. 1984. Maringeologiske undersøgelser på Mejl Flak, Arhus Bugt. Dansk Geologisk Forening, Arsskrift for 1983:73-79.

Nowak, J. 1977. Specyficzna budowa geologiczna form polodowcowych zaleznych od podłoza. (Specific structure of forms of Quaternary age dependent on the substratum.) Studia Geol. Pol. 52:347-360. Warszawa.

Nye, J.F. 1952. The mechanics of glacier flow. Journal Glaciology 2:82-93.

Occhietti, S. 1973. Les structures et deformations engendrees par les glaciers--Essai de mise au point. Revue Geographique de Montreal 27:365-380.

Oldale, R.N. 1980. Pleistocene stratigraphy of Nantucket, Martha's Vineyard, the Elizabeth Islands, and Cape Cod, Massachusetts. In Larson, G.J. and Stone, B.D. (eds.), Late Wisconsin glaciation of New England, p. 1-34. Kendall/Hunt, Dubuque, Iowa.

Oldale, R.N. and O'Hara, C.J. 1984. Glaciotectonic origin of the Massachusetts coastal end moraines and a fluctuating late Wisconsinan ice margin. Geological Society America, Bulletin 95:61-74.

Osterkamp, W.R., Fenton, M.M., Gustavson, T.C., Hadley, R.F., Holliday, V.T., Morrison, R.B. and Toy, T.J. 1987. Great Plains. In Graf, W.L. (ed.), Geomorphic Systems of North America. Geological Society America, Centennial Special Vol. 2:163-210.

Parizek, R.P. 1964. Geology of the Willow Bunch Lake Area (72-H) Saskatchewan. Saskatchewan Research Council, Geology Division, Report 4, 46 p.

Pedersen, S.A.S. 1986. Videregående undersøgelser af sandkiler på Fur. Danmarks Geologiske Undersøgelse, Intern rapport nr. 32 (1986).

Pedersen, S.S. and Petersen, K.S. 1985. Sandkiler i moler på Fur. Danmarks Geologiske Undersøgelse, Intern rapport nr. 32 (1985).

Petersen, K.S. 1978. Applications of glaciotectonic analysis in the geological mapping of Denmark. Danmarks Geologiske Undersøgelse, Arbog 1977:53-61.

Petersen, K.S. and Buch, A. 1974. Dislocated tills with Paleogene and Pleistocene marine beds. Tectonics, lithology, macro- and microfossils. Danmarks Geologiske Undersøgelse, Arbog 1973:63-91.

Petersen, K.S. and Konradi, P.B. 1974. Lithologiske og palaeontologiske beskrivelse af profiler i kvartaeret på Sjaelland. Dansk Geologisk Forening, Arsskrift for 1973:47-56.

Picard, K.-E. 1969. Pleistocene tectonics and glaciation in Schleswig-Holstein, Germany. In Wright, H.E. Jr. (ed.), Quaternary geology and climate, p. 67-71. National Academy Science, Washington, D.C.

Prange, W. 1983. Fabric analyses from Weichselian glacial deposits in Schleswig-Holstein. In Ehlers, J. (ed.), Glacial deposits in North-west Europe, p. 321-324. A.A. Balkema, Rotterdam.

Prange, W. 1985. Glazialtektonik im Weichselglazial Schleswig-Holsteins und ihre Beziehungen zur morphologie. Geological Society Denmark, Bulletin 34:33-45.

Prest, V.K. 1983. Canada's heritage of glacial features. Geological Survey Canada, Miscellaneous Report 28, 119 p.

Prest, V.K., Grant, D.R. and Rampton, V.N. 1967. Glacial Map of Canada. Geological Survey Canada, Map 1253A, map scale = 1:5,000,000.

Puggaard, C. 1851. Møns Klint section. Reproduced in International Geological Congress XXI, Session Norden (1960), Guidebook I.

Rampton, V.N. 1982. Quaternary geology of the Yukon Coastal Plain. Geological Survey Canada, Bulletin 317, 49 p.

Rappol, M. 1987. Saalian till in the Netherlands: A review. In Meer, J.J.M. van der (ed.), Tills and glaciotectonics, p. 3-21. A.A. Balkema, Rotterdam.

Rasmussen, L.Aa. 1974. Om moraenestratigrafi i det nordlige Øresundsområde. Dansk Geologisk Forening, Arsskrift for 1973:132-139.

Rasmussen, L.Aa. 1975. Kineto-stratigraphic glacial drift units on Hindsholm, Denmark. Boreas 4:209-217.

Rasmussen, L.Aa. and Petersen, K.S. 1980. Resultater fra DGU's genoptagne kvartaergeologiske kortlaegning. Dansk Geologisk Forening, Arsskrift for 1979:47-54.

Rau, J.L., Bakken, W.E., Chmelik, J. and Williams, B.J. 1962. Geology and ground water resources of Kidder County, North Dakota. North Dakota Geological Survey, Bulletin 36, part 1 Geology.

Ringberg, B. 1980. Beskrivning till Jordartskartan Malmö SO (Description to the Quaternary map Malmö SO). Sveriges Geologiska Undersökning, Ae38.

Ringberg, B. 1983. Till stratigraphy and glacial rafts of chalk at Kvarnby, southern Sweden. In Ehlers, J. (ed.), Glacial deposits in North-west Europe, p. 151-154. A.A. Balkema, Rotterdam.

Ringberg, B., Holland, B. and Miller, U. 1984. Till stratigraphy and provenance of the glacial chalk rafts at Kvarnby and Angdala, southern Sweden. Striae 20:79-90.

Rosenkrantz, A. 1944. Nye bidrag til forståelsen af Ristinge Klints opbygning. Meddelelser Dansk Geologisk Forening 10:431-435.

Rotnicki, K. 1976. The theoretical basis for and a model of the origin of glaciotectonic deformations. Quaestiones Geographicae 3:103-139.

Rozycki, S.Z. 1970. Dynamiczne uławicenie glin zwałowych i inne procesy w dennej czesci moren ladolodow czwartorzedowych. (Dynamic stratification of tills and other processes in the basal part of moraines of the Quaternary ice sheets.) Acta Geol. Pol. 20:561-586. Warszawa.

Rozycki, S.Z. 1972. Plejstocen Polski srodkowej. Panstw. Wyd. Nauk., Warszawa, 314 p.

Rozycki, S.Z. 1982. Zaburzenia glacitektoniczne w rejonie Julianki. (Glacitectonic deformations nearby Julianka.) Biuletyn Geologiczny Uniwersytetu Warszawskiego 26:161-171. Warszawa.

Ruegg, G.H.J. 1981. Ice-pushed Lower and Middle Pleistocene deposits near Rhenen (Kwintelooijen): sedimentary-structural and lithological/granulometrical investigations. In Ruegg, G.H.J. and Zandstra, J.G. (eds.), Geology and archaeology of Pleistocene deposits in the ice-pushed ridge near Rhenen and Veenendaal. Mededelingen Rijks Geologische Dienst 35-2/7:165-177.

Ruegg, G.H. 1983. Glaciofluvial and glaciolacustrine deposits in the Netherlands. In Ehlers, J. (ed.), Glacial deposits in North-west Europe, p. 379-392. A.A. Balkema, Rotterdam.

Ruegg, G.H.J. and Zandstra, J.G. (eds.) 1981. Geology and archaeology of Pleistocene deposits in the ice-pushed ridge near Rhenen and Veenendaal. Mededelingen Rijks Geologische Dienst 35-2/7:163-268.

Ruszczynska-Szenajch, H. 1973. Kry lodowcowe wycisniete glacitektonicznie na terenie SE Mazowsza i S Podlasia (Glacial floes--bedrock masses--squeezed by ice sheets in mid-eastern Poland). Kwartalnik Geologiczny 17:560-576. Warszawa.

Ruszczynska-Szenajch, H. 1976. Glacitektoniczne depresje i kry lodowcowe na tle budowy geologicznej południowo-wschodniego Mazowsza i południowego Podlasia. (Glacitectonic depressions and glacial rafts in mid-eastern Poland). Studia Geologica Polonica 50:1-106. Warszawa.

Ruszczynska-Szenajch, H. 1978. Glacitectonic origin of some lake-basins in areas of Pleistocene glaciations. Polskie Archiwum Hydrobiologii 25:373-381.

Ruszczynska-Szenajch, H. 1979. Zroznicowanie zaburzen glacitektonicznych w zaleznosci od przewagi oddziaływania ciezaru lodu lub ruchu lodu. (Differentiation of glacitectonic deformations due to prevalence of ice-pressure or ice-movement). Biuletyn Geologiczny Uniwersytetu Warszawskiego 23:131-142.

Ruszczynska-Szenajch, H. 1980. Glacial erosion in contra-distinction to glacial tectonics. In Stankowski, W. (ed.), Tills and glacigene deposits, p. 71-76. Wydawnictwa Uniwersytetu im. Adama Mickiewicza, Poznan.

Ruszczynska-Szenajch, H. 1981. Fossil remnants of up-turned debris bands in Pleistocene glacial deposits of Poland. Sedimentology 28:713-722.

Ruszczynska-Szenajch, H. 1983. Lodgement tills and syndepositional glacitectonic processes related to subglacial thermal and hydrologic conditions. In Evenson, E.B., Schlüchter, C. and Rabassa, J. (eds.), Tills and related deposits, p. 113-117. A.A. Balkema, Rotterdam.

Ruszczynska-Szenajch, H. 1985. Origin and age of the large-scale glaciotectonic structures in central and eastern Poland. Annales Societatis Geologorum Poloniae 55:307-332, Krakow.

Ruszczynska-Szenajch, H. 1987. The origin of glacial rafts: detachment, transport, deposition. Boreas 16:101-112.

Rutten, M.G. 1960. Ice-pushed ridges, permafrost, and drainage. American Journal Science 258:293-297.

Sardeson, F.W. 1905. A particular case of glacial erosion. Journal Geology 13:351-357.

Sardeson, F.W. 1906. The folding of subjacent strata by glacial action. Journal Geology 14:226-232.

Sarnacka, Z. 1965. Struktury glacitektoniczne i marzłociowe w Gorze Kalwarii i Osiecku na południe od Warszawy. (Glacitectonic and frozen ground structures at Gora Kalwaria and Osieck south of Warsaw, central Poland.) Biuletyn Inst. Geol. 187:217-238. Warszawa.

Sauer, E.K. 1978. The engineering significance of glacier ice-thrusting. Canadian Geotechnical Journal 15:457-472.

Schlüchter, C. and Wohlfarth-Meyer, B. 1987. Till facies varieties of the western Swiss Alpine foreland. In Meer, J.J.M. van der (ed.), Tills and glaciotectonics, p. 67-72. A.A. Balkema, Rotterdam.

Schwan, J. and Loon, A.J. van 1979. Structural and sedimentological characteristics of a Weichselian kame terrace at Sønderby Klint, Funen, Denmark. Geologie en Mijnbouw 58:305-319.

Schwan, J. and Loon, A.J. van 1981. Structure and genesis of a buried ice-pushed zone near Rold (Funen, Denmark). Geologie en Mijnbouw 60:385-394.

Shaler, N.S. 1888. Geology of Martha's Vineyard. United States Geological Survey, Report for 1886, vol. 3:297-363.

Shaler, N.S. 1898. Geology of the Cape Cod district. United States Geological Survey, Report for 1896-97, part 2:497-593.

Sirkin, L. 1976. Block Island, Rhode Island: Evidence of fluctuation of the late Pleistocene ice margin. Geological Society America, Bulletin 87:574-580.

Sirkin, L. 1980. Wisconsinan glaciation of Long Island, New York to Block Island, Rhode Island. In Larson, G.J. and Stone, B.D. (eds.), Late Wisconsin glaciation of New England, p. 35-59. Kendall/Hunt, Dubuque, Iowa.

Sjørring, S. 1973. Some problems in the till stratigraphy of the northeastern part of Sjaelland. Bulletin Geol. Instit. Univ. Uppsala, n.s. 5:31-35.

Sjørring, S. 1974. Klinterne ved Hundested Dansk Geologisk Forening, Arsskrift for 1973:108-117.

Sjørring, S. 1977a. Glacialtektonik og istidsgeologi. Dansk Natur-Dansk Skole, Arsskrift 1977:31-44.

Sjørring, S. 1977b. The glacial stratigraphy of the island of Als, southern Denmark. Zeitschrift für Geomorphologie N.F., Suppl.-Bd. 27:1-11.

Sjørring, S. 1981. The Weichselian till stratigraphy in the southern part of Denmark. Quaternary Studies in Poland 3:103-109.

Sjørring, S. 1983a. The glacial history of Denmark. In Ehlers, J. (ed.), Glacial deposits in North-west Europe, p. 163-179. A.A. Balkema, Rotterdam.

Sjørring, S. 1983b. Ristinge Klint. In Ehlers, J. (ed.), Glacial deposits in Northwest Europe, p. 219-226. A.A. Balkema, Rotterdam.

Sjørring, S. (editor) 1985. INQUA and IGCP field meeting in Denmark 1981. Geological Society Denmark, Bulletin 34:1.

Sjørring, S., Nielsen, P.E., Frederiksen, J.K., Hegner, J. Hyde, G., Jensen, J.B., Morgensen, A. and Vortisch, W. 1982. Observationer fra Ristinge Klint, felt- og laboratorie-undersøgelser. Dansk Geologisk Forening, Arsskrift for 1981:135-149.

Slater, G. 1926. Glacial tectonics as reflected in disturbed drift deposits. Geologists' Association Proceedings 37:392-400.

Slater, G. 1927a. The structure of the disturbed deposits in the lower part of the Gipping Valley near Ipswich. Geologists' Association Proceedings 38:157-182.

Slater, G. 1927b. The structure of the disturbed deposits of the Hadleigh Road area, Ipswich. Geologists' Association Proceedings 38:183-261.

Slater, G. 1927c. The structure of the disturbed deposits of Møens Klint, Denmark. Transactions Royal Society Edinburgh 55, part 2:289-302.

Slater, G. 1927d. The disturbed glacial deposits in the neighborhood of Lönstrup, near Hjörring, north Denmark. Transactions Royal Society Edinburgh 55, part 2:303-315.

Slater, G. 1927e. Structure of the Mud Buttes and Tit Hills in Alberta. Geological Society America, Bulletin 38:721-730.

Slater, G. 1929. The structure of the drumlins exposed on the south shore of Lake Ontario. New York State Museum, Bulletin 281:3-19.

Slater, G. 1931. The structure of the Bride Moraine, Isle of Man. Proceedings Liverpool Geological Society 14:184-196.

Smed, P. 1962. Studier over den fynske øgruppes glaciale landskabsformer. Meddelelser Dansk Geologisk Forening 15:1-74.

Stalker, A.MacS. 1973a. Surficial geology of the Drumheller Area, Alberta. Geological Survey Canada, Memoir 370.

Stalker, A.MacS. 1973b. The large interdrift bedrock blocks of the Canadian Prairies. Geological Survey Canada, Paper 75-1A:421-422.

Stalker, A.MacS. 1976. Megablocks, or the enormous erratics of the Albertan Prairies. Geological Survey Canada, Paper 76-1C:185-188.

Stephan, H.-J. 1985. Deformations striking parallel to glacier movement as a problem in reconstructing its direction. Geological Society Denmark, Bulletin 34:47-53.

Stephan, H.-J., Kabel, C. and Schlüter, G. 1983. Stratigraphic problems in the glacial deposits of Schleswig-Holstein. In Ehlers, J. (ed.), Glacial deposits in North-west Europe, p. 305-320. A.A. Balkema, Rotterdam.

Stewart, T.G. and England, J. 1983. Holocene sea-ice variations and paleoenvironmental change, northernmost Ellesmere Island, N.W.T., Canada. Arctic and Alpine Research 15:1-17.

Stone, B.D. and Koteff, C. 1979. A late Wisconsinan ice readvance near Manchester, New Hampshire. American Journal Science 279:590-601.

Sugden, D.E. and John, B.S. 1976. Glaciers and landscape: A geomorphological approach. J. Wiley and Sons, New York, 376 p.

Surlyk, F. 1971. Skrivekridtklinterne pâ Møn. In Hansen, M. and Poulsen, V. (eds.), Geologi pâ øerne. Varv, Ekskursionsfører 2:5-23.

Sønstegaard, E. 1979. Glaciotectonic deformation structures in unconsolidated sediments at Os, south of Bergen. Norsk Geologisk Tidsskrift 59:223-228.

Thomas, G.S.P. 1977. The Quaternary of the Isle of Man. In Kidson, C. and Tooley, M.J. (eds.), The Quaternary history of the Irish Sea. Geology Journal, Special Issue 7:155-178.

Thomas, G.S.P. 1984. The origin of the glacio-dynamic structure of the Bride Moraine, Isle of Man. Boreas 13:355-364.

Thomas, G.S.P. and Summers, A.J. 1984. Glacio-dynamic structures from the Blackwater Formation, Co. Wexford, Ireland. Boreas 13:5-12.

Torell, O. 1872. Undersökningar öfver istiden del I. Aftryck ur Ofversigt af

Kungliga Vetenskapsakademiens Förhandlingar 1872, P.A. Nordstedt & Söner, Stockholm, 44 p.

Torell, O. 1873. Undersökningar öfver istiden del II. Skandinaviska landisens utsräckning under isperioden. Ofversigt af Kungliga Vetenskapsakademiens Förhandlingar 1873, no. 1:47-64.

Upham, W. 1899. Glacial history of the New England islands, Cape Cod, and Long Island. American Geologist 24:79-92.

Ussing, N.V. 1913. Danmarks geologi. Danmarks Geologiske Undersøgelse, III Raekke, nr. 2, 372 p.

Vorren, T.O. 1979. Weichselian ice movements, sediments and stratigraphy on Hardangervidda, South Norway. Norges Geologiske Undersøkelse, Nr. 350 (Bull. 50), 117 p.

Wateren, D.F.M. van der 1981. Glacial tec-R tonics at the Kwintelooijen Sandpit, Rhenen, the Netherlands. In Ruegg, G.H.J. and Zandstra, J.G. (eds.), Geology and archaeology of Pleistocene deposits in the ice-pushed ridge near Rhenen and Veenendaal. Mededelingen Rijks Geologische Dienst 35-2/7:252-268.

Wateren, D.F.M. van der 1985. A model of glacial tectonics, applied to the ice-pushed ridges in the Central Netherlands. Geological Society Denmark, Bulletin 34:55-74.

Wateren, D. van der 1987. Structural geology and sedimentology of the Dammer Berge push moraine, FRG. In Meer, J.J.M. van der (ed.), Tills and glaciotectonics, p. 157-182. A.A. Balkema, Rotterdam.

Wee, M.W. ter 1962. The Saalian glaciation in the Netherlands. Meded. Geol. Stichting NS 15:57-77.

Wee, M.W. ter 1981. The Saalian glaciation in the northern Netherlands. In Ehlers, J. and Zandstra, J.G. (eds.), Glacigenic deposits in the southwest parts of the Scandinavian icesheet. Mededelingen Rijks Geologische Dienst 34-1/11:7-9.

Wee, M.W. ter 1983. The Saalian glaciation in the northern Netherlands. In Ehlers, J. (ed.), Glacial deposits in North-west Europe, p. 405-412. A.A. Balkema, Rotterdam.

Weisse, R. 1978. Die Bedeutung der skandinavischen Vereisungen für die Gestaltung des heutigen Reliefs der Jungund Altmoränenlandschaft. In, 100 Jahre Glazialtheorie im Gebiet der skandinavischen Vereisungen. Schriftenreihe für Geologische Wissenschaften 9:291-308.

Whiteman, C.A. 1987. Till lithology and genesis near the southern margin of the Anglian ice sheet in Essex, England. In Meer, J.J.M. van der (ed.), Tills and glaciotectonics, p. 55-66. A.A. Balkema, Rotterdam.

Whittecar, G.R. and Mickelson, D.M. 1977. Sequence of till deposition and erosion in drumlins. Boreas 6:213-217.

Wilke, H. and Ehlers, J. 1983. The thrust moraine of Hamburg-Blankenese. In Ehlers, J. (ed.), Glacial deposits in North-west Europe, p. 331-333. A.A. Balkema, Rotterdam.

Wolford, J.J. 1932. A record size glacial erratic. American Journal Science 224: 362-367.

Woodworth, J.B. 1897. Unconformities of Martha's Vineyard and of Block Island. Geological Society America, Bulletin 8:197-212.

Woodworth, J.B. and Wigglesworth, E. 1934. Geography and geology of the region including Cape Cod, the Elizabeth Islands, Nantucket, Martha's Vineyard, No Mans Land, and Block Island. Memoirs Museum Comparative Zoology, Harvard College, 322 p.

Zagwijn, W.H. 1974. The palaeogeographic evolution of The Netherlands during the Quaternary. Geologie en Mijnbouw 53:369-385.

Zandstra, J.G. 1981. Petrology and lithostratigraphy of ice-pushed Lower and Middle Pleistocene deposits at Rhenen (Kwintelooijen). In Ruegg, G.H.J. and Zandstra, J.G. (eds.), Geology and archaeology of Pleistocene deposits in the ice-pushed ridge near Rhenen and Veenendaal. Mededelingen Rijks Geologische Dienst 35-2/7:178-191.

Åmark, M. 1984. The deglaciation of the eastern part of Skâne, southern Sweden. Lundqua Thesis 15, Lund University, Sweden, 124 p.

Åmark, M. 1985. Subglacial deposition and deformation of stratified drift at the formation of tills beneath an active glacier--an example from Skâne, Sweden.

Geological Society Denmark, Bulletin 34:75-81.

Âmark, M. 1986a. Clastic dikes formed beneath an active glacier. Geologiska Föreningens i Stockholm, Förhandlingar 108, pt. 1:13-20.

Âmark, M. 1986b. Glacial tectonics and deposition of stratified drift during formation of tills beneath an active glacier--examples from Skåne, southern Sweden. Boreas 15:155-171.

3. GEOGRAPHIC INDEX

3.1A Canada

3.1B Canada

3.2A Denmark

3.2B Denmark

3.3 General

Moran (1971); et al. (1980)
Muller (1983)
Nye (1952)
Occhietti (1973)
Rotnicki (1976)
Ruszcznska-Szenajach (1980, 1987)
Rutten (1960)
Sjørring (1985)
Sugden and John (1976)

3.4A Germany (GFR+GDR)/Switzerland

Carle (1938)
Eybergen (1987)
Figge (1983)
van Gijssel (1987)
Grube and Vollmer (1985)
Höfle and Lade (1983)
Johnstrup (1874)
Kabel-Windloff (1987)
Prange (1983, 1985)
Schlüchter and Wohlfarth-Meyer (1987)
van der Wateren (1987)
Weisse (1978)
Wilke and Ehlers (1983)

3.4B Germany (GFR+GDR)/Switzerland

Ehlers (1982, 1983b)
Lange et al. (1979)
Meyer (1983, 1987)
Picard (1969)
Stephan (1985); et al. (1983)

3.5A Iceland/Greenland/Svalbard

Croot (1987)
Funder and Petersen (1979)
Gripp (1929)
Humlum (1983)
Ingolfsson (1985)
Krüger (1987); and Humlum (1980)

3.5B Iceland/Greenland/Svalbard

Ingolfsson (1987)

3.6A Netherlands

Crommelin and Maarleveld (1949)
de Gans et al. (1987)
de Jong (1952, 1955)
Ruegg (1981); and Zandstra (1981)
van der Wateren (1981)
Zandstra (1981)

3.6B Netherlands

van den Berg and Beets (1987)
Jelgersma and Breeuwer (1975)
de Jong (1967)
Maarleveld (1953, 1981, 1983)
Rappol (1987)
Ruegg (1983)
van der Wateren (1985)
ter Wee (1962, 1981, 1983)
Zagwijn (1974)

3.7A Norway/Sweden

Aber and Aarseth (1988)
Adrielsson (1984)
Erdmann (1873)
Hillefors (1973, 1985)
Holst (1903, 1911)
Königsson and Linde (1977)
Landvik and Hamborg (1987)
Lundqvist (1967, 1987); and Lagerbäck (1976)
Mangerud et al. (1981)
Ringberg (1980, 1983); et al. (1984)
Sønstegaard (1979)
Torell (1872, 1873)
Vorren (1979)
Åmark (1984, 1985, 1986a, 1986b)

3.7B Norway/Sweden

Lagerlund (1987)
Lundqvist (1985)

3.8A Poland

Ber (1987)
Brykczynska and Brykczynski (1974)
Brykczynski (1982)
Drozdowski (1981)
Dylik (1961)
Galon (1961)
Jahn (1950, 1956)
Klatkowa (1972)
Lewinski and Rozycki (1929)
Mojski (1979)
Nowak (1977)
Rozycki (1982)
Ruszcznska-Szenajch (1973, 1976, 1978, 1981)
Sarnacka (1965)

3.8B Poland

Brodzikowski and van Loon (1985)
Dylik (1967)
Rozycki (1970, 1972)
Ruszcznska-Szenajch (1979, 1983, 1985)

3.9A United Kingdom/Ireland

Ehlers et al. (1987)

Slater (1927a, 1927b, 1931)
Thomas (1977, 1984); and Summers (1984)
Whiteman (1987)

3.9B United Kingdom/Ireland

Banham (1975)
Slater (1926)

3.10A United States

Aber (1988c)
Andrews (1980)
Barbour (1913)
Bluemle (1966, 1971, 1973, 1975, 1981)
Carlson and Freers (1975)
Dellwig and Baldwin (1965)
Feldmann (1964)
Fuller (1914)
Hansen, D.E. (1967)
Hollick (1894)
Howe (1968)
Hyyppa (1955)
Kaye (1964a, 1964b, 1980)
Lammerson and Dellwig (1957)
Merrill (1886a)
Mills and Wells (1974)
Rau et al. (1962)
Sardeson (1905, 1906)
Shaler (1888, 1898)
Sirkin (1976)
Slater (1929)
Stone and Koteff (1979)
Whittecar and Mickelson (1977)
Wolford (1932)
Woodworth (1897)

3.10B United States

Aber (1982a, 1985b)
Bluemle and Clayton (1984)
Clayton et al. (1980, 1985)
Merrill (1886b)
Mickelson et al. (1982)
Oldale (1980); and O'Hara (1984)
Sirkin (1980)
Upham (1899)
Woodworth and Wigglesworth (1934)

Glaciotectonics: Forms and Processes, Croot (ed.), © 1988 Balkema, Rotterdam. ISBN 90 6191 848 0

List of contributors

James S Aber
Department of Earth Sciences
Emporia State University
1200 Commercial
EMPORIA
Kansas 66801
USA

Katharine Albino
Department of Geology
University of Western Ontario
London
Ontario
N6A 5B7
Canada

Peter H Banham
Department of Geology
Royal Holloway & Bedford New College
Egham Hill
Egham
Surrey
UK

David G Croot
Department of Geographical Sciences
Plymouth Polytechnic
Drake Circus
PLYMOUTH
PL4 8AA
Devon
UK

Edward Derbyshire
Department of Geography
University of Leicester
Leicester
LH1 7RH
UK

Aleksis Dreimanis
Department of Geology
University of Western Ontario
London
Ontario
N6A 5B7
Canada

Volker Feeser
Geologische U. Palaeontologische Inst.
Olshavsenstrasse 40
Gebande S13A
D-2300 KIEL
Germany

Joanne M R Fernlund
Department of Quaternary Geology
University of Uppsala
Box 555
S-7521 22 Uppsala
Sweden

Michael Houmark-Nielsen
Institute of General Geology
Osterfoldgade 10
DK-1350 Copenhagen-K
Denmark

Olafur Ingolfsson
Kvartargeologiska avd
Lunds Univeristet
Solveg 13
S-223 62 Lund
Sweden

Juha Pekka Lunkka
Sub Department of Quaternary Research
University of Cambridge
Godwin Laboratory
Free School Lane
CAMBRIDGE CB2 3RS
UK

Lewis A Owen
Department of Geography
University of Leicester
LEICESTER LH1 7RH
UK

S A Schack-Pedersen
Geological Survey of Denmark
Thovavej 8
DK-2400
COPENHAGEN NV
Denmark

K S Petersen
Geological Survey of Denmark
Thoravej 8
DK-2400
COPENHAGEN NV
Denmark

Leif A Rasmussen
Geological Survey of Denmark
Thoravej 8
DK-2400
COPENHAGEN NV
Denmark

Hanna Ruszczynska-Szenajch
Department of Geology
Warzaw University
Al Zwirki i Wigary 93
02 089 WARZAW
Poland